学会提问

**原书第12版
中英文对照
学习版**

Asking
the Right Questions

A Guide to
Critical Thinking
12th Edition

[美] 尼尔·布朗（Neil Browne） 著
斯图尔特·基利（Stuart Keeley）

许蔚翰 吴礼敬 译

机械工业出版社
CHINA MACHINE PRESS

图书在版编目（CIP）数据

学会提问：原书第12版：中英文对照学习版／（美）尼尔·布朗（Neil Browne），（美）斯图尔特·基利（Stuart Keeley）著；许蔚翰，吴礼敬译 .-- 北京：机械工业出版社，2021.8（2025.8重印）

书名原文：Asking the Right Questions: A Guide to Critical Thinking (12th Edition)
ISBN 978-7-111-68895-2

I . ①学⋯ II . ①尼⋯ ②斯⋯ ③许⋯ ④吴⋯ III . ①学习方法 – 青少年读物 – 汉、英 IV . ① G791-49

中国版本图书馆 CIP 数据核字（2021）第 174180 号

北京市版权局著作权合同登记　图字：01-2021-3905 号。

Neil Browne, Stuart Keeley. Asking the Right Questions: A Guide to Critical Thinking, 12th Edition.
ISBN: 978-0-13-443199-4
Copyright © 2018, 2015, 2012 by Pearson Education, Inc.

This authorized English-Chinese bilingual edition is jointly published by Pearson Education Inc. and China Machine Press.

No part of this publication may be reproduced or transmitted in any form or by any means, electronic or mechanical, including photocopying, recording, taping, or any information and retrieval system, without the written permission of the publisher. This edition is authorized for sale in the Chinese mainland (excluding Hong Kong SAR, Macao SAR and Taiwan).

All rights reserved.

本书英汉双语版由机械工业出版社和 Pearson Education（培生教育出版集团）合作出版。未经出版者预先书面许可，不得以任何方式复制或抄袭本书的任何部分。此版本仅限在中国大陆地区（不包括香港、澳门特别行政区及台湾地区）销售。

本书封底贴有 Pearson Education（培生教育出版集团）激光防伪标签，无标签者不得销售。

学会提问（原书第 12 版 · 中英文对照学习版）

出版发行：机械工业出版社（北京市西城区百万庄大街 22 号　邮政编码：100037）
责任编辑：邹慧颖　向睿洋　　　　　　　　　　责任校对：殷　虹
印　　刷：涿州市京南印刷厂　　　　　　　　　版　　次：2025 年 8 月第 1 版第 9 次印刷
开　　本：186mm × 240mm　1/16　　　　　　　印　　张：23
书　　号：ISBN 978-7-111-68895-2　　　　　　定　　价：99.00 元

客服电话：（010）88361066　68326294

版权所有·侵权必究
封底无防伪标均为盗版

赞　誉

（按姓氏笔画排序）

在本书的帮助下培养清晰的思维，既能让你获得一种生存能力，也能提升你的人格魅力！

——贝小戎　《三联生活周刊》资深主笔

尽信书，不如无书——信息爆炸时代，专家满天飞，盲信比不信还糟糕。如果你希望成为一个听完别人说的话以后，不是只会说"对对对"，而是会说"你说得真好，我有一个好问题"的人，那么这本书是你的必读书。

——古典　生涯规划师，《跃迁》《拆掉思维里的墙》作者

这是一本关于批判性思维的经典著作。在最新版中，作者与时俱进，参考了认知心理学、逻辑学与思维训练方面的新进展，比如加入了对灰度思维、基本归因错误等的介绍。书中详细介绍了论证的各个组成要素，探讨了提问与独立思考的技巧。这本书值得向每位关心思维成长的读者推荐。

——阳志平　安人心智董事长，心智工具箱公众号作者

或许，正是 2006 年 7 月，初三升高一暑假的某个午后，看到那则"复旦附中学生获哈佛、耶鲁、斯坦福等名校录取"的新闻，我听到了来自灵魂的提问："为什么我不可以？"从此，我成为一名 risk taker（敢于冒险的人），一路上从厦门到耶鲁，从高盛到哈佛，再到回国创业成为 90 后年轻高管，一路上不违本心，勇敢去闯，努力成为更好的自己。我感谢 2006 年的那一个提问。

——李柘远（学长 LEO）　畅销书作家，金影科技联席总裁

本书是我在 20 世纪 90 年代读的第一本关于批判性思维的书，也是我迄今读过的众多关于批判性思维的书中最好的一本。它通过 10 个批判性问题来组织全书，把逻辑分析、认知评价和价值评价熔于一炉，提纲挈领，条理分明，简

明扼要，精心选择例证，分析透彻深入。我至今记得，当年的阅读带给我的震撼：原来我们的思维中隐藏着如此多的谬误和陷阱！它也使我明白了什么是批判性思维，如何去批判性思考。我后来逐渐认识到，批判性思维还是通向创造性思维的重要途径：因为不同凡"想"，所以不同凡响。向各位读者强烈推荐本书：学习批判性思维，从本书开始！

——陈波　北京大学哲学系、外国哲学研究所教授，博士生导师

在这个信息过剩的时代，会问问题的人，运气不会太差。面对海量的信息，只有掌握批判性思维，我们才能"披沙拣金"，成为自己的"信息把关人"。

——陈新焱　"书单"公众号创始人

学术研究的本质就是对社会现象进行不断提问、假设和验证的过程。不盲从于表面现象，深入肌理，发现真问题、提出好问题是学术研究中最关键也最困难的环节，不具备这种能力成为很多学术研究者的"软肋"。《学会提问》是批判性思维训练领域的经典之作，阅读本书，你可以系统学习批判性思考的基本框架和逻辑，不断提升独立思考和创新思维能力。《学会提问》是每个学术人都应该学习的思维宝典。

——学君　"学术志"创始人

这本书是批判性思维领域的经典之作，常读常新。如果一个人不具备批判性思维的能力，那么他就很容易上当受骗，我身边不乏这样的例子。批判性思维是我们每个人都需要上的一课。

——采铜　心理学博士，畅销书《精进》作者

这本《学会提问》是一本非常经典的批判性思维读物，很出色地完成了传授批判性提问的技能这一目标，既简洁又全面，实践指导性强，会为同学们提供很大的帮助。想要去国外留学深造的同学应该学习批判性思维，学会用批判性眼光去看待、评价问题，理性思考，扩展你的思维和眼界，丰富你的内心世界！

——俞敏洪　新东方教育集团董事长

我推荐每一个成年人阅读这本书。我经常说熟练利用批判性思维永远是少数人的武器，拥有批判性思维，就相当于具备真实的自我反省能力。如果你也想学习批判性思维，《学会

提问》这本书你一定要看。

——秋叶　秋叶品牌创始人

当代社会的流动性和信息爆炸，让人们感受到强烈的不确定性，这对心智状态提出了更高要求。我们面对的这一处境，使得批判性思维凸显出重要价值。这本书脉络清晰，层层递进，极佳地展现了如何学习批判性提问、分析和判断，能够帮助我们在诸多情境中去伪存真，做出更好的生活选择。

——夏超　KnowYourself 媒体部总监

提问比告知更能够解决问题，因为告知在很多情况下意味着指责。优秀的提问者可以激发他人的动力，让他人意识到盲点，甚至帮对方找到解决问题的方法。前提是，你得"学会提问"！

——樊登　"樊登读书"创始人

在信息泛滥的时代，我们每天都会遇到许多他人兜售的观点。如何合理"抬杠"，将批判性思维延伸到发言甚至写作中，让自己的提问更为理性、正确，《学会提问》给出了答案和具体的练习方式，让我们得以拨开迷雾，一语中的。

——壹心理

目　录

赞　誉
前　言

第1章　正确提问的益处和方法·001
　　我们生活在嘈杂、混乱的世界·001
　　专家说得再动听，也不一定靠得住·004
　　有必要依靠自己的大脑·006
　　激发你的批判性思维·006
　　海绵式思维和淘金式思维·008
　　弱势批判性思维和强势批判性思维·012
　　　　勤于练习很重要·014
　　批判性思维和与人交往·016
　　　　批判性思维者拥有的主要价值观·018
　　让对话有效进行下去·020
　　　　营造友好的交流氛围·024
　　像批判性思维者那样写作和发言·024

第2章　论题和结论是什么·030
　　论题的种类·032
　　寻找论题·034
　　寻找写作者或发言者的结论·036
　　找到结论的线索·038
　　思维体操·040
　　给个提示·042

第3章　理由是什么·048
　　开始质疑的过程·052
　　理由的提示词·056
　　让理由和结论一目了然·056
　　　　先有理由，后有结论·056
　　像批判性思维者那样写作和发言·060
　　思维体操·064
　　给个提示·068

CONTENTS

Preface

Chapter 1 The Benefit and Manner of Asking the Right Questions 001
    The Noisy, Confused World We Live In 001
    Experts Cannot Rescue Us, Despite What They Say 005
    The Necessity of Relying on Our Mind 007
    Critical Thinking to the Rescue 007
    The Sponge and Panning for Gold: Alternative Thinking Styles 009
    Weak-Sense and Strong-Sense Critical Thinking 013
        The Importance of Practice 015
    Critical Thinking and Other People 017
        Primary Values of a Critical Thinker 019
    Keeping the Conversation Going 021
        Creating a Friendly Environment for Communication 025
    Writing and Speaking as Critical Thinkers 025

Chapter 2 What Are the Issue and the Conclusion? 031
    Kinds of Issues 033
    Searching for the Issue 035
    Searching for the Author's or Speaker's Conclusion 037
    Clues to Discovery: How to Find the Conclusion 039
    Practice Exercises 041
    Sample Responses 043

Chapter 3 What Are the Reasons? 049
    Initiating the Questioning Process 053
    Words That Identify Reasons 057
    Keeping the Reasons and Conclusions Straight 057
        Reasons First, Then Conclusions 057
    Writing and Speaking as Critical Thinkers 061
    Practice Exercises 065
    Sample Responses 069

第 4 章　哪些词语意思不明确 · 072

让人捉摸不透的多义词 · 074
找准关键词 · 076
检查有没有歧义 · 78
判定歧义 · 080
上下文和歧义 · 084
歧义、定义和字典 · 084
澄清歧义不只是你的责任 · 088
像批判性思维者那样写作和发言 · 088
思维体操 · 094
给个提示 · 096

第 5 章　价值观假设和描述性假设是什么 · 102

价值观冲突和价值观假设 · 108
从价值观到价值观假设 · 110
典型的价值观冲突 · 112
把立论者的背景作为寻找价值观假设的线索 · 112
把可能发生的后果作为寻找价值观假设的线索 · 114
寻找价值观假设的其他方法 · 116
了解其他人的价值倾向的价值 · 116
价值观与相对性 · 118
找出并评价描述性假设 · 118
描述性假设举例说明 · 120
常见的描述性假设 · 122
找到假设的线索 · 124
像批判性思维者那样写作和发言 · 126
思维体操 · 134
给个提示 · 136

第 6 章　论证中有没有谬误 · 140

用提问的方法找出论证中的谬误 · 144
从评价假设出发 · 144

Chapter 4 What Words or Phrases Are Ambiguous? 073
   The Confusing Flexibility of Words   075
   Locating Key Terms and Phrases   077
   Checking for Ambiguity   079
   Determining Ambiguity   081
   Context and Ambiguity   085
   Ambiguity, Definitions, and the Dictionary   085
   Limits of Your Responsibility to Clarify Ambiguity   089
   Writing and Speaking as Critical Thinkers   089
   Practice Exercises   095
   Sample Responses   097

Chapter 5 What Are the Value and Descriptive Assumptions?   103
   General Guide for Identifying Assumptions   107
   Value Conflicts and Assumptions   109
   From Values to Value Assumptions   111
   Typical Value Conflicts   113
   The Communicator's Background as a Clue to Value Assumptions   113
   Consequences as Clues to Value Assumptions   115
   More Hints for Finding Value Assumptions   117
   The Value of Knowing the Value Priorities of Others   117
   Values and Relativism   119
   Identifying and Evaluating Descriptive Assumptions   119
   Illustrating Descriptive Assumptions   121
   Common Descriptive Assumptions   123
   Clues for Locating Assumptions   125
   Writing and Speaking as Critical Thinkers   127
   Practice Exercises   135
   Sample Responses   137

Chapter 6 Are There Any Fallacies in the Reasoning?   141
   A Questioning Approach to Finding Reasoning Fallacies   145
   Evaluating Assumptions as a Starting Point   145

发现其他常见的论证谬误 ·150
警惕分散注意力的干扰 ·160
愚弄人的循环论证 ·162
论证谬误小结 ·164
扩展关于谬误的知识 ·166
思维体操 ·166
给个提示 ·170

第7章 证据的效力：个人经历、典型案例、当事人证言和专家陈述 ·174

事实还是见解 ·176
我们需要可靠的证据 ·178
证据的来源 ·180
以个人经历作为证据 ·182
以典型案例作为证据 ·184
以当事人证言作为证据 ·184
以专家意见作为证据 ·188
思维体操 ·192
给个提示 ·196

第8章 证据的效力：个人观察和调查研究 ·198

以个人观察作为证据 ·198
有偏差的调查和问卷 ·200
以研究报告作为证据 ·204
　　研究发现中存在的一般问题 ·206
从研究样本进行概括 ·214
从测量方法进行概括 ·216
专家观点什么时候最可信 ·220
　　科学研究和互联网 ·222
像批判性思维者那样写作和发言 ·224
思维体操 ·230
给个提示 ·232

第9章 有没有替代原因 ·236

何时该寻找替代原因 ·238
替代原因的普遍性 ·240

Discovering Other Common Reasoning Fallacies   151
Looking for Diversions   161
Sleight of Hand: Begging the Question   163
Summary of Reasoning Errors   165
Expanding Your Knowledge of Fallacies   167
Practice Exercises   167
Sample Responses   171

Chapter 7 The Worth of Personal Experience, Case Examples, Testimonials, and Statements of Authority as Evidence   175

Fact or Opinion?   177
The Need for Dependable Evidence   179
Sources of Evidence   181
Personal Experience as Evidence   183
Case Examples as Evidence   185
Testimonials as Evidence   185
Appeals to Authority as Evidence   189
Practice Exercises   193
Sample Responses   197

Chapter 8 How Good Is the Evidence: Personal Observation and Research Studies?   199

Personal Observation as Evidence   199
Biased Surveys and Questionnaires   201
Research Studies as Evidence   205
  General Problems with Research Findings   207
Generalizing From the Research Sample   215
Generalizing From the Research Measures   217
When You Can Most Trust Expert Opinion   221
  Research and the Internet   223
Writing and Speaking as Critical Thinkers   225
Practice Exercises   231
Sample Responses   233

Chapter 9 Are There Rival Causes?   237

When to Look for Rival Causes   239
The Pervasiveness of Rival Causes   241

寻找替代原因 ·242
唯一的原因还是原因之一 ·242
将多种视角作为寻找替代原因的指导 ·244
混淆相关性和因果关系 ·246
混淆"在此之后"与"因此" ·248
解释单个事件或行为 ·250
评价替代原因 ·252
替代原因与你的表达和交流 ·252
　发掘潜在的原因 ·254
思维体操 ·256
给个提示 ·258

第 10 章　数据有没有欺骗性 ·264
不知来历和有偏差的数据 ·266
令人困惑的平均值 ·268
测量误差 ·272
错把一件事的结论用来证明另一件事 ·272
通过省略信息欺骗 ·274
在自己的写作中使用统计数字 ·278
思维体操 ·278
给个提示 ·280

第 11 章　有什么重要信息被省略了 ·284
找到省略信息的益处 ·286
不完整的论证在所难免 ·288
帮你识别省略信息的问题 ·290
我们需要知道确切的数字 ·290
负面视角的重要性 ·294
如果省略的信息还是找不到 ·296
像批判性思维者那样写作和发言 ·298
思维体操 ·304
给个提示 ·306

第 12 章　能得出哪些合理的结论 ·308
二分式思维：妨碍我们考虑多种可能性 ·310

Detecting Rival Causes   243
The Cause or *a* Cause   243
Multiple Perspectives as a Guide to Rival Causes   245
Confusing Causation with Association   247
Confusing "After This" with "Because of This"   249
Explaining Individual Events or Acts   251
Evaluating Rival Causes   253
Rival Causes and Your Own Communication   253
   Exploring Potential Causes   255
Practice Exercises   257
Sample Responses   259

Chapter 10 Are Any Statistics Deceptive?   265
Unknowable and Biased Statistics   267
Confusing Averages   269
Measurement Errors   273
Concluding One Thing, Proving Another   273
Deceiving by Omitting Information   275
Using Statistics in Your Writing   279
Practice Exercises   279
Sample Responses   281

Chapter 11 What Significant Information Is Omitted?   285
The Benefits of Detecting Omitted Information   287
The Certainty of Incomplete Reasoning   289
Questions That Identify Omitted Information   291
But We Need to Know the Numbers   291
The Importance of the Negative View   295
Omitted Information That Remains Missing   297
Writing and Speaking as Critical Thinkers   299
Practice Exercises   305
Sample Responses   307

Chapter 12 What Reasonable Conclusions Are Possible?   309
Dichotomous Thinking: Impediment to Considering Multiple Conclusions   311

灰度思维：两面还是多面 ·312
条件句的生产力 ·314
识别多个结论能解放思维 ·316
小结 ·316
思维体操 ·318
给个提示 ·320

第 13 章　干扰批判性思维的障碍 ·324
正确提问带给人的不快 ·324
思考过快 ·326
刻板印象 ·326
背叛我们的思维习惯 ·328
　　晕轮效应 ·328
　　信念固着 ·330
　　可得性启发法 ·332
　　答非所问 ·332
自我中心 ·334
一厢情愿：批判性思维最大的障碍 ·336
最后的话 ·340

译后记 ·344

Grey Thinking: Two Sides or Many? 313
Productivity of If-Clauses 315
The Liberating Effect of Recognizing Alternative Conclusions 317
Summary 317
Practice Exercises 319
Sample Responses 321

Chapter 13 Speed Bumps Interfering with Your Critical Thinking 325
The Discomfort of Asking the Right Questions 325
Thinking Too Quickly 327
Stereotypes 327
Mental Habits That Betray Us 329
   Halo Effect 329
   Belief Perseverance 331
   Availability Heuristic 333
   Answering the Wrong Question 333
Egocentrism 335
Wishful Thinking: Perhaps the Biggest Single Speed Bump on the Road to Critical Thinking 337
Final Words 339

前　言

"我知道做个慎思明辨的人是很好的，会问很多恰到好处的问题也很棒，可我就是不知道该问哪些问题，不知道该如何提问。"确实有很多人在谈论学习或教授批判性思维，而且这些人知道批判性思维有赖于系统性地提问。但是他们并未言明：该问哪些问题呢？这些问题又是如何促使人们对他人所阐述的内容进行深思熟虑和公平允正的评价的呢？

针对这些疑问，我们专门撰写了本书，目前这已是本书第12次修订了。只有公众都能独立思考和判断，社会才能高效地运转。如果我们经过批判性问题的层层考验后才形成决定和信念，我们就会更加自信。如果任何东西在进入我们的头脑之前都接受特定标准（这些标准被不同文化背景下的思考者所尊重）的检验，我们将为此备感自豪。

本书从一开始就处于不断改进之中，我们不断汲取来自我们的学生以及使用本书的广大教师的意见和建议。一方面，我们为本书的大获成功以及来自很多国家的广大读者的积极反馈而感到无比振奋；另一方面，我们深感任重而道远，觉得我们需要付出前所未有的巨大努力来教导公众如何"提出正确的问题"。

一些本应博闻广识的身居高位者为事实贴上"假消息"的标签，声称他们有其他事实，以此为自己的主张辩护。"事实可能存在争议"或"事实本身是不完整的，还需要得到更多事实支持"这样的言论本来都是有益的，可以提醒我们，对"现实"做断言是很复杂的。然而，仅仅因为我想得出一个特定的结论，就认为我掌握的事实与你掌握的事实等价，这是最低劣的智力欺诈。

我们一直都非常需要批判性思维，尤其是在这个一些人把科学仅仅视作"另一种视角"的历史时刻。

新的意见一多，对哪些需要认真考虑、哪些可以不予考虑的取舍就变得越来越难做。每天都有人想尽一切办法说服我们，跟我们死缠烂打，其中很多人都喜欢钻牛角尖、走极端，他们的论说诉诸情感的成分多，诉诸理智的成分少。记不清在多少次的公开讨论中，我们遇到普遍的极度无视证据、语言草率、错把声高当有理的情况。相信内心而不顾真相，或者说对真相漠不关心的态度正变得越来越普遍。

我们一心追求的是，希望修订本既能保持《学会提问》这本书的主要特色，同时又能适当调整内容以适应我们新的思考重点和读者不断产生的新需求。例如，我们想保留本书简明扼要、清楚易懂以及篇幅短小的特色。经验告诉我们，这本小书出色地完成了它的目

PREFACE

"I know it's good to be a critical thinker and to be able to ask lots of good questions, but I don't know what questions to ask or how to ask them." Indeed, many people speak about learning or teaching critical thinking, and these people know that critical thinking requires the systematic asking of questions. But which questions? And how does each of those questions contribute to a thoughtful and fair evaluation of what someone else is saying?

We are now on our 12th edition of a book that we wrote in response to exactly this situation. Democracy works best with a public capable of critical thinking! We can be more confident of our decisions and beliefs when we have formed them after asking critical questions. We can be proud that before anything gets into our heads, it passes particular standards that both we and thinking people in multiple cultures respect.

From the beginning, our book has been a work in progress, as we continue to incorporate input from our students and from the many teachers using this book. While we continue to be immensely pleased by this book's success and the positive feedback from many readers from many countries, we cannot also help but notice the need for a greater-than-ever expansion of efforts to educate the public in "asking the right questions."

People in high places, who should know better, label facts "fake news" and justify their claims by saying they have alternative facts. That facts may be in dispute, or that facts are incomplete and need to be supported by yet more facts, is all to the good. Such language *could* alert us to the complexity of making claims about "reality." But suggesting that facts I have are equivalent to facts you have just because I want to reach a particular conclusion is intellectual dishonesty of the worst order.

We very much need critical thinking always, but especially at this moment in history where some refer to science as "just another perspective."

Selecting which new suggestions to embrace and which to reject has become increasingly difficult. We are bombarded daily with efforts to persuade us, many of which are highly polarized and appeal much more to the emotional part of the brain than to the reasoning part. We encounter a general, immense disrespect for evidence; the sloppy use of language; and substitution of hollering for reason in so much of our public discussion. "Truthiness," or a lack of concern for the truth, becomes more and more common.

Always uppermost in our mind has been the desire to retain the primary attributes of *Asking the Right Questions*, while adjusting to new emphases in our own thought and the evolving needs of our readers. For instance, we want most of all to keep this book concise, readable, and short. Also, our experience has convinced us that the short book succeeds in the job for which it was intended—the teaching of critical-thinking questioning skills.

标——传授批判性思考和提问的技能。

　　正因如此，本书才能实现其一贯秉持的、其他书无法实现的一系列目标。它将帮助学生掌握一整套提问的技能，并且这些技能可以被广泛应用于各个领域。对这些技能的训练都是在轻松、自然的讨论中展开的。（我们的读者对象是普通大众，而不是专业人士。）

　　本书最为显著的特色之一就是它的适用范围远远超出了你的想象，延伸到生活的方方面面。与批判性思维联系在一起的种种习惯和态度，可以灵活运用到消费、医疗、法律及一般性的伦理和个人选择当中。当外科医生说需要动手术时，寻找本书所倡导的批判性问题的答案就有可能变成生死攸关的事情。此外，坚持提出这些批判性问题也可以巩固我们不断增长的知识，帮助我们更快地发现世界运行的规律，更好地理解这个世界，并理解怎样做才能让世界变得更美好。

　　本书对谁特别有帮助呢？鉴于我们的教学经验涵盖了各种不同水平和层次的学生，我们很难想象出这本书对哪一门专业或课程派不上用场。事实上，本书已被广泛应用于法律、英语、制药学、哲学、教育学、心理学、社会学、宗教学以及其他各类社科课程，同时还被普遍应用于中学课堂里。

　　本书特别适合在以下几个领域中应用。通识课程的老师在第一堂课上，面对学生关于他们能从这门课中学到什么的疑问，可以让学生阅读本书，这将为他们的问题提供清晰明了的答案。英语老师在课堂上训练学生写作说明文时，也可以使用本书，不仅可以将其作为构思文章前评估各种不同论点的范本，还可以将其作为写作过程中应避免的问题的检查清单。本书还可以用作旨在培养学生批判性阅读和思考技能的课程中的核心材料。

　　本书主要生发自我们的课堂教学经验，旨在帮助每个人培养阅读和倾听的习惯。它力求培养的种种技能，是任何一个不盲从盲信的读者都需要的，是理性判断的基础。本书中反复强调的批判性问题可以提高我们的论证能力，不管我们受过什么程度的正规教育。

　　新版的特色主要包括以下几方面：

1. 长期以来，我们一直觉得有必要将批判性思维延伸至写作和发言。为了达到这个目标，我们向一位经验丰富的写作老师寻求帮助，他使我们能更好地帮助那些希望成为批判性思维者的写作者和发言者。
2. 我们的很多学生受益于我们对文字内容的视觉呈现。为帮助读者更好地学习批判性思维，新版中在讨论特别重要的观点的部分增加了一系列图表。

Thus, our book continues to do a number of things that other books have failed to do. This text develops an integrated series of question-asking skills that can be applied widely. These skills are discussed in an informal style. (We have written to a general audience, not to any specialized group.)

One feature that deserves to be highlighted is the applicability of *Asking the Right Questions* to numerous life experiences extending far beyond the classroom. The habits and attitudes associated with critical thinking are transferable to consumer, medical, legal, and general ethical and personal choices. When a surgeon says surgery is needed, it can be life sustaining to seek answers to the critical questions encouraged in *Asking the Right Questions*. In addition, practicing the critical-thinking questions enhances our growth of knowledge in general and helps us better discover the way the world is, how it could be better understood, and how we can make it a better world.

Who would find *Asking the Right Questions* especially beneficial? Because of our teaching experiences with readers representing many different levels of ability, we have difficulty envisioning any academic course or program for which this book would not be useful. In fact, this text been used in law, English, pharmacy, philosophy, education, psychology, sociology, religion, and social science courses, as well as in numerous high school classrooms.

A few uses for the book seem especially appropriate. Teachers in general education programs may want to begin their courses by assigning this book as a coherent response to their students' requests to explain what is expected of them. English courses that emphasize expository writing could use this text both as a format for evaluating arguments prior to constructing an essay and as a checklist of problems that the writer should attempt to avoid as she writes. The text can also be used as the central focus of courses designed specifically to teach critical reading and thinking skills.

While *Asking the Right Questions* stems primarily from our classroom experiences, it is written so that it can guide the reading and listening habits of almost everyone. The skills that it seeks to develop are those that any critical reader needs to serve as a basis for rational decisions. The critical questions stressed in the book can enhance anyone's reasoning, regardless of the extent of her formal education.

The special features of this new edition include the following:

1. We have long felt a need to extend the understanding of critical thinking to writing and speaking. Toward that end we sought the assistance of an experienced writing teacher who has greatly enlarged our attempted assistance to writers and speakers who wish to be critical thinkers while communicating.

2. Many of our students benefit from visual representations of what we have already said in print form. To assist their learning of critical thinking, this edition of our text has added multiple charts at those points where especially important ideas are being discussed.

3. 我们继续吸收丹尼尔·卡尼曼所著的《思考，快与慢》一书里的真知灼见，并使之贯穿全书。我们尤其强调"慢思考"的重要性。

4. 我们用与学生生活更紧密相关的材料替换了 1/3 的练习文章。我们继续在前几篇练习文章里采取自问自答的方式，即呈现思考以得到答案的过程——好比一个人正努力思考如何评价一篇文章，而读者则在这个人的脑海里旁观整个思考的过程。我们认为与其直接让读者看答案，不如让他们切身"感受"一下接受、拒绝、修正和组织答案的点滴过程，这样他们可以更真切地了解到用来获取答案的批判性思维的实际过程。在这一点上，我们认同著名教育家约翰·加德纳的重要比喻，他曾严厉批评一些老师和培训师只向学生展示从知识园地里采摘的缤纷花朵，而不给学生看让那美丽花朵得以绽放的种植、除草、施肥和修剪的整个过程。

5. 我们还强调了批判性思维的社会属性或者说互动属性，强调了一个人提出批判性问题的方式，可能会极大地影响他提问的价值。例如，很多读者在与人交往时刚跃跃欲试，想秀一秀自己批判性提问的能力，就发现并不是每个人都乐于见到别人对自己的信念展开批判性的盘问。有些互动方法能在批判性思维者与发言者/写作者之间激发出更多令人满意的对话，而有些方法则难以奏效。我们建议读者一定要注意提问和倾听的策略，让对话得以顺利进行，而不要迫使对方关闭对话的渠道。例如，常常会有人一听完你的问题就回应道"你怎么单单就盯上了我呢"，批判性提问的进程就此戛然而止，没了下文。

6. 我们添加了许多新的部分，包括测量误差、灰度思维、操控型论证，以及基本归因错误。

7. 新版还提供了 200 多个问题及相应解答，教师可以将其用作评估工具或家庭作业。若有钻研和使用这些问题的需求，请访问学习催化平台（https://learningcatalytics.com/sign_in?login=true）。在登录页创建一个个人账户以获取这些问题。

教师手册

教师手册为使用本书的教师提供综合全面的帮助。可以从 http://pearsonhighered.com/ 网站上下载这一补充材料，也可以从当地的培生代理商那里获取相关渠道。

本书第 12 版的问世要特别感谢很多人。首先我们要对培生集团的以下评论者提出的中肯意见表示感谢：内华达大学雷诺分校的 Diane K. Lofstrom Miniel、普利茅斯州立大学的 Clarissa M. Uttley、亨廷顿学院的 John Saunders、坎伯兰大学的 Joshua Hayden，以及峡谷学院的 Leslie St. Martin。

3. Throughout the book, we have continued to integrate insights from Daniel Kahneman's *Thinking, Fast and Slow*. We especially emphasize the importance of slow thinking.
4. We have replaced one-third of the practice passages with material more closely tied to our students' lives. We continue with think-aloud answers for early practice passages—expressing critical-thinking responses to a passage as if the reader were inside the head of a person struggling with the challenge of evaluating the practice passages. We think that "hearing" the bit-by-bit process of accepting, rejecting, revising, and organizing an answer gives the reader a more realistic picture of the actual critical-thinking process used to achieve an answer than would simply observing an answer. Here we are relying on the important metaphor of John Gardner, who chastised teachers and trainers for showing learners only the cut flowers of knowledge and not the planting, weeding, fertilizing, and pruning that result in a beautiful bouquet.
5. We also emphasize the social or interactive nature of critical thinking and the way in which one asks critical-thinking questions can greatly influence the value of the questioning. For example, many readers initially flexing their critical-questioning muscles with others find that not everyone welcomes the critical questioning of their beliefs. Some interactive approaches stimulate much more satisfactory dialogues between the critical thinker and the speaker or writer than others. We suggest questioning and listening strategies to keep the conversation going rather than shutting it down. For example, critical questioning will often be brought to a quick halt by a listener's response of, "Why are you picking on me?"
6. We have added many new sections, including Measurement Errors, Grey Thinking, Managed Reasoning, and Fundamental Attribution Errors.
7. This edition also makes available over 200 questions and answers that instructors may use as an assessment instrument or as homework assignments. To examine and use these questions, please go to the Learning Catalytics platform at https://learningcatalytics.com/sign_in?login=true. After arriving at the sign-in page, create a personal account to reach the questions.

Instructor's Manual

An Instructor's Manual provides comprehensive assistance for teaching with *Asking the Right Questions*. Instructors may download this supplement at http://www.pearsonhighered.com/ or request access through their local Pearson representative.

This 12th edition owes special debts to many people. We wish to acknowledge the valuable advice of the following Pearson reviewers: Diane K. Lofstrom Miniel, University of Nevada, Reno; Clarissa M. Uttley, Plymouth State University; John Saunders, Huntingdon College; Joshua Hayden, Cumberland University; and Leslie St. Martin, College of the Canyons.

我们的学生一直是本书修订意见的主要来源，我们要特别感谢其中几位。第12版尤其得益于 Alex Jacobs、Joseph Seipel、Caitlyn Reeder 以及 Arataenrique Kaku 等人提供的大量协助。

如果你对批判性问题以及习惯性追问在思维能力的培养中所起的重要作用非常感兴趣，那么你可以关注尼尔·布朗的博客，与他一起探讨问题与答案之间的复杂关系。博客名为：

1. "层层追问，答无止境"，网址为 http://www.celebratequestions.com。
2. "打破脑中坚冰"，网址为 http://www.iceinthehead.com。

尼尔·布朗

斯图尔特·基利

"我的写作实验室"：可用来帮助写作

"我的写作实验室"是一个在线作业、辅导和评价程序，旨在为教师和学生提供一些有趣的体验。通过将写作练习与评估结合在一起，教师可以制定出有意义的学习任务，根据自己设定的测试标准来给学生打分，进而通过高级的报告方式来分析课堂表现。对那些在进入一门课程时准备不足的学生来说，"我的写作实验室"提供诊断性测试和个性化辅导，以便学生了解自己的进步程度，这样教师在课堂上就可以少花一些时间来复习要点。"我的写作实验室"包含大量多媒体资源以吸引学生的注意力，并为教师提供全程支持。想要获取更多信息，可以访问以下链接：http://www.pearsonmylabandmastering.com/northamerica/mywritinglab/。

While our students are always a major source of suggested improvements, a few distinguished themselves in that regard. The 12th edition benefited from the especially valuable assistance of Alex Jacobs, Joseph Seipel, Caitlyn Reeder, and Arataenrique Kaku.

If you are fascinated by critical questions and the significance of habitual questioning for our mental development, please join Neil Browne in discussing the complicated relationship between questions and answers at his blogs:

1. Celebrating Bold Questions and Humble Answers: http://www.celebratequestions.com
2. Breaking the Ice in Our Heads: http://www.iceinthehead.com

M. Neil Browne
Stuart M. Keeley

MyWritingLab: Now Available for Composition
MyWritingLab is an online homework, tutorial, and assessment program that provides engaging experiences to today's instructors and students. By incorporating rubrics into the writing assignments, faculty can create meaningful assignments, grade them based on their desired criteria, and analyze class performance through advanced reporting. For students who enter the course under-prepared, MyWritingLab offers a diagnostic test and personalized remediation so that students see improved results and instructors spend less time in class reviewing the basics. Rich multimedia resources are built in to engage students and support faculty throughout the course. Visit http://www.pearsonmylabandmastering.com/northamerica/mywritinglab/ for more information.

第 1 章

正确提问的益处和方法

| 学 习 目 标 |

1. 对比不需要批判性思维的世界和我们所处的世界之间的异同。
2. 弄清楚专家在我们的信念和推论形成中所起的作用。
3. 阐明批判性思维的内涵。
4. 辨别弱势批判性思维和强势批判性思维。
5. 将价值观的作用与批判性思维结合起来。
6. 掌握用批判性思维激发对话的技巧。

我们生活在嘈杂、混乱的世界

　　本书旨在鼓励你学习一项技能，我们认为它能让你活得更精彩。这项技能不是别的，正是"批判性思维"。但是我们中有些人生活在一个想象的世界中，在那里，批判性思维根本没有存在的必要。这个想象的世界里盛行的是以下这些情况：

1. 想要说服我们相信某件事的人，总是会向我们解释照他们说的那样做的种种弊端。
2. 无论何时，只要我们对人生重大问题产生困惑，总是很快就能找到一位可靠的专家、权威或大师，并且这些充满智慧的人的观点总是若合符契。一句话，我们根本不用操心该怎么办，该相信什么，因为大师们必定会找到答案。我们的任务就是找到这

CHAPTER 1

The Benefit and Manner of Asking the Right Questions

LEARNING OBJECTIVES

1. Contrast a world where no critical thinking is needed with the world we live in.
2. Understand the role of experts in forming our beliefs and conclusions.
3. Clarify the meaning of critical thinking.
4. Distinguish between weak-sense and strong-sense critical thinking.
5. Integrate the role of values with critical thinking.
6. Identify techniques for using critical thinking as a conversation stimulus.

THE NOISY, CONFUSED WORLD WE LIVE IN

This text encourages you to learn something we think can change your life for the better. That something is "critical thinking." But there is an imaginary world that some of us inhabit where there is no need at all for critical thinking. In this imaginary world several conditions prevail:

1. Anyone trying to persuade us of anything will always explain the disadvantages of what he or she wants us to do.
2. Any time we are confused about one of life's important questions, we can quickly find a dependable expert, authority, or wise person. Furthermore, these voices of knowledge will all agree with one another. In short, we need not be anxious about what to do or believe because the

些大师，然后倾听他们的忠告。

3. 我们在面临重要抉择的时候总是能平心静气、心无旁骛、举一反三并且兴致勃勃。

我们希望你能意识到，我们实际生活的这个世界，绝不会和上述想象的世界有任何相似之处。

在真实的世界里，各色人等总是坚称我们一定要按他们所说的那样去做，因为他们最懂行，知道我们该穿什么，该吃什么，该买什么，该相信什么。他们宣称自己掌握的是真理，我们必须要接受。他们口口声声说想要帮助我们。他们不肯对我们放手，让我们独自琢磨出自己到底应该做一个什么样的人。

若想知道练习瑜伽会有哪些负面效果，你只需进行五分钟的网络搜索，就可以发现以下值得警惕的问题：

1. 从潜意识中毫无控制地释放出恐惧、病态、性依恋和可怕的原型形态；
2. 伪死亡、伪精神病、困惑、积聚的焦虑、惊恐发作、自杀行为模式、抑郁、杀人欲望和自残欲望；
3. 躁狂、精神病、幻觉、抑郁和自杀倾向、精神崩溃、心率骤增、慢性疼痛和人格分裂。

你打算现在就终止瑜伽练习吗？

更糟糕的是，那些想说服我们的人，在努力塑造我们的行为习惯时并未做到光明磊落。他们所说的话至多不过是真假参半。保守派的人没有说清美国存在的严重贫富不均的状况，这种状况让很多人根本就难以自力更生、白手起家。那些贩卖最新款牛仔裤的商人没有告诉我们，他们之所以敢号称售价低廉，很可能是因为这是他们剥削亚洲的产业工人所得。药品公司告诉我们，我们需要服用黄色或蓝色药丸才能解决身体问题，但他们没有告诉我们的是，支持这些药物疗效的大部分研究，恰恰正是由售卖这些药物的公司资助的。你应该已经明白我们的意思了吧。

如果我们能依赖那些大师、专家来获得我们需要的答案，那么以上勾勒出的场景也构成不了什么太大的难题。如果他们能够为我们提供正确的解答，我们就能抵制住那些喧嚣的游说者。可惜当我们真正需要那些号称手里掌握了答案的人时，他们却根本就没准备好。他们常常错误百出，并且往往自相矛盾。在后文中我们还会强调这个事实对你和你的思维方式的重要性。

第13章将主要关注我们在应付这个纷繁复杂、令人困惑的世界时，如何常常有余于心，而力所不逮。有时候我们的大脑能够胜任惊人的既充满想象力又复杂艰难的任务，但

wise ones will have the answer. Our task is simply to locate and listen to them.
3. Our minds are calm, engaged, reflective, and curious whenever faced with an important choice.

We hope you realize that the world we actually live in is nothing like the Never-Never Land we just described.

In the real world, we are assaulted on all sides by others who insist that we *must* do what they tell us we should do. They know best. They know what we should wear, eat, buy, and believe. They claim to possess a truth that we must accept. They say they want to help us. They will not leave us alone to form our own understanding of who we should become.

In a five-minute Internet search for the possible negative effects of yoga, we found the following cautionary concerns:

1. uncontrolled release of fears, illnesses, sexual attachment, and terrifying archetypal formations from the unconscious;
2. pseudo death, pseudo psychosis, confusion, increased anxiety, panic attacks, suicidal patterns, depression, homicidal urges, and desire to self-mutilate; and
3. mania, psychosis, hallucination, depression and suicidal tendency, nervous breakdown, sudden surge of heart rate, chronic pain, and split personality.

Planning to end your yoga practice now?

To make matters worse, those trying to persuade us do not play fair as they try to shape us. They tell us half-truths at best. The socialist does not explain the dangers of a large government. The conservative does not explain to us the severe inequality in the United States that makes it very difficult for many of us to pull ourselves up by our bootstraps. The people selling us the latest jeans do not explain to us that the low prices they claim to charge are possible only because they exploit workers in Asia. The drug companies who tell us we need the yellow or blue pill to solve our problems do not explain to us that much of the research that supports the effectiveness of the drugs was paid for by the very companies selling you the drugs. We think you get the picture.

But the scenario we are sketching here would not be much of a problem if we could depend on the wise people, the experts, to have the answers we need. If they could give us the right answers we could resist the noisy persuaders. But when we need those who claim to have the answers, they are not there for us. They are often wrong, and they disagree among themselves. The next section emphasizes the significance of this reality for you and how you think.

Chapter 13 will focus on ways in which our brains often fail us as we try to handle our messy, confusing world. Sometimes our brains perform amazing,

是大脑常常受到丹尼尔·卡尼曼所谓的"快思考"或"系统1思维"原则的引导。我们的大脑常常依赖那些会将我们引入歧途的思维模式。"快思考"来得自然、迅速，并且特别容易受我们的情感掌控。乔纳森·海特把我们对情感的依赖描述为一头发怒的大象沿着乡间一路踩躏肆虐，而我们的理智则如同一个弱小的骑手，竭尽全力想要控制住这头大象狂热的撒野行为。

专家说得再动听，也不一定靠得住

我们已经说得很清楚，如果你指望专家当帮手，助你拨云见日，在纷至沓来的试图说服你的观点中理出一个头绪来，那你就等着大失所望吧。他们只是听起来比实际懂得多。他们自己似乎也深谙其中的道理，那就是只要他们让说出口的话听起来显得信誓旦旦，你就极有可能会相信他们的说法。所以，他们只挑你爱听的话说。

为了让你明白这一点，我们从大卫·弗里曼2010年出版的一本重要著作《大错特错：为什么专家总是靠不住》中举出三个例子。

1. 我们该不该避免阳光直射？美国疾病预防控制中心说，太阳紫外线的照射可能是诱发皮肤癌的一个最重要的因素。简单地说，就是要避免阳光直射。但是，且慢！世界卫生组织说，在世界范围内，紫外线的照射只是诱发疾病的一个微不足道的原因。然后，让我们更加困惑的是，他们又说在世界范围内，晒太阳少的人比经常晒太阳的人患病的概率更大。

2. 买个宠物可以让自己活得更健康一点，这种说法有没有道理？美国心脏协会说，很多研究显示，养宠物有助于促进主人的健康。但是，芬兰开展的一项可靠的研究发现，饲养宠物和健康状况恶化有着直接的联系。

3. 手机会不会产生有害的辐射？国际流行病研究院的院长说，"手机会产生有害的辐射"这种说法没有任何根据。但是南卡罗来纳州医院的一位专家对这个问题却有着截然相反的回答。他宣称，有足够的证据可以证明手机和癌症之间的关系，足以对手机使用者发出健康预警。

专家们多多少少给我们提供了合情合理的观点，为我们做出深思熟虑的决定提供了材料。但是我们自己必须要做能工巧匠，对那些观点和主张加以剪裁、取舍，将它们整合成自己的决定，这样才能做个有主见的人。

在这里我们务必要小心谨慎。毫无疑问，我们并不是说专家们根本就无关紧要。实际

imaginative, and complex tasks. But the human brain is frequently guided by what Daniel Kahneman calls "fast thinking" or "System 1 thinking." Our brains often rely on patterns of thinking that get us into trouble. Fast thinking is automatic, immediate, and typically controlled by our emotions. Jonathan Haidt describes our reliance on emotion as a raging elephant tearing through the countryside with our rational tendencies taking the form of a tiny rider trying desperately to control the elephant's passionate rampage.

EXPERTS CANNOT RESCUE US, DESPITE WHAT THEY SAY

We already made the point that if you expect to lean on experts as the tool with which to wade through the multitude of people wishing to own your mind, you are in for a big disappointment. They often sound as if they know far more than they do. They probably understand at some level that you are much more likely to listen to them when they sound certain about what they claim to be true. So, they give you what you want to hear.

But we want to drive this point home to you by providing three examples from David Freedman's important 2010 book, *Wrong: Why Experts Keep Failing Us*:

1. Should you stay out of the sun? The U.S. Centers for Disease Control and Prevention says that exposure to the sun's ultraviolet rays may be the most important factor influencing the development of skin cancer. In short, stay out of the sun. But wait. The World Health Organization says exposure to ultraviolet light is a minor contributor to disease in the world. Then to confuse us all the more they add that too little exposure to the sun may cause more disease in the world than does exposure to the sun.
2. Does it make sense to buy a pet as a means of having better health? The American Heart Association says that many studies have demonstrated the positive effect of pet ownership on the owner's health. However, a reliable study in Finland found that pet ownership is linked to poor health.
3. Do cell phones emit harmful radiation? The director of the International Epidemiology Institute says there is no basis for believing that cell phones produce harmful emissions. But an expert linked to a South Carolina hospital has a quite different response to this question. He claims there is sufficient evidence to justify a health advisory warning about the link between cell phones and cancer.

Experts provide us with more or less reasonable assertions. They give us the materials for a thoughtful decision. But we are the carpenters who must measure and construct those assertions into a decision that is ours.

We need to be very careful here. We are definitely not saying that experts cannot be helpful. Indeed, we cannot function without depending on people

上，如果不依靠那些知识比我们丰富、可以助我们一臂之力的人，我们将寸步难行。从某种意义上说，不管你现在多大程度上依赖专家学者，我们都鼓励你加倍重视他们的意见。只不过，我们要倾听来自不同领域的很多专家学者的意见，一边听一边评估判断，对这些意见进行整理，做出取舍，这一点你很快就会明白。我们倾听他们，是为了构建出自己的答案，而不是听了他们的话以后，马上就按他们说的去做，好像自己是只无助的羔羊，或者是个牵线的木偶。

有必要依靠自己的大脑

一旦我们对自己做决定时所处的环境有了清晰的把握，我们就要直面一个重大的责任：

我们必须要理性地掌控自己的信念和结论。与之相对的选择是，谁是我们最后遇到的人，我们就甘愿做谁的精神奴隶。

批判性思维会教你很多技能和态度，让你理性地找到对自己有意义的答案并为此感到自豪。批判性思维鼓励你倾听他人，向别人学习，同时掂量别人所说的话，看看它们的分量如何。如此，你将了解到我们必须要依赖他人，但又不能来者不拒，必须做出选择。由此，批判性思维会解放你的身心，让你有能力成为自己迈向未来道路上的指导者。

激发你的批判性思维

所谓批判性地倾听和阅读，即对自己耳闻目见的一切加以系统评价，然后做出回应。这需要一整套技能和态度。这些技能和态度都建立在一系列环环相扣的批判性问题的基础之上。我们会一个一个地学习这些问题，而我们的最终目标是将这些问题融会贯通，从而找到可以做出的最佳决断。理想的效果是，提出这些问题将会成为你不可或缺的一部分，而不仅是你在书上学会的一套本领。

我们在本书中使用的批判性思维这一术语，指由以下三个维度激活的评估技能：

1. 意识到一整套环环相扣的批判性问题；
2. 有能力在适当时机以适当方式提出并回答这些问题；
3. 有积极主动地使用这些批判性问题的强烈渴望。

本书的目标就是激发你朝这三个维度全面发展。

who we think might have knowledge we can use. In a sense, we are encouraging you to pay even more attention to experts than you might already give them. But, as will be clear soon, we need to listen to experts of many different kinds, sorting and discarding as we listen and evaluate. We listen to them to construct *our* answer. We do not listen to them to follow their advice, as if we were but a helpless lamb or a puppet on the expert's string.

THE NECESSITY OF RELYING ON OUR MIND

Once we have a clear grasp of where we live in the sense of the environment in which we make decisions, we come face to face with a heavy responsibility:

> WE MUST ASSERT RATIONAL CONTROL OF OUR BELIEFS AND CONCLUSIONS. THE ALTERNATIVE IS BEING THE MENTAL SLAVE OF THE LAST PERSON WE ENCOUNTERED.

Critical thinking teaches you skills and attitudes that make you proud to have rationally discovered answers that make sense to you. Critical thinking encourages you to listen to and learn from others, while at the same time weighing the quality of what others say. In this regard, you are learning that we must depend on others, but only selectively. Critical thinking thereby liberates you, empowering you to be the supervisor of who you are becoming.

CRITICAL THINKING TO THE RESCUE

Listening and reading critically—that is, reacting with systematic evaluation to what you have heard and read—requires a set of skills and attitudes. These skills and attitudes are built around a series of related critical questions. While we will learn these questions one by one, our goal is to be able to use them as a unit to identify the best decision available. Ideally, asking these questions will become part of who you are, not just something you studied in a book.

Critical thinking, as we will use the term, refers to evaluation skills activated by the following:

1. awareness of a set of interrelated critical questions;
2. ability to ask and answer these critical questions in an appropriate manner; and
3. desire to actively use the critical questions.

The goal of this text is to encourage you in all three of these dimensions.
Questions require the person being asked the question to do something in response. By our questions, we are saying to the person, "I am curious"; "I want to know more"; "help me." This request shows respect for the other person. Critical questions exist to inform and provide direction for all who hear them. In that respect, critical thinking begins with the desire to improve what

提出的问题需要被问对象做出回应。通过提问，我们传达给被问对象的是"我很好奇""我想多了解一点这方面的知识""请帮帮我"。这样的请求体现出我们对他们的尊重。批判性问题的提出是为了让所有听到问题的人明白要点，掌握方向。从这方面来说，批判性思维始于个体完善思想的强烈愿望。批判性问题同样有助于提高我们的书面和口头表达能力，因为它们能在以下几个方面助我们一臂之力：

1. 客观评价一篇文章或图书、杂志中以及网站上提供的证据，不盲从盲信；
2. 评判一场讲座或演说的水平高低；
3. 形成自己的论点；
4. 完成指定阅读任务后，撰写有理有据的论文；
5. 积极参与课堂讨论。

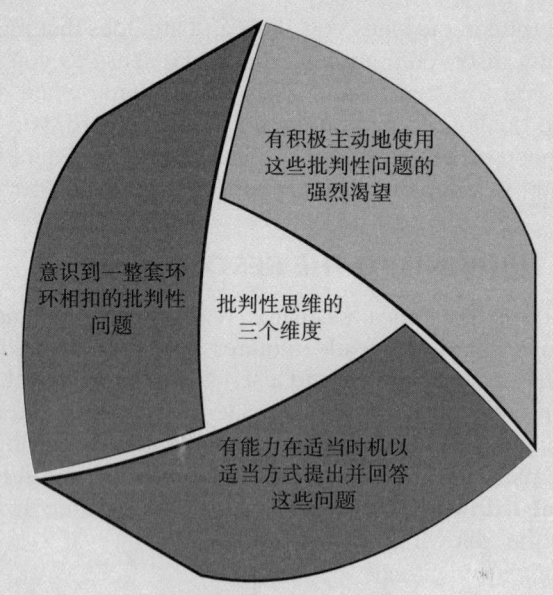

批判性思维的三个维度

⚠ **注意**：批判性思维包括意识到一整套环环相扣的评价性问题，加上在适当时机提出和回答这些问题的能力和意愿。

海绵式思维和淘金式思维

有种常见的思维方式因为类似于海绵放到水中的反应——充分吸收水分，而被称为

we think. The critical questions are also useful in improving your own writing and speaking because they will assist you when you

1. react critically to an essay or to evidence presented in a textbook, in a periodical, or on a Web site;
2. judge the quality of a lecture or a speech;
3. form a persuasive argument;
4. write an essay based on a reading assignment; or
5. participate in class.

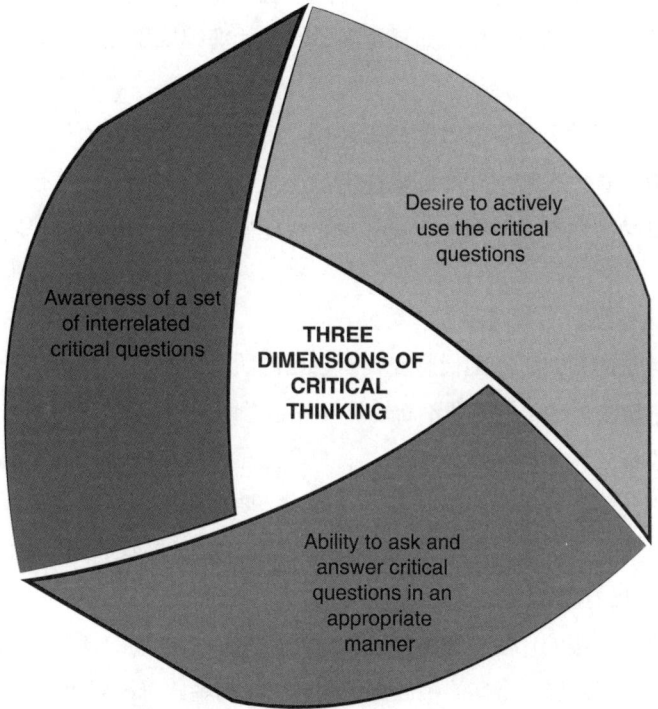

Three Dimensions of Critical Thinking

 ***Attention:** Critical thinking consists of an awareness of a set of interrelated evaluative questions, plus the ability and willingness to ask and answer them at appropriate times.*

THE SPONGE AND PANNING FOR GOLD: ALTERNATIVE THINKING STYLES

One common approach to thinking is similar to the way in which a sponge reacts to water: by absorbing. This popular approach has some clear advantages.

"海绵式思维"。这种流行的思维方式有以下两个显著优点：

第一，吸收外部世界的信息越多，你就越能体会到这个世界的千头万绪。你获取的知识将会为今后展开更复杂的思考奠定坚实的基础。

第二，海绵式思维相对而言比较被动，它并不需要你绞尽脑汁、冥思苦想，往往既轻松又快捷，特别是在你看到的材料组织得井井有条又生动有趣时，运用这种思维更是毫不费力。虽然大量吸收外部信息可以为你成为一个有思想的人提供一个有效的起点，但海绵式思维有个严重而又致命的缺陷：对于哪些信息和观点可以相信，哪些信息和观点应该摒弃，它提供不了任何判断方法。如果读者始终依赖海绵式思维，那么最后读到的内容是什么，他就会相信什么。

我们相信，你一定愿意自己掌握主动权，来选择该吸收什么和忽略什么。要做出这个取舍，你就得带着一种特别的态度去读书，即不断提问的态度。这种思维方式需要你积极主动地参与互动。作者不断向你兜售观点，而你应该随时准备与之辩论，尽管作者本人并不在场。

我们把这种互动的方式称为"淘金式思维"。淘金的过程为积极主动的读者和听众提供了一种可效法的模式，在他们想要判断自己的所见所闻有多大价值时可以借鉴。要想在对话的过程中披沙拣金，就需要不断地提问并思考问题的答案。

海绵式思维强调知识获取的结果，而淘金式思维则重视在获取知识的过程中积极和它展开互动。因此这两种思维方式其实可以互补。要想淘出智慧的金子，首先你的淘金盘里得有东西供你掂量才行。此外，要评价各种论点，我们必须掌握足够的知识，也就是说，得有一些信得过的见解才行。

采取淘金式思维的读者会怎样做呢？像采取海绵式思维的读者一样，他也希望通过阅读来获取新的知识，但两者间的相似点仅此而已。淘金式思维要求读者问自己一系列问题，这些问题旨在找出最佳的决定或最合理的信念。

采用淘金式思维的读者常常质疑作者为什么要提出各种各样的主张。他在书页的边缘写批注，提醒自己注意论证中存在的问题。他时刻和自己的阅读材料进行互动，意在批判性地评价所读的材料，并在客观评价的基础上得出自己的结论。

淘金式思维最重要的特点就是互动式参与，即作者和读者之间、演讲者和听众之间展开的对话。一个会思考和判断的人愿意赞同别人的观点，但他首先得为自己的问题找到一些令人信服的答案。

First, the more information you absorb about the world, the more capable you are of understanding its complexities. Knowledge you have acquired provides a foundation for more complicated thinking later.

A second advantage of the sponge approach is that it is relatively passive. Rather than requiring strenuous mental effort, it tends to be rather quick and easy, especially when the material is presented in a clear and interesting fashion. Though absorbing information provides a productive start toward becoming a thoughtful person, the sponge approach also has a serious and devastating disadvantage: It provides no method for deciding which information and opinions to believe and which to reject. If a reader relied on the sponge approach all the time, he would believe whatever he read last.

We think you would rather choose for yourself what to absorb and what to ignore. To make this choice, you must read with a special attitude—a question-asking attitude. Such a thinking style requires active participation. The writer is trying to speak to you, and you should try to talk back to him, even though he is not physically present.

We call this interactive approach the panning-for-gold style of thinking. The process of panning for gold provides a model for active readers and listeners as they try to determine the worth of what they read and hear. Distinguishing the gold from the gravel in a conversation requires you to ask frequent questions and to reflect on the answers.

The sponge approach emphasizes knowledge acquisition; the panning-for-gold approach stresses active interaction with knowledge as it is being acquired. Thus, the two approaches complement each other. To pan for intellectual gold, there must be something in your pan to evaluate. In addition, to evaluate arguments, we must possess knowledge, that is, dependable opinions.

What does the reader who takes the panning-for-gold approach do? Like the person using the sponge approach, she approaches her reading with the hope that she will acquire new knowledge. There the similarity ends. The panning-for-gold approach requires that the reader ask herself a number of questions designed to uncover the best available decisions or beliefs.

The reader who uses the panning-for-gold approach frequently questions why the author makes various claims. She writes notes to herself in the margins indicating problems with the reasoning. She continually interacts with the material. Her intent is to critically evaluate the material and formulate personal conclusions based on the evaluation.

The most important characteristic of the panning-for-gold approach is interactive involvement—a dialogue between the writer and the reader, or the speaker and the listener. As a critical thinker, you are willing to agree with others, but first you need some convincing answers to your questions.

- 我有没有问"为什么"别人要我相信某件事？
- 我在想到别人所说的话可能存在的问题时，有没有把它们记下来？
- 我有没有客观评价别人说过的话？
- 对于某一主题，我有没有在别人所说的合理内容的基础上形成自己的结论？

淘金式思维的心理检视表

别人话中的不合理之处并不会自动跳到你的眼前。作为读者或听众，你必须要积极主动地去查究才行。要做到这一点，你就需要不断地提问。最好的查缺补漏方法就是批判性地提问。这些问题的一个巨大好处是，即使你对当前讨论的问题知识有限，你仍然可以打破砂锅问到底。例如，即使你不是育儿专家，你也一样可以对日托中心的管理完善与否提出一些批判性问题。

弱势批判性思维和强势批判性思维

现在，在个人和社会的很多重大问题上，你都有了自己的观点。那么对以下这些问题，你自然也愿意确定自己的立场："酗酒是难于自控的疾病，还是有意为之的恶习？""贝拉克·奥巴马在任时是不是个成功的总统？"你会带着这些初始观点，对你听到和读到的内容加以思考。

Chapter 1 The Benefit and Manner of Asking the Right Questions

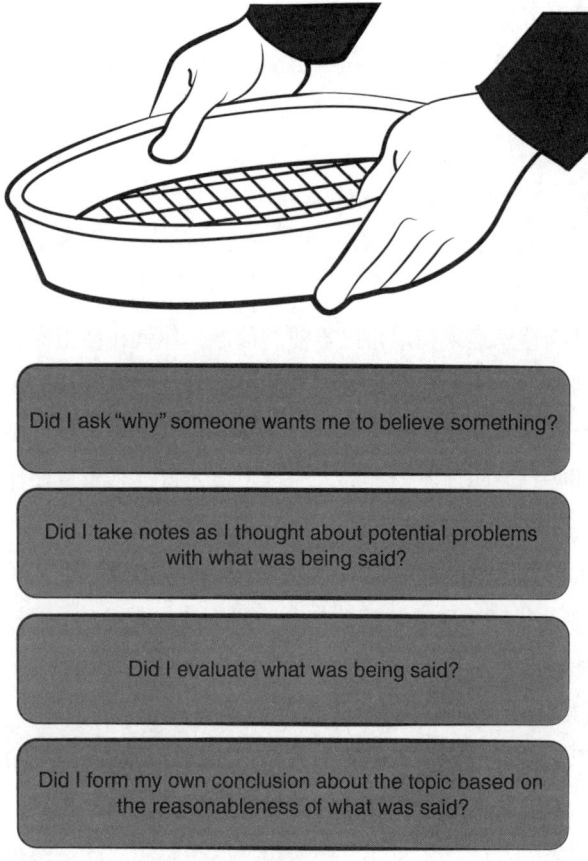

Mental Checklist for Panning for Gold

The inadequacies in what someone says will not always leap out at you. You must be an *active* reader and listener. You can do this by *asking questions*. The best search strategy is a critical-questioning strategy. A powerful advantage of these questions is that they permit you to ask probing questions even when you know very little about the topic being discussed. For example, you do not need to be an expert on child care to ask critical questions about the adequacy of day-care centers.

WEAK-SENSE AND STRONG-SENSE CRITICAL THINKING

Previous sections mentioned that you already have opinions about many personal and social issues. You are willing right now to take a position on such questions as, Is alcoholism a disease or willful misconduct? Was Barack Obama a successful president? You bring these initial opinions to what you hear and read.

批判性思维可以用来捍卫自己的信念，也可以用来对自己初始的信念加以评价和修正。理查德·保罗教授对弱势批判性思维和强势批判性思维的区分有助于我们理解批判性思维这两个对立的用途。

⚠ **注意：** 弱势批判性思维是利用批判性思维来捍卫自己当前的信念。强势批判性思维是利用批判性思维来评价所有的断言和信念，尤其是对自己的信念加以评价。

如果你利用批判性思维来捍卫自己当前的信念，你就在使用弱势批判性思维。为什么说这种思维是弱势的呢？因为以这种方式来使用批判性思维的技能，就意味着你根本不关心自己是否在接近真理和美德。弱势批判性思维的目的就是坚决抵制和驳倒那些不同的观点和论证。将把那些意见和你不一样的人驳得哑口无言、乖乖认输作为批判性思维的最终目标，会毁掉批判性思维潜在的人道和进步性的特征。

相反，强势批判性思维要求我们对所有的主张都提出批判性问题，包括对我们自己的主张。只有强迫自己批判性地看待初始信念，我们才能保证自己不自欺欺人和人云亦云。抱着初始信念死死不放自然容易，特别是很多人都持这样的信念时更是如此，可一旦我们选择走这条容易的道路，我们就极有可能犯下原本可以避免的错误。

强势批判性思维并不一定会迫使我们放弃初始信念。它有可能会让我们变得对初始信念更加坚定，因为对这些信念进行批判性的检视有时会让我们相信得更坚决。区分强势和弱势批判性思维的另一方法就是对比一下思想开放和思想封闭这两种状态。当我们思想开放的时候，我们欢迎别人对我们的信念展开批评，而当我们思想封闭的时候，我们就会坚决维护自己的各种信念，容不得别人说三道四。

只有当某个观点是我们千挑万选出来的——我们理解并且仔细评估过其他的观点，最终选择了这个观点，我们才会对所持的这个观点感到自豪。

勤于练习很重要

我们的目标就是要让你的学习过程变得尽可能简单。不过，批判性思维习惯的养成，首先依赖于大量的练习。

除了本章以外，后面每章末尾的练习题和参考答案是本书不可或缺的重要内容。我们所提供的答案并不是唯一正确的解答，但它们确实提供了一些例证，教我们怎样将各种定义和提问的技能在实践中加以运用。对于每章结尾部分的第三篇文章，我们有意不提供参考答案，目的就是让读者有机会运用在该章中所学的知识自己去找出答案。

Critical thinking can be used to either (1) defend or (2) evaluate and reconsider your initial beliefs. Professor Richard Paul's distinction between weak-sense and strong-sense critical thinking helps us appreciate these two antagonistic uses of critical thinking.

 ***Attention:** Weak-sense critical thinking is the use of critical thinking to defend your current beliefs. Strong-sense critical thinking is the use of the same skills to evaluate all claims and beliefs, especially your own.*

If you approach critical thinking as a method for defending your present beliefs, you are engaged in *weak-sense critical thinking*. Why is it weak? To use critical-thinking skills in this manner is to be unconcerned with moving toward truth or virtue. The purpose of weak-sense critical thinking is to resist and annihilate opinions and reasoning different from yours. To see domination and victory over those who disagree with you as the objective of critical thinking is to ruin the potentially humane and progressive aspects of critical thinking.

In contrast, *strong-sense critical thinking* requires us to apply the critical questions to all claims, including our own. By forcing ourselves to look critically at our initial beliefs, we help protect ourselves against self-deception and conformity. It is easy to just stick with current beliefs, particularly when many people share them. But when we take this easy road, we run the strong risk of making mistakes we could otherwise avoid.

Strong-sense critical thinking does not necessarily force us to give up our initial beliefs. It can provide a basis for strengthening them because critical examination of those beliefs will sometimes reinforce our original commitment to them. Another way of thinking about this distinction is to contrast open- and closed-mindedness. When my mind is open, it welcomes criticism of my own beliefs. But when my mind is closed, the beliefs I have are going to be the ones I keep.

To feel proud of a particular opinion, it should be one we have selected—selected from alternative opinions that we have understood and evaluated.

The Importance of Practice

Our goal is to make your learning as simple as possible. However, the habit of critical thinking will initially take a lot of practice.

The practice exercises and sample responses at the end of subsequent chapters, except this introductory chapter, are an important part of this text. Our answers are not necessarily the only correct ones, but they do provide illustrations of how to apply the definitions and question-asking skills. We intentionally failed to provide sample answers for the third passage at the end of each chapter. Our objective is to give you the opportunity to struggle with the answer using your knowledge of the chapter we have just studied.

批判性思维和与人交往

你可以把其他人当成自己最有价值的资源，当成自己最终拥有的事实、观点和结论的基础。在孕育出你的众多结论的"大家庭"中，其他人也是其中的一部分，他们不可或缺并持续发挥影响。这个大家庭的主题就是相互联系。

与其他人的互动所起的作用，由你的价值观以及你从那些与你交往之人身上体会到的价值观所塑造。在你发现价值观对形成结论的重要性之前，最好先对价值观是什么有一个初步的了解。本书所谓的价值观，是指人们认为有价值的观念。你会发现，正是人们重视的抽象观念对其选择和行为产生了重大影响。

通常，我们偏好的目标、体验和行动正是来自我们重视的观念。例如，我们也许会选择性地去做一些事情，让自己有机会接触到重要人物。我们之所以重视"重要人物"（此为具体观念），是因为我们重视"身份地位"（此为抽象观念）。我们在本章中使用"价值观"这个词时，指的是抽象观念，代表某人心中重要和美好的事物。

⚠ **注意**：价值观指的是人们认为较有价值的观念（常常不曾言明）。它们树立了一定的行为准则，我们据此来衡量人类行为的品质高下。

为了让你更熟悉价值观这个概念，请你先写下自己的一些价值观。注意尽量不要写一些具体的人名，也不要写一些可以触摸到的物体或具体的行动。比萨饼或网球对你来说也许很重要，但最能影响你在面对有争议的社会问题时所做出的抉择和行为的，是你重视的抽象观念。比如说，你对于"安乐死"的反对态度离不开你对"人的生命的神圣性"这个抽象概念的重视，"人的生命的神圣性"这个抽象观念就是你的价值观。在你列举种种价值观的时候，请注意那些能在很多方面影响到你的观点和行为的重要价值观。

你在列举价值观的时候是否遇到了困难？我们可以为你提供一些帮助。价值观就是我们自己所认可的行为准则，我们希望他人的言行能与之相一致。当我们期望政治代表们能"实话实说"时，我们向他们和我们自身暗示的信息就是：诚实是我们最重视的价值观之一。你可以问问自己，你希望自己的朋友都是些什么样的人，你希望自己的孩子被培养出什么样的行为准则。对诸如此类问题的求解可以帮助你拓宽对价值观的理解。

让我们提醒一下自己，关于价值观的知识是怎样和批判性思维的社会属性产生联系的。虽然我们时常要求自己必须耐心倾听那些和我们价值观倾向不同的人的观点，但是价值观所带来的最常见的社会联系是相似性。那些把个人责任当成重要价值观的人，必然会去寻

CRITICAL THINKING AND OTHER PEOPLE

Think of other people as your most valuable resource, the basis for the facts, opinions, and conclusions that you will eventually have. In an important and ongoing manner, other people are part of your extended family, those who nurture your conclusions. The theme here is connectedness.

How these interactions work is shaped by your values and the values you perceive in those with whom you interact. Before you can discover the importance of values in shaping conclusions, you must have some understanding of what a value is. *Values*, as we will use the term, are ideas that someone thinks are worthwhile. You will find that it is the importance one assigns to *abstract ideas* that has the major influence on one's choices and behavior.

Usually objects, experiences, and actions are desired because of some idea we value. For example, we may choose to do things that provide us with contacts with important people. We value "important people" (concrete idea) because we value "status" (abstract idea). When we use the word *value* in this chapter, we will be referring to an (abstract) idea representing what someone thinks is important and good.

 Attention: *Values are ideas, often unstated, that people see as worthwhile. They provide standards of conduct by which we measure the quality of human behavior.*

To better familiarize yourself with values, write down some of your own values. Try to avoid writing down the names of people, tangible objects, or actions. Pizza and tennis may be important to you, but it is the importance you assign to abstract ideas that most influences your choices and behavior concerning controversial public issues. Your willingness to argue against assisted suicide, for instance, is strongly related to the importance you assign to the sanctity of human life—an abstract idea. As you create your list of values, focus on those that are so significant that they affect your opinions and behavior in many ways.

Do you have problems making your list? We can provide some help. Values are *standards of conduct* that we endorse and expect people to meet. When we expect our political representatives to "tell the truth," we are indicating to them and to ourselves that honesty is one of our most cherished values. Ask yourself how you expect your friends to be. What standards of conduct would you want your children to develop? Answers to these questions should help you enlarge your understanding of values.

Let's remind ourselves how knowledge about values relates to the social nature of critical thinking. While we must require ourselves to listen carefully to those who have different value priorities than our own, the most obvious social link established by values is similarity. Those of us who see individual

找同样相信"更好的个人选择可以解决人类的大多数问题"的人为伴，和他们惺惺相惜。可见，我们许多最有价值的社会交往或学习经历都始于与拥有类似价值观的人之间的交流。鉴于此，我们面对的一个巨大挑战，就是尽量去理解那些价值观倾向和我们不同的人的论证方式。

敢冒险、有志气、自主决断、让人放心、出类拔萃、有正义心、有理性、能包容、自发主动，这些价值观也许对我们而言都很重要，但其他通情达理之人很可能也有一套重要的价值观，并且他们的价值观与上述价值观中的很多条都存在冲突。我们最常见的倾向就是只愿听那些价值观倾向和我们相似的人的话，我们需要下大力气去抵制这种倾向。我们必须要与这种倾向做斗争。

批判性思维者拥有的主要价值观

本书的主要目标是帮你成为批判性思维者。而作为批判性思维者，你的目标就是不断追求更好的结论、更好的信念以及更好的决定。有些价值观能让你加速迈向这些目标，而另一些价值观则难以奏效。理解和欣赏那些批判性思维者持有的主要价值观，可以让你获得一些心智力量，你可以用这些力量来不断提醒自己：密切关注那些与你价值观倾向不同的人的观点很有必要。下面我们就一起看看这些主要价值观。

1. 自主决断。乍一看，这个价值观好像和鼓励人们注意不同的视角有点扯不上关系。一方面它要我们形成自己的结论，另一方面它又鼓励我们去寻找或倾听那些不属于我们自己的见解，这可能吗？当然可能！这两者并不矛盾！在追求自主决断的过程中到底该使用哪些原材料呢？我们当然都想从最多、最全的可能性当中进行选择，否则就有可能遗漏掉某个本该选中的重要选项，仔细听听那些和我们价值倾向不同的人的话，也许就不会有此遗憾。举例来说，美国民主党人如果只听其他民主党人的意见，势必会铸成大错。

2. 好奇心。要想充分利用淘金式思维来立身处世，你就需要多听、多看，真正用心去听、用心去看。其他人有力量驱使你不断向前，将你从当前狭小的知识面里解放出来。要想成为一个会思考和判断的人，你就需要对自己遇到的一切不断提问。你从别人那里能获得一些真知灼见——符合本书倡导的严密论证的一系列标准。

3. 谦恭有礼。世界上最聪明的人每周都会犯下很多错误，认识到这一点，我们就能更积极主动地和其他人交流。我们当中自然有人具备其他人没有的真知灼见，但我们每个人能做到的事情都很有限，当我们真诚地自我反省的时候，就会印证苏格拉底

responsibility as an extremely important value tend to be comfortable with and to seek out those who similarly believe that *improved personal choices* are the solution to most human problems. Hence, many of our most valuable social interactions or learning experiences start with communications with those who have similar value priorities. Our huge challenge in this regard is to make ourselves work hard to understand the reasoning of those whose value priorities *differ* from ours.

While adventure, ambition, autonomy, comfort, excellence, justice, rationality, tolerance, and spontaneity may be important values to us, it is quite likely that other reasonable people will have important values that conflict with many of these. Our normal tendency to listen to only those with similar value priorities needs our active resistance. We have to fight against the tendency.

Primary Values of a Critical Thinker

This text is dedicated to helping you become a critical thinker. As a critical thinker, you will be pursuing *better* conclusions, *better* beliefs, and *better* decisions. Certain values advance your effort to do so; others do not. By knowing and appreciating the primary values of a critical thinker, you have some mental muscle that you can use to remind yourself of the necessity of your paying close attention to those who do not share your value priorities. Let's examine these primary values.

1. Autonomy. At first this value may seem as if it has little to do with encouraging people to pay attention to those with different perspectives. How does a drive to form one's own conclusions encourage us in any fashion to seek and listen to views that are not our own? Aha! And what raw material should you use in pursuing this autonomy? Surely, we all want to pick and choose from the widest possible array of possibilities; otherwise, we may miss the one decision or option that we *would have chosen* if only we had paid attention to those who did not share our value priorities. For example, Democrats make a huge mistake if they listen only to other Democrats.
2. Curiosity. To take advantage of the panning-for-gold method of living your life, you need to listen and read, *really* listen and read. Other people have the power to move you forward, to liberate you from your current condition of partial knowledge. To be a critical thinker requires you to then ask questions about what you have encountered. Part of what you gain from other people are their insights and understanding, **when what they have to offer meets the standards of good reasoning** that you will learn in this text.
3. Humility. Recognizing that even the smartest person in the world makes many mistakes each week provides the ideal platform for engaging actively with other people. Certainly some of us have insights that others

说过的那句话："我唯一知道的事就是我一无所知。"一旦接受这个现实，我们就能更好地认识到，我们与他人之间的交往，至少能填补一部分我们当前认识的空缺。而且，谦恭有礼可以让我们避免一个常见的批判性思维的障碍：相信凡是和自己意见不同的人都心存偏见，只有自己才客观公允。

4. 发自内心地尊重严密的论证。虽然我们想尊重和倾听其他人的观点，但这并不是说所有的观点和结论都不分轩轾，都有价值。你从本书中学到的那些批判性问题，将会为你提供一个基本框架，帮你精挑细选，从所有想对你施加影响的人里择优、汰劣。一旦你发现说理透彻、论证严密的人，不论其肤色、年龄、财富、国籍如何，都要信赖他的观点，直到更加透彻、严谨的论证出现为止。

总而言之，你可以基于自己的观点自信行事，对自己的结论不要犹豫不定，但也要留有一点余地，让自己时不时地反省一下：我有没有可能是错的呢？

因为这些都是你自己的观点，所以你自然会对它们另眼相看，加倍呵护。我们可以听一听政治讽刺家斯蒂芬·科尔伯特怎样嘲笑我们的这个态度："我并不是什么事实的拥趸，你看，事实还会发生改变，可是我的观点从来都不会变，不论它面对的是什么样的事实。"

任何下决心死抱自己的结论不放手的人，都有可能找到许多理由来证明自己的观点。这种论证方式被称作"操控型论证"，意思就是这种论证方式经过事先挑选，以便得出某个特定结论。

让对话有效进行下去

批判性思维是一种社会性活动，当我们对别人的信念和结论不断提问时，要考虑到他们可能会有什么样的反应。只要和我们交流的人也认同批判性思维的主要价值观，他们就会接受我们的提问，因为这些提问证明大家都是同道中人，都在为共同的问题寻找更好的答案。不过，除了这种大家一起成长的天赐良机，你还可能遇到其他类型的社交互动。

很多人并不喜欢别人对他们的思考过程逐一提问，他们常把别人的提问当成不怀好意，没事找碴。有的人可能会这样想：他为什么老是问我这些难应付的问题？他为什么就是不肯接受我的观点？如果你迫切地想要多了解一点知识，而其他人的反应却是责问你为什么要这样刻薄、小气，那么你千万不要觉得惊讶。很多人就是不习惯这样的情形，不习惯一些人对某个观点的来龙去脉特别好奇、一定要打听清楚。

do not have, but each of us is very limited in what we can do, and at honest moments, we echo Socrates when he said that he knew that he did not know. Once we accept this reality, we can better recognize that our experiences with other people can fill in at least a few of the gaps in our present understanding. Also, a sense of humility keeps us from avoiding a very common obstacle to critical thinking, the belief that "those who disagree with me are biased, but I am not."

4. Respect for good reasoning wherever you find it. While we want to respect and listen to other voices, all conclusions or opinions are not equally worthwhile. The critical questions you will learn as you study this text will provide a framework to assist you in picking and choosing from among all the people trying to influence you. When you find strong reasoning, regardless of the race, age, wealth, or citizenship of the speaker or writer, rely on it until a better set of reasoning comes along.

By all means, act with confidence based on your beliefs, but hold your conclusions with only that degree of firmness that permits you to still wonder to yourself, "Might I be wrong?"

They are *your* opinions, and you quite understandably feel protective of them. Listen as political satirist Stephen Colbert mocks this attitude of ours: "I'm not a fan of facts. You see, the facts can change, but my opinion will never change, no matter what the facts are."

Anyone determined to keep the conclusions he already has may well use reasons to justify his opinion. However, this kind of reasoning is called "managed reasoning," meaning that the reasoning is being selected so as to reach a particular conclusion.

KEEPING THE CONVERSATION GOING

Because critical thinking is a social activity, we need to consider how other people are likely to react to us when we ask them questions about their beliefs and conclusions. As long as we are interacting with others who share the primary values of critical thinking, our questions will be received as evidence that we are a partner in the search for better answers to the questions we share. But that terrific opportunity to grow together is not going to be the only kind of social interaction you will have.

Many people are not eager to have their thinking questioned; often, they experience questioning as annoying and unfriendly. Some may wonder, "Why is she asking me all these challenging questions? Why does she not just agree with me?" Don't be surprised if someone reacts to your quest to learn more by asking you why you are being so mean. Many people are unaccustomed to situations where someone is so excited to know more about why a particular viewpoint is held.

人们对于论证的惯常理解 © Viorel Sima/Shutterstock

以实现批判性思维为目标，论证就会大不一样。因为我们将论证当成一种机制，依靠这种机制，我们可以充实和提炼当前的结论，所以我们使用的"论证"这个概念的含义也就变得迥然不同。一个论证是由两种形式的陈述结合在一起构成的：一个结论和貌似支持这一结论的各种理由。正是各种理由和结论之间的紧密结合构建起一个人的论证。我们之所以提供这样的论证，是因为我们关心人们的生活，关心人们的信念。我们能不断取得进步，也得益于其他人足够关心我们，提供各种论证，并且客观评价我们所做的论证。只有这样，我们才能成为有思想的人。

最重要的是，当你使用批判性思维的各种技能时，一定要让别人明白你抱着学习的态度。此外，你要让他们相信你的本意是好的，如果你和他们的意见有分歧，不管这些分歧多么重大和严肃，都不必发展成口诛笔伐来收场。下面我们列举一些谈话策略，使用这些策略，可以让谈话一直继续下去：

1. 不妨问一下"你是不是说……"，确保你完全明白对方所说的话是什么意思。
2. 问一下对方，有没有什么证据可以让他改变想法。
3. 提议大家休息一会儿，这样双方都可借此找出支撑自己结论的最佳证据。
4. 问一下对方，为什么他认为你所倚赖的证据这么薄弱。
5. 尽量弥合分歧。如果你把对方最好的理由拿过来跟你最好的理由放到一起，能不能得出一个双方都可以接受的新结论呢？

Common Understanding of an Argument © Viorel Sima/Shutterstock

For purposes of critical thinking, an argument is altogether something else. Because we see argument as the mechanism whereby we fertilize and prune our current conclusions, we will use the concept in a very different manner. An *argument* is a combination of two forms of statements: a conclusion and the reasons allegedly supporting it. The partnership between reasons and conclusion establishes a person's argument. It is something we provide because we care about how people live their lives and what they believe. Our continual improvement depends on someone's caring enough about us to offer us arguments and to evaluate the ones we make. Only then will we be able to develop as thoughtful people.

Above all else, when you use your critical-thinking skills, make it clear to other people that you want to learn. Furthermore, give them assurances that you wish them well and that any disagreement you have with them, as serious and important as that disagreement might be, need not result in a verbal bloodbath. What follows are a few verbal strategies that you can use to keep the conversation moving:

1. Try to clarify your understanding of what the other person intends by asking, "Did I hear you say?"
2. Ask the other person whether there is any evidence that would cause him to change his mind.
3. Suggest a time-out in which each of you will try to find the very best evidence for the belief you hold.
4. Ask why the person thinks the evidence on which you are relying is so weak.
5. Try to come together. If you take that person's best reasons and put them together with your best reasons, is there some conclusion that both of you could embrace?

6. 寻找一些共同的价值观，或者双方都认可的其他结论，以此为基础，找出双方谈话中产生分歧的起始点。

7. 好奇心再强，也要表现得体贴他人且平静随和，一旦讨论的语气升温，你就要不断提醒自己，你的目的是虚心学习，而不是舌战群英。

8. 表情和动作都要表现出谦恭的样子，不要摆出一副全知全能、目空一切的架势。

营造友好的交流氛围

在写作和演讲的时候，你会面临一个重要的选择，那就是决定为自己的读者和听众营造一种什么样的氛围。你会不会做出这样一种选择：一旦有人不同意你的观点，气氛马上就变得剑拔弩张？在当前这种极化的环境下，这样一触即发的诱惑是非常大的。我们可以看看美国大选季里大家采用的策略，也就是《每日秀》主持人乔恩·斯图尔特在节目里揶揄过的策略，他说："我不同意你的观点，但我相信你肯定不是希特勒。"

本着斯图尔特这句话的精神，你可以选择营造这样一种氛围，让理智的人能够体面而大方地表达不同的看法和见解——一种欢迎讨论和提问的氛围。我们当然提倡这样的方式，但实话实说，确实有一些难以抗拒的理由，让我们写作的语气变得不容置疑，让我们把批判性的思考者拒之门外，甚至将他们一竿子全部撂倒。

首先，遇到一个难回答的问题，将它直接推翻确实比仔细斟酌之后再谨慎回答要容易得多。而且，这样做还能让你显得一言九鼎，让听众没胆量挑战你的权威判断。此外，这种写作风格有时还显得轻松有趣。不知你有没有读过"毒舌"的影评、书评以及对唱片或游戏的评论，并为之击节称赏？

我们可以看看下面这段对 2009 年票房大获成功的影片《变形金刚 2：卷土重来》的影评，注意其中的语气和措辞。流行电影评论家罗杰·埃伯特这样写道：

> 要是你想省下这张电影票的钱，那就干脆走进厨房，播放一个男子合唱团演唱的地狱之歌，让一个孩子胡乱敲打锅碗瓢盆。然后闭上眼睛，动用点想象力就完事了。⊖

面对一个做出这样的评价的人，你可以尝试去说服他，让他冷静下来重新考虑一下。

像批判性思维者那样写作和发言

本章已接近尾声，我们希望大家都能清醒地认识到，轻松而又简单地回答复杂问题的

⊖ 罗杰·埃伯特，《变形金刚：卷土重来》，2009 年 6 月 23 日。

6. Search for common values or other shared conclusions to serve as a basis for determining where the disagreement first appeared in your conversation.
7. Try to present a model of caring and calm curiosity; as soon as the verbal heat turns up, try to remind yourselves that you are learners, not warriors.
8. Make certain that your face and body suggest humility, rather than the demeanor of a know-it-all.

Creating a Friendly Environment for Communication

As a writer or speaker, you are faced with an important choice. You have to decide the type of environment you'll create for your audience. Will you choose one that is hostile to people who disagree with your conclusions? In the current polarized climate, the temptation is great. Just look at the tactics employed during the American election season—the tactics the *Daily Show's* Jon Stewart mocked when he said, "I disagree with you, but I am pretty sure you're not Hitler."

In the spirit of this Jon Stewart quote, you could choose to create an environment in which reasonable people can productively and respectfully disagree—an environment that welcomes discussion and question-asking. Of course we prefer this approach, but let's be honest: There are some compelling reasons to write in a tone that excludes, even shoots down, critical thinkers.

First, it's easier to shoot down a hard question than to consider and respond to it. Plus, you'll surely sound authoritative, daring your audience to challenge your expert judgment. Not to mention that this writing style can even be fun. Have you ever read and enjoyed a vicious review of a movie, book, album, or video game?

Take a look at the tone and word choice in this review of the 2009 box office success *Transformers: Revenge of the Fallen*. Popular film critic Roger Ebert suggests:

> If you want to save yourself the ticket price, go into the kitchen, cue up a male choir singing the music of hell, and get a kid to start banging pots and pans together. Then close your eyes and use your imagination.*

Just try to convince a person who makes that kind of assessment that he should calm down and reconsider!

WRITING AND SPEAKING AS CRITICAL THINKERS

As we come to the end of this chapter, we hope it is clear that a Never-Never Land of easy answers to complex questions does not exist. Rather, we are under persistent assault by others who want to tell us what to believe, what choices to make, and how to live. Critical thinking is our defense against this

---

*Roger Ebert. Transformers: Revenge of the Fallen. June 23, 2009.

理想方式是不存在的。恰恰相反，我们持续不断地受到他人的抨击，他们想告诉我们应该相信什么，应该选择什么，应该怎样生活。批判性思维是我们难得的防御手段，它能帮助我们抵御那些老奸巨猾、固执己见、花言巧语的敌人。最有效的防御手段之一就是提出批判性问题。

记住：批判性思维通常是对别人所述或所著的一种训练有素的反应。

作为批判性思维者，我们当然希望自己在阐述论点时，能做到平允公正、深思熟虑。在我们试图说服别人相信我们的观点是正确的时，我们不希望写出推理不当的论文或大声喊叫未经检验的观点。我们也不希望只是鹦鹉学舌地重复别人的论点，无论它论证得多么完美无瑕。

让我们来看看本章建议的两种不同的策略，借助这两种策略，我们可以在日常的发言和写作中提高批判性思维的有效性。

1. 在演讲和写作中识别和使用语气；
2. 避免"过度依赖"专家。

倘若我们能有效运用这些策略，就能增加别人倾听我们的机会，同时也能鼓励我们发展出自己的观点和想法。

说话和写作过程中，语气对于吸引观众至关重要。咄咄逼人、狭隘偏执、武断专横的语气很容易使彼此疏远、产生隔阂；相反，好的语气会在我们的听众周围编织起一张网，通过深思熟虑和明智理性的论证把他们和我们紧密地联系在一起。我们很容易对自己拥有的信仰充满激情，但也很容易在努力分享我们的激情时，疏远了读者或听众。举个例子，还记得埃伯特把电影《变形金刚》比作地狱里的合唱团吗？当然，强大的视觉语言可以为其书评招徕读者，但也有可能疏远那些欣赏（和尊重）更加深思熟虑和平衡淡然语气的读者。

我们说话或写作的方式会对听众和读者如何"听"产生巨大的影响。如果埃伯特说"你倘若喜欢震撼、嘈杂的电影，《变形金刚》那种电影正中你的下怀"，相比于他把《变形金刚》比作来自地狱的合唱团，我们会有不同的反应吗？同样，在更重要的论证中，我们可以换一种方式来提出我们的批评，表明我们对不同的观点和价值观给予了周密的考虑。

当我们说起或写到某人"愚蠢而又无知"时，语气会让人不舒服，甚至可能会让我们的听众或读者，特别是持有不同意见者，把我们拒之千里之外。一旦开骂，我们就降低了自己主张的档次，把它等同于"校园呐喊节"上的叫骂：谁喊得声音最大，谁想出最粗鲁的骂名，谁就赢了。相反，让我们试着用强势批判性思维的语言来塑造我们的论点。这种方式会让听众或读者更有可能"听见"我们。如果我们说话或写作没有接受对象，那我们这样做又有什么意义呢？

cunning, constant, and incredibly persuasive enemy. And asking critical questions is one of the most effective tools in our defensive arsenal.

Remember: Critical thinking is ordinarily a trained reaction to what others are saying or writing.

As critical thinkers, we certainly want to be fair and thoughtful when making our own arguments. Once we are in the role of trying to persuade others that our views are correct, we don't want to go around writing poorly reasoned papers or loudly shouting out unexamined opinions. Neither do we want to simply parrot the arguments of others, no matter how well argued.

Let's look at two different strategies suggested in this chapter through which we can increase our critical-thinking effectiveness in our writing and speaking:

1. recognizing and using tone in our speaking and writing and
2. avoiding "sponging" off the experts.

If these strategies are used effectively, we increase our chances that someone will listen to us while also encouraging the development of our own opinions and ideas.

Tone is crucial in engaging our audience when we speak and write. Tone can easily alienate when it is aggressive, intolerant, and judgmental, or it can weave a web around our listeners, binding them to us with a combination of thoughtful and rational reasoning. It's easy to become passionate about a belief we have and, in our effort to share our passion, to alienate readers or listeners. For example, remember the quote from Ebert at the end of this chapter comparing the movie *Transformers* to a choir in hell? Certainly the powerful visual language sells reviews, but it also has the potential to alienate readers who would appreciate (and respect) a more considered and balanced tone.

How we say or write something can make a big difference in how our listeners "hear" us. If Ebert had said, "If you like jarring and noisy movies, *Transformers* will be your kind of film," might we have reacted differently than when he compared it to a choir from hell? Similarly, in more important arguments we can present our criticisms in a way that shows we have given extensive consideration to different perspectives and values.

When we say or write that someone is "stupid and uninformed," the tone is off-putting and will probably go far in making sure that our audience, particularly if they disagree with the claim, tune us out. Once we start calling people names, we've reduced our claim to a schoolyard shouting fest: whoever screams loudest or thinks up the rudest names wins. Instead, let's try to shape our arguments in language that suggests strong-sense critical thinking. This approach will make our listeners much more likely to "hear" us. If no one is listening when we speak or write, what point is there in our doing it?

Tone also can suggest that we are intolerant of the other side's claims, rather than showing that we have asked questions regarding all relevant conclusions and used those questions to evaluate our own beliefs. Remind yourself

有时，某种语气会表明我们只是对对方的主张不能容忍，而非我们对所有相关结论都提出了问题，并用这些问题来评估我们自己的信念。提醒你自己一下，理查德·费曼是20世纪最著名的物理学家之一，他曾说过，当一个科学家有了一个想法时，他的第一步不是走出去为自己的想法寻找证据，而是思考想法中每一个可能的错误。

在写作和发言中，娴熟运用强势批判性思维的一个好方法就是在得出论点之前，充分论证自己的主张并从反面对这个主张加以辩驳。如果我们认为论据站不住脚，那么在我们为自己的主张论证之前，就需要寻找其他观点来支撑。现在，持有不同观点的读者或听众会聚精会神地关注我们，也许会想："这个人已经考虑过我的立场了。也许他的话值得一听。"

在写作和发言中运用批判性思维的另一个步骤就是有效地将专家的意见融入我们自己的声音中。由于缺乏自信、害怕犯错、慵懒随性甚至蒙昧无知，我们常常"过度依赖"专家。我们背诵并引用他人的作品，而不是自己辛勤工作，加深理解并清晰表达自己的观点。想一想，我们是否常常读到别人说过的话，被他们分析问题的那种鞭辟入里、入木三分的神来之笔折服、震撼，可能会对自己说："我要表达的就是这个意思！"我们发现的精彩名言也被揉进了我们的写作中。或者，与人争论时，我们是否常常把别人的观点当成自己的而不假思索地说出来？这么做很简单，而且听起来也不错。但我们必须记住，虽然鹦鹉学舌并宣称那些观点是我们自己的确实易如反掌，但这样做就标志着我们不是一个好的批判性思维者。

那么，我们如何有效地防止自己在写作和发言中"依赖"专家呢？如果我们发现自己在反复引用专家的话或者频繁地转述别人的观点，我们可以退后一步，问自己："我是怎么想的？为什么？"引用和转述在推理中占有一席之地，但仅能用来作为自己主张的佐证。例如，读完埃伯特对《变形金刚》的评论后，我们可能会忍不住在发言中说："《变形金刚》是有史以来最差的电影。"我们可以引用埃伯特的话作为佐证：他的地狱合唱团的例子非常鲜活、生动。

现在让我们把埃伯特的观点放在一边，转而问问自己："我不喜欢《变形金刚》的哪一点？"也许是主演马克·沃尔伯格单一的表演，或者是永无止境的动作场面，或者是落入窠臼的对话。在我们的发言和写作中，我们可以使用埃伯特作为精彩的后援来深化我们的观点。作为批判性思维者，我们不仅要培养成为主动读者或听众的技能，还要经历成为主动发言者和写作者的激动人心的过程。

将他人的观点有机融入我们自己的立场中，并有分寸地把控我们的语气，这两种方法都可以让我们将批判性思维纳入自己的写作和思考中。如果我们两者都做到了，那么我们更有可能被其他人"听见"。此外，我们还可以展开高效的讨论，这种讨论能让我们有机会学习到更多的东西，推动我们的探索和论证向前发展，让我们保持对话而不终止。

that one of the most famous physicists of the 20th century, Richard Feynman, argued that when a scientist has an idea, one of her first steps is not to go out and find proof for her idea but to consider every conceivable way she might be wrong.

In our writing and speech, a good way to approach strong-sense critical thinking is to develop both the claims and the arguments against the claim (counterarguments) fully before developing a thesis. If we believe that tanning beds rock, then we need to hunt for support for other perspectives even before we provide the support for our own claim. Now our reader or listener who disagrees with us will sit up and take notice, perhaps thinking, "This person has given my position some thought. Maybe he is worth hearing out."

Another step toward using critical thinking in our speaking and writing is effectively incorporating expert opinions into our own voice. Out of a lack of confidence, fear of being wrong, laziness, and even ignorance, we often "sponge" off the experts. We memorize and quote the work of others rather than doing the hard work of understanding and voicing clearly our own opinions. How often have we read something someone else said and been blown away by how well written it is, saying to ourselves, "That's exactly what I meant!" And that excellent quote we discovered then gets inserted into our own writing. Or, how frequently do we find ourselves stating someone else's opinion as our own when arguing with others? It's easy, and it sounds good too. But we have to remember, while it's certainly easier to parrot those opinions and claim them as our own, doing so is not a sign of a good critical thinker.

So, how do we effectively resist "sponging" off experts in our own writing and speaking? If we find ourselves quoting experts repeatedly or frequently paraphrasing other people's opinions, we can take a step back and ask, "What do I think and why?" Quotes and paraphrases have a place in our reasoning, but only as support for our own claims. For example, after reading Ebert's review of *Transformers*, we might be tempted to state in a speech that "*Transformers* is the worst movie ever." And we might want to quote Ebert as support: His choir from hell example is pretty vivid.

But let's now put aside Ebert's arguments and ask instead, "What is it that I don't like about *Transformers*?" Maybe it's Mark Wahlberg's one-dimensional acting, or the never-ending action scenes, or the horribly clichéd dialogue. For our speech and writing, we want to develop our ideas using Ebert as colorful backup. As critical thinkers, we aren't just developing the skills of an active reader or listener; we are also going through the exciting process of becoming an active speaker and writer.

Both incorporating others' ideas into our own positions and carefully considering tone are ways we can bring critical thinking into our own writing and thinking. If we do both, we are much more likely to be heard. Moreover, we might open up a discussion that gives us an opportunity to learn more, that moves our exploration and reasoning forward, and that asks for dialogue and doesn't put an end to the discussion.

第 2 章

论题和结论是什么

| 学 习 目 标 |

1. 区分论题的类型。
2. 发现论题并找到结论。
3. 把对论题和结论的识别融入写作中。

在评价一个人的论证之前,我们先要找到他的论证。这听起来很简单,其实不然。要成为一个批判性思维者,首先我们必须练习找准论题和结论的能力。可以把论题看作激发讨论的问题,把结论看作作者对这个问题所持的立场。阅读下面的段落,看看你能不能在其中找到论题和结论:

> 尽管手机给人们带来了种种便利,但使用手机也存在可怕的弊端。例如,一边开车一边发短信容易引发交通事故,美国有些州觉得有必要通过法律,强制规定人们在开车时不准使用手机,违者将面临巨额罚款,以此来减少交通事故发生率。面对日渐增长的滥用手机人口,我们需要施加更加严厉的惩戒才行。

写这篇文章来评价手机的人很想让你去相信某个观点。但他到底想让你相信什么观点,我们又为什么要相信这样的观点呢?

一般来说,那些建网页、开博客、写社论、出书、给杂志写文章或者做演讲的人,

CHAPTER 2

What Are the Issue and the Conclusion?

LEARNING OBJECTIVES

1. Distinguish between types of issues.
2. Discover the issue and conclusion.
3. Integrate identification of the issue and conclusion into your own writing.

Before we evaluate someone's reasoning, we must first find it. Doing so sounds simple; it isn't. To get started as a critical thinker, you must practice the identification of the issue and the conclusion. Think of the issue as the question stimulating the discussion, and the conclusion as the position the writer is taking with respect to that question. See whether you can find the issue and conclusion in the following paragraph:

> Even though cell phones have multiple benefits, there are horrible drawbacks in using them. Some states have found it necessary to try to reduce accidents caused by texting while driving by imposing large fines for violating their law against the use of a cell phone. We need stronger penalties associated with abuse of the growing population of cell phones.

The person who wrote this assessment of cell phones very much wants you to believe something. But what is that something and why are we supposed to believe any such thing?

In general, those who create Web pages, blogs, editorials, books, magazine articles, or speeches are trying to influence your perceptions or beliefs.

都在竭力影响你的观念或是信仰。要对他们循循善诱的说法做出合理的回应，首先就得找出其中的争议之处，或者说论题之所在，然后再找到作者想要推销给你的论点或者说结论。如果找不准作者的结论，那你就会曲解其意图，这样做出的回应往往也会显得驴唇不对马嘴。

读完本章以后，你应该能比较自如地回答我们提出的第一个批判性问题：

⚠ **批判性问题**：论题和结论是什么？

⚠ **注意**：论题就是引起对话或讨论的问题或争议。它是后续所有讨论的原动力。

论题的种类

此时此刻，识别出你通常遇到的两类论题将大有裨益。以下这些问题展示了其中的第一类论题：

> 音乐学习是不是有助于提高一个人的数学能力？
> 导致家庭暴力的最常见的诱因是什么？
> 服用帕罗西汀是不是治疗抑郁症的有效方法？

以上问题都有一个共同点：对它们的回答无一例外地都要描述某一事物过去、现在或将来客观是什么样的。比如说，前两个问题的答案有可能是"一般来说，受过音乐训练的孩子比没有受过音乐训练的孩子学起数学来更得心应手"以及"长期酗酒是引发家庭暴力最常见的诱因"。

这些论题都属于"描述性论题"。它们在图书、杂志、互联网、电视节目中较为常见，反映了我们对世间万物的存在方式及其秩序的好奇。请注意上述问题中的黑体部分：以这些方式提出的问题极有可能属于描述性论题。

⚠ **注意**：描述性论题是指关于各种对过去、现在或将来的描述准确与否的问题。

现在，让我们来看看第二类论题的一些例子：

> 公立学校里应不应该教授智能设计？
> 对骗取医疗补助的行为应该采取什么措施？
> 我们应不应该禁止运动型多用途汽车的使用以缓解哮喘病发病率的上升势头？

For you to form a reasonable reaction to their persuasive effort, you must first identify the controversy or *issue* as well as the thesis or conclusion being pushed onto you. Fail to identify the author's conclusion, and you will be reacting to a distorted version of the attempted communication.

When you have completed this chapter, you should be able to answer the first of our critical questions successfully:

 Critical Question: **What are the issue and the conclusion?**

 Attention: An issue is a question or controversy responsible for the conversation or discussion. It is the stimulus for what is being said.

KINDS OF ISSUES

It will be helpful at this point to identify two kinds of issues you will typically encounter. The following questions illustrate one of these:

Does musical training improve a person's ability to learn math?

What is the most common cause of domestic abuse?

Is Paxil an effective way to treat depression?

All these questions have one thing in common. They require answers attempting to describe the way the world was, is, or is going to be. For example, answers to the first two questions might be, "In general, children who are musically trained learn math more easily than nonmusical children," and "Chronic alcohol use is the most common cause of domestic abuse."

Such issues are *descriptive issues*. They are commonly found in textbooks, magazines, the Internet, and television. Such issues reflect our curiosity about patterns or order in the world. Note the boldfaced words that begin each question above; when questions begin with these words, they will probably be descriptive questions.

 Attention: Descriptive issues are those that raise questions about the accuracy of descriptions of the past, present, or future.

Now let's look at examples of a second kind of question:

Should intelligent design be taught in the public schools?
What ought to be done about Medicaid fraud?
Must we outlaw SUVs to reduce increasing rates of asthma?

All these questions require answers suggesting the way the world *ought to be.* For example, answers to the first two questions might be, "Intelligent

上述问题都需要关于世间万物应该是什么样的答案。举例来说，前两个问题的答案可能是"公立学校里应该教授智能设计"以及"我们应该采取更严厉的惩罚措施来应对骗取医疗补助的行为"。

这些论题都属于伦理或道德范畴，它们提出的都是关于什么是对、什么是错、什么该有、什么不该有、什么是好、什么是坏的问题。它们需要的是规定性的答案。因此，我们将这些论题称为"规定性论题"。社会论争通常都属于规定性论题。

这样说多少有点过度简化，有时候，我们很难判断所讨论的到底属于哪一类论题。不过，记住这两种不同的论题是很有用的，因为你最终做出的判断和评价的类型，将取决于你要回答的论题的类型。

> ⚠ **注意**：规定性论题是指关于什么该做、什么不该做、什么是对、什么是错、什么是好、什么是坏的问题。

寻找论题

我们该怎样着手确定基本的问题或者说论题呢？有时这很简单：写作者或发言者会直接告诉你论题是什么。或者，你可以在文章的正文中找到论题，通常在文章的开头甚至标题中就可以找到它。如果论题直截了当地说出来，会有一些提示性的话语。

> 我想问的问题是：我们为什么一定要有规范烟草制品的各种法律呢？
> 降低法定饮酒年龄：这样做到底对不对？
> 学校到底该不该提供性教育？

可惜的是，论题中涉及的问题并不总是直截了当地提出来，很多时候需要我们从其他线索中去推断。例如，很多写作者或发言者会对他们关心的一些热点事件进行回应，例如发生在校园里的一系列暴力事件。问一问"这个人是在对什么事进行回应"常常能帮你找出文章或发言的中心论题。还有一个比较好的线索是作者的背景，例如他所属的组织机构。所以在你想要确定论题的时候，有必要去查一查写作者的背景信息。

在识别论题的时候，千万不要以为存在且只存在一种正确的陈述论题的方法。一旦你找到整篇文章或整个发言都在论述的问题，以及该问题和该文章（或该发言）间的联系，你就已经找到了一个引发了争论的论题。只要确定你所谓的论题符合"论题"的界定标准即可。

在论题没有直截了当地陈述出来时，最有效的方法是先找准结论。很多情况下，必须

design *should* be taught in the public schools," and "We *ought* to impose more severe penalties for Medicaid fraud."

These issues are ethical or moral; they raise questions about what is right or wrong, desirable or undesirable, good or bad. They demand prescriptive answers. Thus, we will refer to these issues as *prescriptive issues*. Social controversies are often prescriptive issues.

We have somewhat oversimplified. Sometimes, it will be difficult to decide what kind of issue is being discussed. Keeping these distinctions in mind, however, is useful because the kinds of critical evaluations you eventually make will differ depending on the kind of issue to which you are responding.

 Attention: *Prescriptive issues are those that raise questions about what we should do or what is right or wrong, good or bad.*

SEARCHING FOR THE ISSUE

How does one go about determining the basic question or issue? Sometimes, it is very simple: The writer or speaker will tell you what it is. Alternatively, the issue may be identified in the body of the text, usually right at the beginning, or it may even be found in the title. When the issue is explicitly stated, it will be indicated by phrases such as the following:

> *The question I am raising is*: Why must we have laws regulating tobacco products?
>
> Lowering the legal drinking age: *Is it the right thing to do?*
>
> *Should* sex education be taught in the schools?

Unfortunately, the question is not always explicitly stated and instead must be inferred from other clues in the communication. For example, many writers or speakers react to some current event that concerns them, such as a series of violent acts in schools. Asking, "What is the individual reacting to?" will often suggest the central issue of a communication. Another good clue is knowledge of the author's background, such as organizations to which she belongs. So check for background information about the author as you try to determine the issue.

When you are identifying the issue, try to resist the idea that there is one and only one correct way to state the issue. Once you have identified a question that the entire essay or speech is addressing and the link between that question and the essay or speech, *you have found an issue that stimulated the argument in question*. Just make certain that what you are calling an issue meets the definitional criteria that define an "issue."

The surest way to detect an issue when it is not explicitly stated, however, is to locate the conclusion. In many cases, the conclusion must be found

先找到结论，才能确认论题。因此，在这种情况下，批判性评价的第一步就是找出结论之所在——这通常是一个比较难的步骤。

> 只有找到结论，我们才能做出批判性的评价！

我们一起看看如何着手寻找结论这一至关重要的结构要素。

⚠ **注意**：所谓结论，即写作者或发言者希望你接收的信息。

寻找写作者或发言者的结论

要找到结论，批判性思维者首先就要问"作者或演说者想要证明什么"或者"他们表达的要点是什么"。这两个问题的答案都是结论。同样，写作者或发言者对所提问题进行的任何解答也都是结论。

寻找结论，其实就是寻找写作者或发言者希望你相信的一个陈述或一系列陈述。他希望你相信他在其他陈述的基础上所得出的结论。简言之，有说服力的交流或论证的基本结构是：甲之所以成立是因为乙。"甲"指结论，"乙"指结论的支撑材料。这个结构代表了推论的过程。

结论是推论出来的，它们来源于论证。结论是需要其他观点来支撑的观点。因此，如果有人断言某事是真的或应该去做某事，却没有提供相应陈述来支撑他的这一断言，那么这一断言就不能被称为结论，因为提出此断言的人并没有提供任何能让人相信的基础信息。我们把没有证据支撑的断言称为纯观点。

要想学会批判性地阅读和倾听，理解结论的本质是个必不可少的步骤。我们来仔细看看一个结论。阅读下面这一小段内容，看看你能否找出其中的结论，以及支撑这一结论的陈述：

> 转基因食品不应该合法化。我们无从知晓食用部分由化学家制造的食物会对我们的健康产生什么样的影响。

"转基因食品不应该合法化"是作者对"转基因食品应不应该合法化"这个问题的回答，这就是作者的结论。作者通过警告人们食用经过化学家改性的食品会对健康造成危害来支撑这个观点。

before you can identify the issue. Thus, in such cases, the first step in critical evaluation is to find the conclusion—a frequently difficult step.

WE CANNOT CRITICALLY EVALUATE UNTIL WE FIND THE CONCLUSION!

Let's see how we go about looking for that very important structural element.

 Attention: *A conclusion is the message that the speaker or writer wishes you to accept.*

SEARCHING FOR THE AUTHOR'S OR SPEAKER'S CONCLUSION

To identify the conclusion, the critical thinker must ask, "What is the writer or speaker trying to prove?" or "What is the communicator's main point?" The answer to either of these questions will be the conclusion. Also, any answer to the question provided by the speaker or writer will be the conclusion.

In searching for a conclusion, you will be looking for a statement or set of statements that the writer or speaker wants you to believe. He wants you to believe the conclusion on the basis of his other statements. In short, the basic structure of persuasive communication or argument is *this, because of that.* *This* refers to the conclusion; *that* refers to the support for the conclusion. This structure represents the process of *inference*.

Conclusions are *inferred*; they are derived from reasoning. Conclusions are ideas that require other ideas to support them. Thus, whenever someone claims something is true or ought to be done and provides no statements to support her claim, that claim is not a conclusion because that person has not offered any basis for belief. In contrast, unsupported claims are what we refer to as *mere* opinions.

Understanding the nature of a conclusion is an essential step toward critical reading and listening. Let's look closely at a conclusion. Here is a brief paragraph; see whether you can identify the conclusion, then the statements that support it.

> Genetically modified foods should not be legal. We have no way of knowing what the effects on our health will be from consuming food that is created in part by chemists.

"Genetically modified foods should not be legal" is the author's answer to the question, "Should genetically modified foods be legal?" It is her conclusion. The author supports this belief by warning of health dangers from foods that have been modified by chemists.

你是否明白为什么用作支撑材料的观点并不是一个结论？它之所以不是结论，是因为它被用来证明别的观点。记住：你相信一个陈述（结论），因为你认为它得到其他观点支撑，这就是推论。当人们从事这一思维活动时，他们就是在进行论证，而论证得出的结果就是结论。

有时候，和我们交流的人并不直截了当地说出他们的结论。在这种情况下，你就得依靠推断来找到结论：根据作者摆出的种种观点，推断他想证明什么。

使用这个批判性问题

一旦你找到结论，就要将结论作为评价的重点。结论是写作者或发言者希望你选择的"终点站"。接下来，你关心的问题是："支撑这一断言的材料是否足以让我接受这个结论？"

找到结论的线索

线索一：问问论题是什么。因为结论总是对论题的回应，知道论题是什么将有助于找到结论。我们已经讨论过怎样确定论题。首先，看看文章的标题；其次，看看文章的开头几段。如果这些技巧都不管用，那就有必要接着往下浏览几页。

线索二：寻找指示词。结论前面常有指示词引导，指示词表明结论即将出现。看到指示词的时候务必提高警惕，接下来出现的很可能就是结论。下面，我们为你列举一些该类指示词：

- 因此
- 表明
- 由此可知
- 由此
- 因此得出
- 我要说的重点是
- 显示出
- 证明
- 告诉我们
- 事情的真相是

很可惜，很多书面和口头材料并不会用指示词来引出结论。不过，如果你与人交流的目的在于让读者和听众明白你的结论，那么你就应该凸显指示词以吸引他们注意你的论点。这些词的作用就像霓虹灯一样，吸引读者和听众注意你想要他们接受的重点。

线索三：在可能的位置查看一下。结论一般都在特定的位置出现。首先要注意的两个地方是文章的开头和结尾。有些写作者写文章喜欢开门见山，一开始就说出自己想要证明的东西。另一些写作者则喜欢在文章的结尾部分概括出结论。如果你在读一篇冗长而又复

Do you see why the supporting belief is not a conclusion? It is not the conclusion because it is used to prove something else. *Remember:* To believe *one statement* (the conclusion) because you think it is supported by *other* beliefs is to make an inference. When people engage in this process, they are reasoning; the conclusion is the outcome of this reasoning.

Sometimes, communicators will not make their conclusions explicit; in such cases, you will have to infer the conclusion from what you believe the author is trying to prove by the set of ideas she has presented.

USING THIS CRITICAL QUESTION

Once you have found the conclusion, use it as the focus of your evaluation. It is the destination that the writer or speaker wants you to choose. Your ongoing concern is, Should I accept that conclusion on the basis of what is supporting it?

CLUES TO DISCOVERY: HOW TO FIND THE CONCLUSION

Clue No. 1: **Ask what the issue is.** Because a conclusion is always a response to an issue, it will help you find the conclusion if you know the issue. We discussed earlier how to identify the issue. First, look at the title. Next, look at the opening paragraphs. If this technique does not help, skimming several pages may be necessary.

Clue No. 2: **Look for indicator words.** The conclusion will frequently be preceded by indicator words that announce a conclusion is coming. When you see these indicator words, take note of them. They tell you that a conclusion may follow. A list of such indicator words follows:

consequently	suggests that
therefore	thus
it follows that	the point I'm trying to make is
shows that	proves that
indicates that	the truth of the matter is

Unfortunately, many written and spoken communications do not introduce the conclusion with indicator words. However, when *you* communicate with the goal of making your conclusion clear to your audience, you should draw attention to your thesis by highlighting indicator words. Those words act as a neon sign, drawing attention to the point you want the reader to accept.

Clue No. 3: **Look in likely locations.** Conclusions tend to occupy certain locations. The first two places to look are at the beginning and at the end. Many writers begin with a statement of purpose,

杂的文章，不明白作者到底在说些什么，那么不如直接跳到结尾去看结论。

线索四：记住结论不是什么。以下这些都不可能作为结论出现：

- 例证
- 数据
- 定义
- 背景材料
- 证据

线索五：检查一下交流的语境和作者的背景。写作者、发言者或者网站常会在某些论题上持一种可以预测的立场。如果有些文章的结论不太明显，那么其来源可能存在的偏见和作者的背景信息就可能是特别有价值的线索。我们尤其要留心那些和写作者或发言者可能有关系的组织机构的信息。

思维体操

⚠ **批判性问题**：论题和结论是什么？

请找出下面几篇文章里的论题和结论。注意寻找其中的指示词。我们在第一篇文章后提供了自问自答的范例，以展现批判性思维的过程。通过示范怎样分析这篇文章，我们希望能为你提供一些帮助，使你在将来能更加自如地提出和回答批判性问题。我们为第二篇文章提供了一个较扼要的参考答案，第三篇文章则留给你，供你自行去寻找其中的论题和结论。

第一篇

如果父母将教育子女当成自己的全职工作，并且具备一定的眼光、知识和耐心来从事这一职业，那么家庭教学不失为一个合理的概念。但是，事情的真相是，父母在家教育子女往往是一种错误。

一些父母之所以选择将子女从公立学校接回家来自行教育，也许是找错了理由。有时候，孩子在学校里不守纪律，父母往往宁愿选择让孩子辍学，也不愿容忍学校为处罚违纪学生而设立的种种规矩。这样的动机并不能保证接下来的家庭教学就能带来不一样的结果。此外，当家里没有其他成人来监控发生的一切时，可能会出现这样的情况：父母虐待子女，

containing what they are trying to prove. Others summarize their conclusions at the end. If you are reading a long, complex passage and are having difficulty seeing where it is going, skip ahead to the end.

Clue No. 4: **Remember what a conclusion is not.** Conclusions will not be any of the following:

- examples
- statistics
- definitions
- background information
- evidence

Clue No. 5: **Check the context of the communication and the author's background.** Often writers, speakers, or Internet sites take predictable positions on issues. Knowing probable biases of the source and the background of the authors can be especially valuable clues when the conclusion is not explicit. Be particularly alert to information about organizations with which writers or speakers may be associated.

PRACTICE EXERCISES

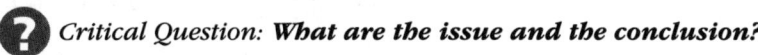
Critical Question: *What are the issue and the conclusion?*

In the following passages, locate the issue and the conclusion. As you search, be sure to look for indicator words. Notice that a self-talk model of this critical-thinking process follows the first passage. By thinking aloud about how we would approach this passage, we hope to make it easier for you to ask and answer the critical questions in the future. We provide a more condensed version of a sample response for passage 2 and leave you on your own to find the issue and conclusion for the third practice passage.

Passage 1

Homeschooling is a valid concept if the parent makes teaching a full-time job and has the insight, knowledge, and patience to do so. However, the truth of the matter is that it is usually a mistake for parents to homeschool their child.

Parents may choose to pull their child out of public schools for the wrong reasons. Sometimes, when children have a discipline problem, the parents will pull them out of school rather than tolerate the rules associated with the punishment. Such a motivation does not speak well for the probable results of the homeschooling that follows. In addition, when there are no other adults to monitor what is going on at home, it

却根本就没人知道。社会必须要了解这些孩子在家里是否接受了应受的教育，是否得到了正常的待遇。

第二篇

　　社交媒体兴起于 21 世纪初，其受欢迎程度与日俱增。随着社交媒体越来越受欢迎，儿童和青少年也越来越多地使用社交媒体。研究表明，过度使用社交媒体会导致学生学业荒疏、成绩下降。社交媒体的使用也与注意缺陷多动障碍和成瘾行为等疾病有关，还有可能使儿童面临抑郁、自卑和饮食失调的风险。此外，社交媒体减少了学生与他人面对面交流的时间。缺乏人际交往会导致年轻人沟通能力和人际交往技能发展滞后，这些技能对年轻人的未来成功至关重要。缺乏面对面交流还与网络欺凌的不断增加有关联，带来一系列新的问题。犯罪分子也可以利用社交媒体来肆意侵扰毫无戒心的年轻人。如果让孩子少接触社交媒体，这些问题大多都可以避免。家长应该主动限制孩子访问社交媒体网站。

第三篇

　　该不该让孩子玩暴力电子游戏？玩暴力游戏似乎会增强孩子在现实生活中的暴力倾向。暴力游戏会根据暴力行为的等级发给玩家一定的分数和奖品，通过条件作用，让玩家沉迷于虚拟的暴力行为。这样玩家会习惯暴力行为，因为当他沉浸在这些虚拟世界的暴力行为中时，他获得的分数或奖品会让他获得快乐和成就感。由此，他们将快乐和暴力联系在一起。如果一个孩子花大量的时间玩暴力游戏，那么从这些游戏中获得的暴力习惯就会渗透到这个孩子在现实生活里的行为中。研究还显示，这些游戏会让儿童玩家变得对暴力行为无动于衷，让孩子变得不太可能去谴责或厌恶现实生活中发生的暴力行为。

给个提示

第一篇

- 有时候，论题很容易找到，因为它会在论证中直截了当地说出来。不过，我认为这篇论述文章并没有直接告诉我论题是什么，因为作者根本没有提及引起争议的问题

is likely that if there is a case of abuse in the home, it will go unnoticed. Society needs to know whether these children are getting the education and treatment they deserve.

Passage 2

Social media has been growing in popularity since its beginnings in the early 2000s. With this rise in popularity comes more social media use by children and teenagers. Studies have shown that heavy social media use can lead to less studying and lower grades. Social media use has also been associated with disorders such as attention-deficit hyperactivity disorder (ADHD) and addictive behaviors, and can also put children at risk for depression, low self-esteem, and eating disorders. Also, social media causes schoolchildren to spend less time interacting with others face-to-face. This lack of personal interaction leads to stunted development of communication and interpersonal skills vital to a young person's future success. The lack of face-to-face communication has also been connected to the rise of cyberbullying, bringing with it a new set of problems. Criminals can also use social media to prey on unsuspecting young people. These problems could be mostly avoided if children had less access to social media. Parents should take the initiative to restrict their children's use of social media sites.

Passage 3

Should children be allowed to play violent video games? It seems as though playing violent video games could increase a child's tendency to be violent in real life. Video games give players points and rewards for acting on violence, and thus condition the players to engage in virtual violent behavior. The player is conditioned to be violent because he experiences feelings of happiness and achievement when receiving points or rewards that are achieved by engaging in these virtual violent acts, and thus associates happiness with violence. If the child spends a significant amount of time playing violent video games, the violent conditioning from these games could filter into the real-life behavior of the child. Studies have also shown that video games desensitize child players to violence, making the child less likely to be disapproving of, or turned off by, violent behavior in real life.

Sample Responses

Passage 1

- *Sometimes, the issue is easy to find because it's explicitly stated in an argument. I don't think that this argument explicitly mentions the issue*

所在。接下来就应该去找结论,这样我就能更容易地找到论题。本书明确说,要找到文中没有明说的论题,最有效方法就是先找出结论。

- 找到指示词将有助于我发现结论。"事情的真相是"被列为引出结论的指示词,这篇论述中恰好用到这个词。也许这篇文章的结论就是"父母在家教育子女往往是一种错误"。这一陈述确实有可能就是结论。另一个找到结论的方法是在文章的开头和结尾部分寻找,而这句话出现在文章的开头部分。

- 本书列举了论证中一些可以排除出结论范围的组成部分,我应该核对一下,以确定"父母在家教育子女往往是一种错误"这一陈述不是数据、例证、定义、背景资料或其他类型的证据。很显然,它不属于这些范畴。

- 到了这一步,我基本可以确定这篇文章的结论就是"父母在家教育子女往往是一种错误"。指示词提示它是结论,它所处的位置确定了它是结论,而它又不属于那类有时会让人误认为是结论的论证组成部分。

- 接下来我要弄明白,是什么问题引出了作者的这一讨论,或者说说论题是什么。如果结论是"父母在家教育子女往往是一种错误",那么引起这个讨论的论题有可能是"父母在家教育子女的行为是否可取"。这个论题可以从结论反推出来,而接下来的所有句子都在讨论家庭教学的种种潜在问题,这一点也可辅证。

- 在得出结论之前,我还想确定这一论题到底属于规定性论题还是描述性论题。要做到这一点,我先要问自己,作者到底是在描述一种情况还是在规定一种什么是对、什么是错、什么该有、什么不该有、什么是好、什么是坏的立场。作者详细描述了家庭教学存在的种种问题,并且建议社会需要了解这些孩子在家"是否接受了应受的教育,是否得到了正常的待遇"。这些陈述都在对家庭教学的可取性提出质疑。因此,这个论题肯定属于规定性论题。

第二篇

这篇文章中没有指向结论的指示词。找到结论的好地方是文章的开头或结尾。结论可能是"社交媒体兴起于21世纪初,其受欢迎程度与日俱增"?或者是"家长应该主动限制孩子访问社交媒体网站"?让我们逐字逐句分析一下这篇文章,看看结论在哪里。

because the author never mentions the question that sparked the argument. My next move should be to find the conclusion. Then I'll be able to more easily find the issue. The text said that the surest way to find an issue that is not explicitly mentioned in the text is to find the conclusion.

- *Looking for indicator words may help me find the conclusion. "The truth of the matter" was listed as an indication of a conclusion and is used in the argument. Maybe the conclusion is, "It is usually a mistake for parents to homeschool their child." This statement really could be the conclusion. Another suggestion for finding the conclusion was to look in the introduction and conclusion. And the sentence is in the introduction.*

- *The text provided me with a list of components of arguments that are not the conclusion. I should check to make sure that the statement "Few parents who homeschool their child are capable of doing so" is not a statistic, an example, a definition, background information, or other evidence. Clearly, it is not.*

- *At this point, I am mostly certain that the conclusion is that "it is usually a mistake for parents to homeschool their child." The indicator words suggested it, the location confirmed this belief, and it did not fall into the list of components of arguments sometimes mistaken for the conclusion.*

- *Next, I need to figure out what question stimulated this discussion, or the issue. If the conclusion is that "it is usually a mistake for parents to homeschool their child," the issue that stimulated this discussion might be, "Is it desirable for parents to homeschool their child?" This issue can be inferred from the conclusion and all the subsequent sentences that discuss potential problems with homeschooling.*

- *Before I conclude, I want to figure out whether this issue is prescriptive or descriptive. To do so, I need to ask myself whether the author is describing a situation or prescribing a position about right and wrong, desirable and undesirable, good and bad. The author details some problems with homeschooling and suggests that society needs to know that these children are receiving "the education and treatment they deserve." These statements raise questions about whether a situation—homeschooling—is desirable. The issue, therefore, must be a prescriptive issue.*

Passage 2

There are no indicator words to point toward the conclusion, but a good place to look for the conclusion is at either the beginning or the end of the excerpt. Could the conclusion be, "Social media has been growing in popularity since its beginnings in the early 2000s"? Or, "Parents should take the initiative to restrict their children's use of social media sites"? Let's break down the passage a bit to find out where the conclusion lies.

在前两句中，作者声称社交媒体受欢迎程度与日俱增，尤其受儿童和青少年的欢迎。然后作者开始举例并给出证据，表明社交媒体的使用会伤害年轻人。接下来，作者声称，如果孩子们尽量少地接触社交媒体，那么列举的伤害的例子大多是可以避免的。作者在文章结尾指出，父母应该限制孩子使用社交媒体。

你可以运用的策略是问这样一个问题："这些证据试图支持哪种说法？"因为这种说法很可能是结论。我们可以通过询问"为什么"，将我们潜在的结论与证据联系起来。让我们试着用这个策略来简化开场白。"社交媒体在年轻人中变得越来越受欢迎。"为什么？"因为使用社交媒体会对年轻人造成伤害。"这个结论和证据似乎并不契合。事实上，在这种情况下，开场白可以被视为背景信息。

结束语是否与证据契合呢？"家长应该主动限制孩子访问社交媒体网站。"为什么？"因为使用社交媒体会对年轻人造成伤害。"证据支持文章的最后一句陈述。另外，看看倒数第二句，它也支持最后一句陈述。作者声称，如果儿童较少接触社交媒体，那么儿童因社交媒体而受到的伤害是可以避免的。假如父母限制孩子使用社交媒体，社交媒体带来的种种问题就不会出现。因此，父母应该限制孩子使用社交媒体。

文章通篇的叙述在某种程度上都在支持最后一句陈述，所以我们可以断定最后一句陈述就是结论。从结论倒推，很容易找到论题。结论是肯定的，父母应该限制孩子使用社交媒体，那么论题可能是"父母应该限制孩子使用社交媒体吗"。这个论题并不告诉我们事情是什么样的，而是问父母是否应该做些什么，这表明这个论题是一个规定性论题。

论题：父母应该限制孩子使用社交媒体吗？
结论：父母应该限制孩子使用社交媒体。

In the first two sentences, the author claims that social media has been growing more popular, particularly among children and teenagers. The author then begins to give examples and evidence showing that social media use can harm young people. Next, the author claims that the examples of harm she listed above can be mostly avoided if children had less access to social media. The author finishes the passage by stating that parents should restrict their children's use of social media.

One strategy you can use is to ask the question, "Which statement does this evidence seek to support?" as this statement is very likely to be the conclusion. We can apply this question by linking our potential conclusions with the evidence by using the question, "Why?" Let's try using this strategy with a simplified version of the opening statements. "Social media has been becoming more popular among young people." Why? "Because social media use can cause harm to young people." This conclusion doesn't seem to fit the evidence. In fact, the opening statements would be considered background information in this situation.

How about the closing statement? "Parents should restrict their children's use of social media." Why? "Because social media use can cause harm to young people." The evidence supports the final statement of the passage. Also, take a look at the second-to-last statement. It too supports the final statement. The author claims that the harm children face due to social media could be avoided if children had less access to social media. If parents were to restrict their children's use of social media, they would avoid the problems social media brings. Therefore, parents should restrict their children's use of social media.

The statements in the passage all support the final statement in some way, so we can conclude that the final statement is the conclusion. The issue can be easily discovered by working backwards from the conclusion. The conclusion is yes, parents should restrict their children's use of social media, and the issue could be, "Should parents restrict their children's use of social media?" As opposed to telling us about how things are, the issue is asking whether parents should do something, indicating that the issue is a prescriptive issue.

Issue: *Should parents restrict their children's use of social media?*
Conclusion: *Parents should restrict their children's use of social media.*

第 3 章

理由是什么

| 学 习 目 标 |

1. 认识论证中理由和证据的作用。
2. 理解论证的属性。
3. 区别理由和结论。

为什么有人会做出某个决定？为什么有人会持有某个观点？我们总是感到好奇，"理由"为我们的好奇心提供了答案。看看下面的几个陈述：

1. 强加给大学生的债务被转化成银行的巨大利润。
2. 蜈蚣的蜇伤比大多数蛇的咬伤更危险。
3. 音乐的力量巨大无比，它改变世界的力量比所有政治领导人加起来还要强。

以上三个断言都缺少了点什么。我们既可以同意，也可以不同意，以它们现在的形式，我们既不能说它们经不起推敲，也不能说它们经得起考验。这些断言都不包含解释说明或逻辑依据，不能证明我们为什么应该同意它们。因此，如果我们听到有人提出以上任何一个主张，都只有干瞪眼，迫切想要知道其所以然。我们没有依据来决定是否同意这些陈述。

以上陈述所缺少的正是支撑断言的理由或者证据。所谓理由，就是用来支撑或证明结论的信念、隐喻和其他陈述。这些陈述是结论可信度的基础，是结论成立背后的逻辑。证

CHAPTER 3

What Are the Reasons?

LEARNING OBJECTIVES

1. Recognize the role of reasons and evidence in an argument.
2. Comprehend the attributes of an argument.
3. Distinguish between reasons and conclusions.

Reasons and evidence provide answers for our human curiosity about why someone makes a particular decision or holds a particular opinion. Consider the following statements:

1. The debt imposed on college students translates into huge profits for the banks that lend the money to them.
2. A centipede sting is more dangerous than the bite of most snakes.
3. Music has more power to change the world than is possessed by the combined effect of all political leaders.

Those three claims are each missing something. We may or may not agree with them, but in their current form, they are neither weak nor strong. None of the claims contains an explanation or rationale for *why* we should agree. Thus, if we heard someone make one of those three assertions, we would be left hungry for more. We would have no basis for deciding whether to agree with the statements.

What is missing is the reason or evidence responsible for the claims. *Reasons* are beliefs, metaphors, and other statements offered to support or

据则是支持断言的另一个基础，它由一系列事实组成，这些事实有助于使听众或读者相信你的理由是可靠的。

我们在第2章里已经介绍了一些必要的方法，教你怎样找到论证"大厦"的两块柱石——论题和结论。本章主要介绍找到论证大厦的第三块柱石——理由的各种技巧。第7章和第8章将重点讨论不同形式的证据，它们辅助理由以支撑一个强有力的结论。一个经得起考验的论证的支撑结构被称为论证的依据，即理由和证据。

如果写作者有个结论希望你接受，那么他不但要提供论证的依据来说服你相信他的结论是正确的，还要说明为什么他的论证的依据能够证明这个结论。

判断一个人是否理性，其主要标志就是看他的各种信念是否都有适当的证据来支撑，当这些信念有一定的争议时更是如此。举例来说，如果有人声称中国在不久之后将取代美国成为世界超级大国，这样的言论应该受到以下挑战："你为何有这样的想法？"这个人的依据可能充分，也可能不充分，但只有你提出以上问题并确定他的理由之后，才能见分晓。如果他的回答是"因为我就是相信这是事实"，那么你肯定会对这样的论证极不满意，因为这个人的"理由"不过是结论的翻版而已。但是，如果答案是关于两国的军事和教育支出的证据，那么你在评价这一结论的时候就要认真考虑这些证据。请记住：只有在你找到支撑一个结论的各种理由和证据以后，你才能评判这个结论的价值。

⚠ **批判性问题**：理由是什么？

理由和结论相结合，就构成我们在第2章中所定义的论证。有时候，一个论证只由一个理由和一个结论构成；但更常见的情形是，很多理由会被用来支撑一个结论。所以当我们谈到某个人的论证时，我们可能指的是一个理由以及相关的结论，也可能指一整套的理由以及它们想要证明的结论。

⚠ **注意**：理由是指我们相信某个结论的原因或原理。

⚠ **注意**：证据是指证明理由的真实性的事实。

justify conclusions. They are the statements that form the basis for creating the credibility of a conclusion. They are the logic for why the conclusion makes sense. *Evidence*, an additional basis for support, consists of the facts that assist in convincing the listener or reader that your reasons are true.

Chapter 2 gave you some guidelines for locating two very important parts of the structure of an argument—the issue and the conclusion. This chapter focuses on techniques for identifying a third essential element of an argument—the reasons. Chapters 7 and 8 will focus on alternative forms of evidence that assist reasons in supporting a strong conclusion. The support structure for a strong argument is called the *warrants* for the argument, meaning both the reasons and the evidence.

When a writer has a conclusion she wants you to accept, she must present warrants for that conclusion to persuade you that she is right and to show you *why*.

It is the mark of a rational person to support her beliefs with adequate proof, especially when the beliefs are of a controversial nature. For example, when someone asserts that China will soon overtake the United States as the dominant country in the world, this assertion should be met with the challenge, "What makes you think so?" The person's warrants may be either strong or weak, but you will not know until you have asked the question and identified the reasons. If the answer is "because I believe it is true," you should be dissatisfied with the argument because the "reason" is a mere restatement of the conclusion. However, if the answer is evidence concerning the projected military and educational expenditures of the two countries, you will want to consider such evidence when you evaluate the conclusion. *Remember:* You cannot determine the worth of a conclusion until you identify the reasons and evidence.

 Critical Question: **What are the reasons?**

The combination of the reasons and the conclusion results in what we defined in Chapter 2 as the "argument." Sometimes, an argument will consist of a single reason and a conclusion; often, however, several reasons will be offered to support the conclusion. So when we refer to someone's argument, we might be referring to a single reason and its related conclusion or to the entire group of reasons and the conclusion it is intended to substantiate.

 Attention: *Reasons are explanations or rationales for why we should believe a particular conclusion.*

Attention: *Evidence refers to the facts that demonstrate the truth of the reasons.*

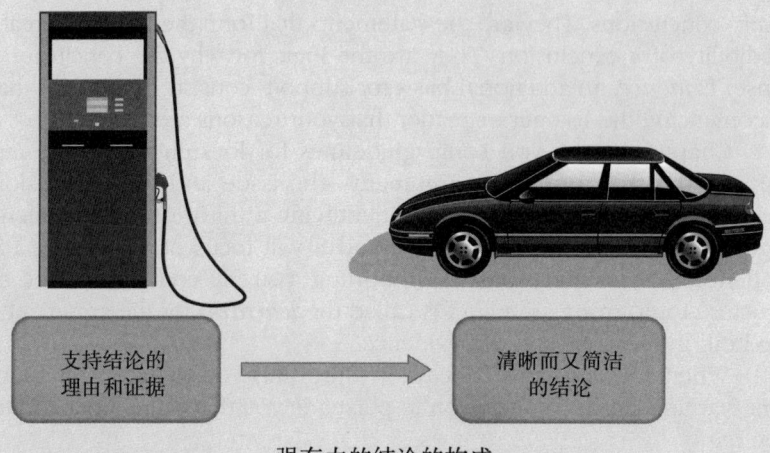

强有力的结论的构成

论证本身有几个特点值得我们加以注意:

- 论证必有其目的。人们之所以展开论证,是因为希望说服我们相信某些事或按照某些特定的方式行事。因而论证需要别人对其做出回应。不论我们的反应类似于海绵还是淘金者,我们一般总会做出回应。
- 论证的质量有高低之分。我们需要依赖批判性思维来判定一个论证的质量高低。
- 论证有两个明显的必要组成部分,一个结论及支撑它的理由。两者当中如果有一个我们找不到,也就意味着我们失去了评价这一论证的机会。我们自然无法对找不到的东西做出评价。

最后一点值得进一步强调和说明。心急火燎地展开批判性思考并没有什么意义。实际上,哲学家维特根斯坦曾说过,在一个聪明人和另一个聪明人说话时,他们都会总是先说"等一等"。花点时间找准论证之所在,然后再去评价我们听到或看到的那些话,这样对提供论证的那个人才够公平。

开始质疑的过程

找出理由的第一步,就是在接触每个论证时都采取怀疑的态度,你要问的第一个问题就是"为什么"。你已经找到了结论,现在你想知道为什么这个结论有道理。如果一个陈述回答不了"为什么写作者或发言者相信这个结论?"这个问题,那么它就不属于理由。一个陈述(或一系列陈述)必须被立论者用作结论的支撑或依据,才能起到理由的作用。

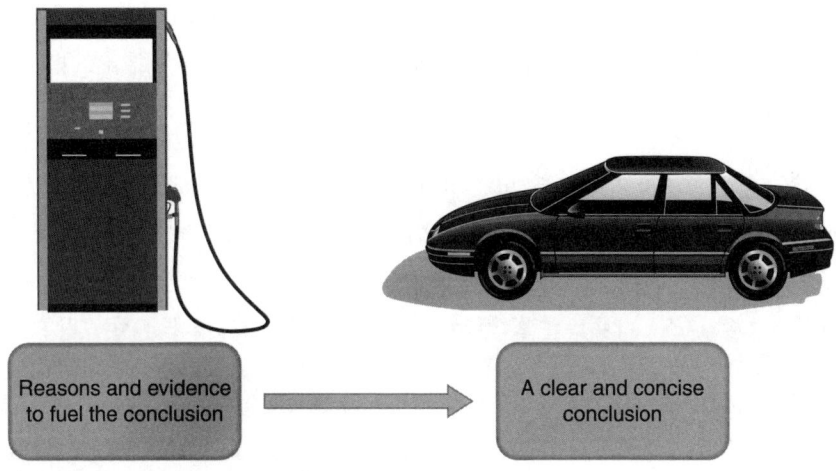

The Components of a Strong Argument

Several characteristics of arguments grab our attention:

- They have intent. Those who provide arguments hope to convince us to believe certain things or to act in certain ways. Consequently, they call for a reaction. We can imitate the sponge or the gold prospector, but we ordinarily must respond somehow.

- Their quality varies. Critical thinking is required to determine the extent of quality in an argument.

- They have two essential visible components—a conclusion and reasons. Failure to identify either component destroys the opportunity to evaluate the argument. We cannot evaluate what we cannot identify.

That last point deserves some repetition and explanation. There is little purpose in rushing critical thinking. In fact, the philosopher Wittgenstein suggests that when one bright person addresses another, each should first say, "Wait!" Taking the time to locate arguments before we assess what we think might have been said is only fair to the person providing the argument.

INITIATING THE QUESTIONING PROCESS

The first step in identifying reasons is to approach the argument with a questioning attitude, and the first question you should ask is *why*. You have identified the conclusion; now you wish to know why the conclusion makes sense. If a statement does not answer the question, "Why does the writer or speaker believe that?" then it is not a reason. To function as a reason, a statement (or a group of statements) must be used by a communicator as support or grounds for a conclusion.

让我们把这种怀疑的态度应用到下面这篇文章中。首先我们必须要找出结论,然后再适当地问问"为什么"的问题。请回想一下找准结论的方法(引导结论的指示词已经用黑体标出来)。

(1)学生评教分数是否应该与教授的薪水挂钩?(2)教授们接受了关于他们的意见的调查。(3)许多人表示,他们认为学生并不熟悉关于年轻人如何学习的不同观点。(4)80%的教授说,他们担心如此运用学生的评教,恐有妨害高等教育质量之虞。(5)**因此**,学生的评教不应该成为一个决定教授工资的因素。

紧跟在"因此"后面的句子回答了第(1)句陈述所提出的问题。因此,结论就是第(5)句陈述:"……学生的评教不应该成为一个决定教授工资的因素"。请给这一结论做上记号。

⚠ **注意**:一个论证由一个结论以及支撑这一结论的各种理由组成。

然后,我们可以问这个问题:为什么写作者或发言者相信这个结论?用于回答这个问题的陈述就是理由。在以上的例子中,作者为我们提供了调查得来的证据作为理由。第(3)句和第(4)句陈述共同提供了证据,也就是说,它们一起为结论提供了依据,因而成为支撑结论的理由。因此,我们可以这样复述上述理由:大部分教员都相信若将学生的评教用来决定教员的薪水,那么恐有妨害高等教育质量之虞。

现在,请试着找出下面这篇文章中的理由。同上面的步骤一样,我们还是要先找出结论,做上标记,然后问"为什么"的问题。

(1)不应该允许对人类胚胎进行基因筛查。(2)人们无权扮演上帝的角色并终止一个生命来到人间,人类无权仅仅因为胚胎的性别不如人意或可能存在某种生理缺陷就这么干。(3)我有两个患自闭症的孩子,他们都生活得很快乐。(4)不能说一个人的生活质量会因为天生的缺陷而发生重大改变。

第(1)句当中的指示词"应该"显示出文章的结论:作者反对对胚胎进行基因筛查。作者为什么相信这一点呢?他给出的主要理由是"人们无权扮演上帝的角色,根据一套自己偏爱的标准来决定终止一个生命来到人间。"第(3)句和第(4)句一起提供了另一条理由来支撑作者的看法:作者本人就有患自闭症的孩子,其正面经验显示,一个人的生活质量并不会因为天生的缺陷而发生重大改变。

在确定论证结构的时候,你应该把每一个被论证提出者拿来支持其结论的想法都当成

Let us apply the questioning attitude to the following paragraph. First we will find the conclusion; then we will ask the appropriate *why* question. Remember your guidelines for finding the conclusion. (The indicator word for the conclusion has been italicized.)

> (1) Should student evaluation scores help determine the salaries of professors? (2) Professors were surveyed about their opinions. (3) Many indicated that they believed that students are not familiar with alternative views about how young people learn. (4) Eighty percent of faculty members said they fear that student evaluations used in this manner would harm the quality of higher education. (5) *Therefore*, student evaluations should not be a contributing factor in determining faculty salaries.

What follows *"Therefore"* answers the question raised in statement (1). Thus, the conclusion is statement (5): "… student evaluations should not be a contributing factor in determining faculty salaries." *Highlight the conclusion!*

 Attention: *An argument consists of a conclusion and the reasons meant to support it.*

We then ask the question, "Why does the writer or speaker believe the conclusion?" The statements that answer that question are the reasons. In this particular case, the writer provides us with survey evidence as reasons. Statements (3) and (4) jointly provide the evidence; that is, together they provide support for the conclusion, thus serving as the reason for it. Thus, we can paraphrase the reason as follows: Most faculty believe that student evaluations used to help determine faculty salaries would damage higher education.

Now, try to find the reasons in the following paragraph. Again, first find the conclusion, highlight it, and then ask the *why* question.

> (1) Genetic screening of embryos should not be allowed. (2) People do not have the right to play God and terminate a potential life just because it might not be the right sex or may have a defect of some kind. (3) I've had two autistic children and they are both happy. (4) It cannot be said that a person's quality of life is severely changed by a birth defect.

The indicator word *should* in the first sentence signals the conclusion: The author is against genetic screening of embryos. Why does the author believe this? The main reason given is "People don't have the right to play God and decide to terminate a potential life based on a set of their preferred criteria." Sentences (3) and (4) together provide an additional reason for the author's belief: Personal positive experience with autistic children demonstrates that a person's quality of life is not severely changed by a birth defect.

As you determine a communicator's reasoning structure, you should treat any idea that seems to be used to support her conclusion as a reason, even if

理由，即使你并不相信它真的能为结论提供支持。在批判性思维的这一阶段，你正在设法找出论证之所在。因为你想对提供论证的人尽量公平，所以你要利用施惠原则。如果写作者或发言者相信他正在摆事实、讲道理以支持自己的结论，那么我们至少应该考虑一下他的论证过程。之后还有大把的时间来细细评价这个论证。

切记！评价之前要先等一等。

理由的提示词

和寻找结论的情形一样，有些特定的词常会指示紧随其后的就是理由。请记住：论证的基本结构是"甲之成立是因为乙"。因此，"因为"这个词，以及与其意义相同或功能相近的词，都可能提示我们，用于支持结论的理由会紧随其后出现。以下是一些理由的提示词：

- 由于
- 研究显示
- 因为这个事实
- 因为这个原因
- 由以下材料支撑
- 因为证据显示

让理由和结论一目了然

有很多论证篇幅冗长、结构松散。有时候，成套的理由只是用来支持一个结论，而这个结论又被当成另一个结论的主要理由。理由可能会由其他理由来支持。面对复杂的论证，当你想批判性地评价自己看到或听到的内容时，你常常发现自己很难在脑海中将论证的结构清晰展现出来。要想克服这一难题，你就要设法组织材料，将理由和结论分开，以逻辑性的方式重新编排。

我们已经提到过许多技巧，它们能帮助你构建一幅有关论证结构的清晰画面。如果你有其他更好的技巧，那就毫不犹豫地加以利用吧。关键是，在你打算评价一个论证以前，一定要让理由和结论一目了然。

使用这个批判性问题

一旦你找到理由，随着阅读或倾听的不断深入，你就需要一遍又一遍地重温这些理由。结论能不能站住脚主要取决于给出的相应理由扎不扎实。薄弱的理由必然导致薄弱的论证！

先有理由，后有结论

我们在第 1 章里警告过弱势批判性思维带来的危险。当你注意到有人似乎在编造理由

you do not believe that it actually provides such support. At this stage of critical thinking, you are trying to identify the argument. Because you want to be fair to the person who made the argument, you want to use the principle of charity. If the writer or speaker believes she is providing support for the conclusion with some evidence or logic, then we should at least consider the argument. There will be plenty of time later to evaluate the argument carefully.

Remember! Wait before evaluating.

WORDS THAT IDENTIFY REASONS

As was the case with conclusions, certain words will typically indicate that a reason will follow. *Remember*: The structure of reasoning is *this, because of that*. Thus, the word *because*, as well as words synonymous with and similar in function to it, will frequently signal the presence of reasons. A list of indicator words for reasons follows:

as a result of	studies show that
because of the fact that	for the reason that
is supported by	because the evidence is

KEEPING THE REASONS AND CONCLUSIONS STRAIGHT

Much reasoning is long and not very well organized. Sometimes, a set of reasons will support one conclusion, and that conclusion will function as the main reason for another conclusion. Reasons may be supported by other reasons. In complicated arguments, it is frequently difficult to keep the structure straight in your mind as you attempt to critically evaluate what you have read. To overcome this problem, try to develop your own organizing procedure for keeping the reasons and conclusions separate and in a logical pattern.

We have mentioned a number of techniques for you to use in developing a clear picture of the reasoning structure. If some other technique works better for you, by all means use it. The important point is to keep the reasons and conclusions straight as you prepare to evaluate.

USING THIS CRITICAL QUESTION

Once you have found the reasons, you need to come back to them again and again as you read or listen further. The conclusion depends on the merit of the reasons. *Weak reasons create weak reasoning!*

Reasons First, Then Conclusions

Chapter 1 warned you about the danger of weak-sense critical thinking. A warning signal that can alert you to weak-sense critical thinking should go off

（甚至是当场编造理由），仅仅因为他们想以此来捍卫自己所持的观点，这时你就应该在心里不断发出警示信号，提醒自己注意弱势批判性思维的局限性。如果有人急于和你分享他的观点，好像其观点是确凿无疑的结论，可一旦被问及有哪些理由，他就变得一脸茫然或恼羞成怒，那么这很可能是弱势批判性思维惹的祸。

操控型论证就是指以一个结论开始的论证。然后提供论证者选择理由和证据，它们发挥着特定的作用。这种作用并不在于这些证据具有很强的真实性，而在于它们有助于听众或读者在脑海中勾勒出一个故事，使这个故事朝着预先确定的结论行进。你可以这样理解操控型论证：想象一个人态度坚定，持有先入为主的结论，他在"信息超市"中"采办"理由和证据的目的仅在于增加分享预期结论的人数。

操控型论证的一个很好的例证就是律师在对抗制的司法体系中的行为。当客户跨进大门的时候，律师接收到的信息是一个特定的结论，他被请求必须接受这个结论。一旦建立了代理人–委托人关系，律师的道德任务就是积极地提出有利于委托人的论证。这与科学家的行事风格截然不同，科学家可能对一个实验将得出什么样的结果有一种预感，但他最终的结论由这样的规则所决定：先有理由和证据，后有结论。

操控型论证

when you notice that reasons seem to be created (on the spot, even) only because they defend a previously held opinion. When someone is eager to share an opinion as if it were a conclusion but looks puzzled or angry when asked for reasons, weak-sense critical thinking is the probable culprit.

Managed reasoning refers to an argument that has begun with a conclusion. Then the reasons and evidence are chosen because they perform a particular function. That function is not because the warrants are strong in the sense of being true, but rather because they help create a story in the mind of the listener or reader that moves toward the predetermined conclusion. When you think of managed reasoning, you can imagine someone with a firmly entrenched conclusion going shopping for whatever reasons and evidence will enlarge the number of people sharing the desired conclusion.

A good illustration of managed reasoning is the behavior of an attorney in an adversarial system of justice. When the client walks in the door, the attorney is receiving a delivery of the conclusion that he is being asked to embrace. If an attorney–client relationship is created, the attorney's ethical task is to vigorously present an argument that will assist that client. Contrast that situation with the actions of a scientist who may have a hunch about how an experiment will play out. But her eventual conclusion is shaped by the rule: Reasons and evidence first—conclusions next.

Managed Reasoning

在这一点上,你必须要自我监控。我们都知道,我们倾向于匆匆下结论。要设法避免"逆向逻辑"或"反向论证",在这种论证中,理由不过是一记马后炮,会随着你的结论而不断变化。理想的做法是,将理由和证据作为模具,结论据此得以成型和修饰。

像批判性思维者那样写作和发言

在我们写作或发言时,我们会用自己的结论来解决问题。我们希望其他人能够分享或至少留意到我们结论中所蕴含的智慧。

论证伊始,我们得界定好论题:我们要探究的具体问题到底是什么?针对这个问题,我们想要得出一个结论,尽管这个结论可能是试验性的。听起来很简单:找到问题然后回答它!读到这里,你可能已经发现这其实并不简单。

在我们自己的写作和发言中,什么是一个好的论题?什么是一个好的结论?

- 清晰而有趣的问题;
- 重点突出、阐述明确的结论;
- 有理由和证据支撑的结论,而且大多数人会觉得理由和证据有说服力;
- 回应论题的结论;
- 容易让读者或听众找到的结论。

很多学生和其他写作者、发言者没有花足够的时间来讲清楚自己的论题,这样做是要自己承担风险的。含糊其词、不清不楚、散漫芜杂的结论往往是没有清楚地确定我们希望回答的问题而导致的结果。当我们接到一个写作或发言的任务,而且这个任务不是以问题的形式来陈述的时候,最好的办法是把它重新组织成一个问题。把论题当成一个问题来思考,你就能集中精力,以通俗易懂的形式让读者或听众明白,你是被某一个问题所困扰,与其他问题无关。

例如,如果我们得到这样的提示:"就你最喜欢的电子游戏,写一篇论述文",我们有可能倾向于描述、分析、总结和评价一款电子游戏,这样的写法只能让读者在寻找我们的观点时感到费心费力。相反,我们可以明确地将提示信息表述成一个问题:"是什么让我最喜欢的电子游戏《魔兽世界》成为一款了不起的游戏?"这样一个特定的问题将引导我们得出一个特定的结论,并使我们的论证更加严密细致、聚焦中心。通常,我们写作时的含糊其词源于我们不知道要就什么论题展开论述。首先确定一个清楚陈述的论题将有助于我们得出一个清楚陈述的结论。

Be your own monitor in this regard. We all know we have a tendency to form conclusions too quickly. Try to avoid "reverse logic" or "backward reasoning," whereby reasons are an afterthought, following the selection of your conclusion. Ideally, reasons and evidence are the tool by which conclusions are shaped and modified.

WRITING AND SPEAKING AS CRITICAL THINKERS

As a writer or speaker, we will be addressing issues with conclusions of our own making. We would like others to share or at least consider the wisdom of our conclusions.

Before we get started, we want to define our issue: Just what is the specific question that we are going to explore? And, in response to that question, we want to put forward a conclusion, however tentative. Sounds pretty easy: Find a question and answer it! At this point in the text, you probably are not surprised that it is not that simple.

What makes for a good issue and conclusion in our own writing and speaking?

- a clearly articulated and interesting question;
- a focused and explicit conclusion;
- a conclusion propped up by reasons and evidence that most people would find compelling;
- a conclusion that addresses the issue; and
- a conclusion that is easy for a reader or listener to locate.

Many, many students and other writers and speakers do not spend enough time clarifying their issue, but they do so at their peril. Fuzzy, unclear, and rambling conclusions are often the result of not clearly identifying the question we hope to answer. When we get a writing or speaking assignment that is not stated as a question, it's a good idea to reframe it as one. By thinking about the issue as a question, you focus your effort and make it clear to your reader or listener that you are agitated by a certain question and not others.

For example, if we are given the prompt, "Write an argument about your favorite video game," our tendency might be to describe, analyze, summarize, and evaluate a video game, leading our reader on a meandering and elusive search for our point. Instead, we can specifically articulate the prompt into a question such as, "What makes my favorite video game, *World of Warcraft*, a great game?" Such a specific question will lead us to a specific conclusion and a much tighter and more focused argument. Often, our fuzziness in our own writing is because we don't really know what the issue is that we are addressing. Starting with a clearly stated issue will help us to create a clearly stated conclusion.

其次，这个论题必须是一个能够激发受众思考的论题。你可能会对《魔兽世界》中哪些级别的玩家被允许复活的问题非常感兴趣，但是受众可能会觉得这个问题非常无聊（如果他们不玩这个游戏，甚至可能感到费解）。作为写作者，我们必须找到一个能让受众感兴趣的论题，或者让我们的论题趣味盎然，以至于任何受众都会觉得它有趣味（后者显然要难得多）。

再次，我们一旦确定了希望论证的结论，就要确保这一结论切合主题，陈述明确，并且易于找到。读者或听众应该能够毫不费力地识别出我们的论点（结论的另一种表述方式）。我们在大学里读到的许多文章都有隐含的结论，但在大多数情况下，作为初学写作和发言的人，我们想让结论清晰明确。指示词（见第 2 章的简表）不仅非常有助于使结论明确，而且可以为读者指明我们的理由和支撑。许多学生脑海中都清楚想要传达什么结论，但对于读者或听众来说，写作者或发言者头脑中的内容并不总能清晰地传递到纸上或他们的耳朵里。再做一次检查是个不错的主意。找个人来读你的作品，确保他能轻而易举地看出你在论证什么，并在文章或发言中找到结论所处的位置。

最后，对论题的回应（即我们的结论）不仅要回答我们提出的问题，还要有强有力的理由和证据作为支撑。我们对上面提到的问题的结论可能是在《魔兽世界》中，只有 5 级到 10 级之间的玩家应该被允许复活。但现在，我们必须超越我们的观点，为我们为什么会有这种看法加上理由和证据。不可能只是因为我们喜欢它或者它对我们有意义。如果想要别人相信我们的话，我们就要提供最强有力的理由。

切记，受众评价我们的结论质量的唯一依据是我们写在纸上或在发言中包含的内容。读者和听众无法了解我们的思维过程，也无法了解当有人问我们到底在说什么时我们可能会给出的解释。我们必须从一开始就在发言或写作中明确给出我们的支撑材料。这是我们唯一一次吸引受众注意力的机会。

要实现清晰晓畅，得具有几个要素：

1. 指示词；
2. 经过研究并聚焦主旨的话题；
3. 明确支撑结论的理由；
4. 文章合理的结构和清晰的布局。

那么，身为写作者或发言者的我们该如何把论证结构组织起来呢？

首先，要确保我们的确对所选主题言之有物，而且要言之凿凿，能够满足任务的要求。动手写作之前，我们可以通过头脑风暴法或自由写作法来评估我们对这个主题的知识储备。

Second, the issue has to be one that activates the mind of our audience. You might be really passionate about the question of which levels players should be allowed to be resurrected in while playing *World of Warcraft*, but your audience might find that question immensely boring (not to mention incomprehensible, if they don't play the game!). As writers, we have to find an issue that interests our audience, *or* make our topic so interesting that any audience would find it interesting (much harder).

Third, once we've determined the conclusion we hope to argue, we want to make sure it's focused, clearly stated, and easily identifiable. Our reader or listener should be able to identify our thesis (another word for "conclusion") without any trouble. Many of the essays we read in college have implied conclusions; but as beginning writers and speakers, in most cases, we will want to make our conclusions explicit. Indicator words (see the brief list in Chapter 2) can be very helpful in making a conclusion explicit but also in identifying for our readers our reasons and support. Many students have a clear idea in their head of what conclusion they want to convey, but for a reader or listener, what's in the writer's or speaker's head doesn't always transfer clearly to paper or ear. It's a good idea to do a double-check. Get someone to read your work to make sure she can easily identify just what it is you are arguing and where in the essay or speech that conclusion is located.

Fourth, the response to our issue, our conclusion, must not only answer the question we raised but also have strong reasons and evidence that we can use as support for our conclusion. Our conclusion to the above question might be that players in *World of Warcraft* should be allowed to be resurrected only in levels 5 to 10. But now, we have to go beyond just our opinion and add the reasons and evidence for why we feel this way. And, it can't be just because we like it or it makes sense to us. We want to offer the strongest reasons possible if we want to be believed.

Remember, the only basis a reader has for evaluating the quality of our conclusion is what we've written on paper or included in our speech. Readers and listeners have no access to our thought processes or the explanation we might provide if someone were to ask us just what the heck it was we were trying to say. We have to make our support clear in the speech or in the writing from the get-go. It's the only time we know we have our audience's attention.

Clarity comes through several processes:

1. indicator words;
2. researched and focused topic;
3. reasons that explicitly support our conclusion; and
4. an organized structure to our essay.

So, how do we go about putting this argument structure together as a writer or speaker?

First, we want to make sure we actually have something to say about the topic and enough to say that it will fulfill the requirements of our assignment. We can evaluate our knowledge by doing some brainstorming or freewriting

这个过程也可以帮助我们认真思考论题，而不是脑海中刚闪现过第一个念头，就匆匆上阵。不能仅仅因为我们对某一主题见解强烈，就认为我们可以很好地论证它。请记住，一个论点，一个好的论点，需要强有力的理由和证据。

其次，我们可以事先做研究。其实我们不愿意使用强烈的措辞，但在这种情况下似乎这样的措辞是恰如其分的：我们必须做好研究。一旦我们认为找到了一个很好的论题，并确定了结论是什么，我们就应该查看为得到这个结论所需要的证据和支撑。人们很容易认为自己已经知之甚多，而忽略了研究的步骤。但是我们当中能称得上即将论述的话题方面专家的人少之又少。事实上，在很多情况下，就算我们对某个话题有见解，这些见解可能也不是通过推理得出的。如果我们找不到强有力的理由，或者我们能找到的理由是毫无说服力的，甚或是道听途说得来的，我们就需要回到论题起点，得出不同的结论。

一旦我们有了一个论题，有了支撑结论的强有力的理由，我们就必须搭建一个条理明晰、结构合理的框架，使读者对我们论证充分的结论一目了然。切记，一篇文章主要由支撑结论的理由和证据所组成。如果我们没有几个令人信服的、强有力的理由来支撑我们的论点，如果这些理由没有按照逻辑关系组织起来，那么我们的写作或发言就很难进行下去，读者或听众也很难理解我们的论点。

虽然组织一篇文章或一篇发言的方法有很多，但是一些一般规则能为我们提供帮助。

1. 把论点放在引言里，用指示词突出，并在文章的结尾部分重复一下这个论点，稍微修改一下论点的措辞。
2. 确保每条理由都能回答"为什么"这个问题。前文中我们曾经指出，身为读者的我们应该不停地问"为什么写作者或发言者相信那个观点"，而作为写作者或发言者，我们可以对自己的作品提出这样的问题：为什么我们相信自己的论点？如果我们所陈述的理由无法回答这个问题，我们就应该舍弃它，或者我们需要更充分地展开陈述，这样理由就能回答这个问题了。
3. 使用指示词来介绍我们的理由和证据（本章的前面提供了一个列表）。

可以把论题、结论和理由看作论点的骨架。骨架越坚固，论据越有力。

思维体操

⚠ **批判性问题**：理由是什么？

首先浏览一下文章，找出结论并做好标记。然后问"为什么"，找出理由。利用指示词

on a topic before we get started. This process can also help us think through an issue instead of simply jumping on board with the first thought that pops into our heads. Just because we have a strong opinion on a topic does not mean we have a good argument. Keep in mind that an argument, a good argument, needs strong reasons and evidence.

Second, we can do research; in fact, we hesitate to use such strong language, but in this case it seems appropriate: we *must* do research. Once we think we've found a good issue and determined what our conclusion is, we should check to see the evidence and support for that conclusion. It is tempting to think we already know enough and skip the research step. But very few of us are experts on the topics we will be writing about. And, in fact, in many cases if we have an opinion about the topic, it may not have come through reasoning. If we can't find strong reasons, or the reasons we can find are weak or in obscure blogs, we need to go back to our issue and develop a different conclusion.

Once we've got an issue and what seem to be strong reasons that support our conclusion, we have to create an organizational structure that makes our conclusion clear to our readers and fully develops our reasons. Remember, an essay is primarily composed of the reasons and evidence that support the conclusion. If we don't have several good, strong reasons for our argument and if they are not organized in a logical way, it will be very difficult to develop our essay or speech or for a reader or listener to follow our argument.

While there are many different ways to organize an essay or a speech, some general rules can be helpful.

1. Put our thesis in the introduction highlighted by an indicator phrase *and* repeat that thesis, perhaps reworded slightly, in the conclusion of our essay.
2. Make sure each of our reasons responds to the "why" question. Remember, earlier we indicated that as readers we should be asking, "Why does the writer or speaker believe that?" As writers or speakers, we can ask that question of our own work: Why do we believe our thesis? If our reason doesn't answer that question, we need to get rid of it or we need to develop it more fully so it does answer that question.
3. Use indicator words to introduce our reasons and evidence (we provided a list earlier in the chapter).

Think of the issue, conclusion, and reasons as the skeleton of your argument. The stronger the skeleton, the stronger the argument.

PRACTICE EXERCISES

 Critical Question: ***What are the reasons?***

First survey the passage and highlight its conclusion. Then ask the question, "Why?" and locate the reasons. Use indicator words to help. Keep the conclusions

来获得帮助。将结论和理由区分开。试试用自己的话复述理由，用自己的话复述有助于澄清它们的含义和作用。

第一篇

　　流行的女性杂志塑造出人们对何为女性美不公正的期待。几乎所有刊登在这些杂志封面上的女性照片都经过了数码技术的处理，通过调整照片的色彩亮度，对照片中女性的身材比例进行处理，让她们看起来更迷人。研究显示，女性根本不可能模仿杂志封面女郎的一些身材特征，因为这些特征都是电脑制造出来的。我们在杂志封面上看到的"美女"都是不现实的，相反，她们都是电脑制作出来的。

第二篇

　　随着获得大学学位所需的费用不断上涨，关于学生是否应该上大学的争论越来越激烈。大学还值得上吗？其实获得大学学位依然有几项益处。

　　首先，越来越多的工作需要大学文凭。这一事实不仅意味着没有大学学历会妨碍一个人的职业选择，还意味着拥有大学学历会让雇主在招新时对他另眼相看，把他归类于更高层次的人员范畴。

　　其次，通常需要大学学历的工作岗位给予的薪酬高于平均水平。与只有高中文凭的求职者相比，这些高薪酬的工作能把生活水平提高到更高的层次，偿付当年上大学所支付的成本更是不在话下。显然，上大学是一个非常不错的选择，关乎一个人的事业前程。

第三篇

　　中学阶段，男篮和男足比赛通常会独占周五晚上的日程安排。到底应不应该这样？这些比赛对学生的中学经历来说，其意义自不待言，可也不至于为此而牺牲掉其他所有运动项目。它们确实有着悠久的传统，但这并不意味着这种模式就一定要一成不变、世代因循。

　　周五晚上腾出时间，对大多数父母和球迷而言都是容易做到的。因此，周五晚上观看男篮和男足的比赛对他们而言都很容易。

　　那么女子篮球队或游泳队该怎么办呢？她们的比赛不应该总是被安排在周一到周五的下午和傍晚进行。她们的家人经常不能腾出时间来观看比赛，因为大多数父母下午都要上班。那些参加这些"二流"体育活动的学生没有获得平等的关注。男篮和男足的日程表应该做出相应改变，为其他的体育活动腾出一点空间。

and the reasons separate. Try to paraphrase the reasons; putting them in your own words helps clarify their meaning and function.

Passage 1

Popular women's magazines create unfair expectations of beauty for women. Almost all the photos of women featured on the covers of magazines have been digitally enhanced by manipulating the lighting and measurements of the bodies of the women in the photo to look more attractive. Studies have shown that it is impossible for women to mimic some of the physical features represented on magazine covers because these features are literally manufactured by computers. The "beauty" that we see on the cover of a magazine is not realistic, but instead, computerized.

Passage 2

With the growing costs of obtaining a college degree, a debate rages about whether students should attend a university. Is it still worth going to college? Obtaining a college degree still holds several benefits.

First, more and more jobs are requiring a college degree. This fact not only means that lacking a college degree will hinder someone's occupational options, but also means that having a college degree will put a person into a separate and higher category when employers are looking for new hires.

Second, the types of jobs that require a college degree are often ones with above-average wages. Not only will these higher wages lead to a higher standard of living than what would be achieved with only a high school diploma, but they will also help mitigate the costs of attending college. It's obvious that attending college is a great choice for securing one's future.

Passage 3

In high school, men's basketball and men's football usually dominate the Friday-night schedule. Should it be that way? These games are significant to the high school experience, but not at the cost of the other sports in the school. Just because it has been a tradition does not mean that the format has to remain that way.

It is easier for most parents and other fans to make it out to the game on Friday nights. Therefore, it is easier for them to come see the men's basketball or men's football games.

What about the girl's basketball team or the swim team? Their games should not always be stuck on weekday afternoons and evenings. Their families often are not able to make it out to see them because most parents are working during the afternoons. The students who play these "secondary" sports are not getting a fair share of the spotlight; the schedule should change to accommodate these other sports.

给个提示

第一篇

  论题：女性杂志有没有塑造出人们对女性美不公正的期待？

  结论：是的，确实如此。

  理由：我们在杂志封面上看到的美女不是真实的，相反，她们都是电脑制作出来的。

（支持性理由）

  a. 杂志封面上的女郎照片都经过数码技术处理，通过调整色彩亮度，调整照片中女性的身材曲线，让她们看起来更加迷人。

  b. 女性要想模仿杂志封面女郎的一些身材特点是不可能的，因为封面女郎实际上是电脑制造出来的美女。

请回想一下，我们是在寻找结论的支撑体系。

我们问自己：为什么写作者宣称女性杂志塑造出人们对女性美不公正的期待？写作者提供了两个理由来证明这个描述性结论：一是宣称杂志封面上的女性照片都经过数码技术处理，改变了女性的身材特征，二是宣称女性不可能去模仿这些经过数码技术处理过的身材特征，它们都是电脑制造出来的。引出支持理由的提示词是"研究显示"。

第二篇

  论题：人们应不应该继续上大学？

  结论：应该，人们都应该继续上大学。

  理由：大学学历能够确保工作岗位。

（支持性理由）

  a. 现在很多工作都需要大学学历，有了大学学历，在求职时就能把自己和其他人区别开来。

  b. 需要大学文凭的工作往往薪酬更高。

   1. 更高的薪酬会带来更高品质的生活和更高的生活水平。

   2. 更高的薪酬还能偿付当年上大学所支付的成本。

为什么我们被告知大家应该继续上大学？这个问题的答案将是写作者的理由。第一个

Sample Responses

Passage 1

ISSUE: *Do women's magazines create unfair expectations of female beauty?*
CONCLUSION: *Yes, they do.*
REASONS: 1. *The beauty that we see on the cover of a magazine is not realistic, but instead, computerized.*

(SUPPORTING REASONS)

   a. *The photos of women featured on the covers of magazines have been digitally enhanced by manipulating the lighting and measurements of the bodies of the women in the photo to look more attractive.*

   b. *It is impossible for women to mimic some of the physical features represented on magazine covers because they are literally manufactured by computers.*

Recall that we are looking for the support system for the conclusion.

We ask ourselves, Why does this person claim that women's magazines create unfair expectations of female beauty? The descriptive conclusion is justified by two reasons: an assertion that the photos of women are digitally enhanced on magazine covers by changing the women's body measurements, and an assertion that it is impossible for women to mimic these digitally enhanced features manufactured by computers. Indicator words for the supporting reason are "studies have shown."

Passage 2

ISSUE: *Should people continue to attend college?*
CONCLUSION: *Yes, people should continue to attend college.*
REASONS: 1. *A college degree can help in securing a job.*

(SUPPORTING REASONS)

   a. *Many jobs now require a college degree, and having a degree sets a person apart when applying for a job.*

   b. *Jobs that require a college degree tend to pay more.*

   1. *The higher pay will lead to a better life and higher standard of living.*

   2. *The higher pay will also help to mitigate the cost of attending college.*

Why are we told that people should continue to attend college? The answer to that question will be the author's reasons. The first reason is

理由是越来越多的工作需要大学学历,这有助于说明大学学历将有助于一个人找到工作。"其次"是指示词,提醒我们注意第二个支持理由。请注意,我们在某种程度上改写了(或用我们自己的话复述)主要理由。你会发现,一个理由越长、越复杂,转述就越有用,它能帮助你准确地找到理由。

supported by claims that more and more jobs are requiring a college degree, and these claims help to illustrate that a college degree will help a person to find a job. *Second* is the indicator word calling our attention to the second supporting reason. Note that we paraphrased (or put into our own words) the major reasons to some extent. You will find that the longer and more complex a reason, the more useful paraphrasing will be to your accurately identifying the reasons.

第 4 章

哪些词语意思不明确

| 学 习 目 标 |

1. 意识到大多数字词有多重含义。
2. 说明在批判性地评价一个论点之前,发现字词意图表达的精确意思的重要性。
3. 展示如何说明模糊性亟待澄清。

第 2 章和第 3 章带你认识了任何一个信息的基本构成要素。读到这儿,如果你能找到一位写作者或发言者的结论和理由,那么你就在快速走向我们的终极目标——形成你自己的理性决策。接下来要做的就是将这幅"论证结构图"放到更清晰的焦点下来审视。

虽然找到结论和理由为你提供了基本清晰的论证结构,但你还需要进一步检查这些部分的确切含义,然后才能对呈现在你面前的观点做出公正的回应。这时你就要特别注意语言的细枝末节。

准确辨认出关键词或短语的确切含义,是你在决定是否同意别人观点时的一个必要步骤。如果你没有仔细核对起决定性作用的关键词或短语的意思,那么也许你回应的观点根本就不是写作者想要表达的内容。

CHAPTER 4

What Words or Phrases Are Ambiguous?

LEARNING OBJECTIVES

1. Demonstrate an awareness of the multiple meaning of most words.
2. Provide illustrations of the importance of discovering the precise intended meaning of a word prior to critically evaluating an argument.
3. Demonstrate how to explain that the ambiguity needs to be clarified.

Chapters 2 and 3 of this text help you identify the basic structural elements in any message. At this point, if you can locate a writer's or speaker's conclusion and reasons, you are progressing rapidly toward the ultimate goal of forming your own rational decisions. Your next step is to put this structural picture into even clearer focus.

While identifying the conclusion and reasons gives you the basic visible structure, you still need to examine the precise *meaning* of these parts before you can react fairly to the ideas being presented. Now you need to pay special attention to the details of the language.

Identifying the precise meaning of key words or phrases is an essential step in deciding whether to agree with someone's opinion. If you fail to check for the meaning of crucial terms and phrases, you may react to an opinion the author never intended.

我们一起看看下面这个例子，来了解一下为什么把握写作者使用的术语的准确含义如此重要。

> 近期一项对大学教练记录的分析表明，教练的经验越丰富，球队获胜的百分比就越高。显而易见，多年从事篮球事业带来了回报，这方面的知识转化为球队的成功。

请注意，如果我们不了解写作者所谓的"经验"具体指哪些内容，我们就很难对这个论证做出评价。打篮球有多少年头了？作为教练执教过多少场比赛？在哪一个特定级别比赛中执教了多少年？在我们对作者声称导致赢得更多篮球比赛的"经验"是什么有更多了解之前，我们不知道该从哪个方面考虑。

这个例子说明了重要的一点：只有理解了关键词和短语的意思（不论是外显的还是内含的意思），我们才能合理地评价一个论证。对这些关键词和短语的解释方式，常常会关系到论证能否被接受。因此，在你决定自己应在多大程度上接受这样那样的结论以前，首先你必须要设法找出结论和理由的准确含义。它们的意思看起来似乎明显，其实不明显。

发现并阐明准确的含义，需要我们自觉遵循一套按部就班的步骤。本章为读者指出这样一套步骤。我们主要关注下面这个问题。

⚠ **批判性问题**：哪些词语是模棱两可（有歧义）的？

让人捉摸不透的多义词

我们使用的语言极其复杂。如果每个词都只有一种潜在的含义，并且大家都认同这个含义，那么迅捷、有效的交流就更有可能实现。可惜的是，论证中许多词语都有不止一种含义。

想一想"自由""淫秽"和"公平"这些词的多重含义吧。这些多重含义在判断一个论证的价值时可能会带来很大的问题。例如，如果有人指出某本杂志根本就不该出版，因为它内容淫秽，只有在你了解到写作者所指的"淫秽"的准确含义之后，你才能评价他的论证是否可信。在这个简短的论证中，结论和支撑的理由都很容易找到，但我们很难判断论证的质量，因为"淫秽"这个词的具体含义不明确。这里，我们要警示大家：我们常常会误解所读到的文章或所听到的言论的含义，因为我们总以为词语的含义都显而易见。

我们在读书、听讲的时候，一定要不断推动自己去寻找那些意思不明确的词或短语，否则就可能对发言者或写作者要表达的内容不得要领。我们说一个词语或短语意思不明确，是指它的意思在我们所考察的论证的上下文语境里让人捉摸不透，需要有进一步的解释，

Let's see why knowing the meaning of a communicator's terms is so important.

> A recent analysis of the records of college coaches demonstrated that the more experienced a coach is, the higher the percentage of games that a team wins. Apparently years of involvement with basketball pay off with the kind of knowledge that translates into basketball success.

Notice that it is very hard to know what to think about this argument until we know more about the kinds of experience that the writer has in mind. Years playing basketball? Number of games coached by this particular coach? Years of coaching at a particular level of competition? We have no idea what to think until we know more about this "experience" the writer is claiming will result in more basketball victories.

This example illustrates an important point: You cannot react to an argument unless you understand the meanings (explicit or implied) of crucial terms and phrases. How these are interpreted will often affect the acceptability of the reasoning. Consequently, before you can determine the extent to which you wish to accept one conclusion or another, you must first attempt to discover the precise meaning of the conclusion and the reasons. While their meaning typically *appears* obvious, it often is not.

The discovery and clarification of meaning require conscious, step-by-step procedures. This chapter suggests one set of such procedures. It focuses on the following question:

 Critical Question: **What words or phrases are ambiguous?**

THE CONFUSING FLEXIBILITY OF WORDS

Our language is highly complex. If each word had only one potential meaning about which we all agreed, effective communication would be more likely. However, many key words in an argument will have more than one meaning.

Consider the multiple meanings of such words as *freedom*, *obscenity*, and *fairness*. These multiple meanings can create serious problems in determining the worth of an argument. For example, when someone argues that a magazine should not be published because it is *obscene*, you cannot evaluate the argument until you know what the writer means by *obscene*. In this brief argument, it is easy to find the conclusion and the supporting reason, but the quality of the reasoning is difficult to judge because of the ambiguous use of *obscene*. A warning: *We often misunderstand what we read or hear because we presume that the meaning of words is obvious.*

Whenever you are reading or listening, force yourself to *search for ambiguity*; otherwise, you may simply miss the point the speaker or writer intended. A term or phrase is ambiguous when its meaning is so uncertain in the context

我们才能判断论证是否恰当。

如果我们当中有人说话总是含糊其词，这个人并不一定做了什么错事或亏心事。事实上，有很多文献，比如美国的宪法，都有意要显得模棱两可，这样这份文献才能与时俱进，其中的关键词如"自由""携带武器"的含义才能适应不断发展的现实需求。实际上，因为我们与人交流时主要依靠词语来将自己的意思传达给他人，所以根本就没有办法完全避免意思表达不明确的情况。但是我们能够而且应该避免在论证当中出现意思表达不明确的情况。如果有人想要说服我们相信某件事或是去做某件事，在我们考虑他的说法有多大价值之前，这个人首先就有责任澄清任何可能存在的歧义。

找准关键词

要确定哪些关键词或短语意思不清楚，第一步就要以表述出来的论题为线索，来寻找可能的关键词。这里所说的关键词或短语，是指在论题的语境里有不止一层潜在含义的词或短语。在你决定是否同意发言者的论证之前，你觉得首先得让他澄清这些词的含义。仔细核对论题中专门术语的含义，这种做法大有好处，为了说明这种好处，我们可以看看下面两个论题：

1. 高收入是否能带来幸福感？
2. 真人秀节目里展示的画面是不是对现实生活的一种歪曲？

⚠ **注意**：歧义是指一个词或短语可能存在多重含义的现象。

以上两个论题中都含有一些需要写作者或发言者进一步加以解释，然后你才能评价他们对这些论题的回答的短语。下列每个短语的意思可能都不太清楚："高收入""幸福感"和"歪曲"。当你读到一篇针对这些论题的回应文章时，你要特别注意写作者如何定义这些术语。

确定哪些词或短语意思不清楚的第二步是要找到在对写作者的理由能否支撑其结论的判断中，哪些词或短语起到关键作用。也就是说，要找出论证结构中的关键词。找到这些关键词后，你就能判断它们的意思是否含糊不清。

在寻找关键词或短语的时候，你应该牢记寻找它们的原因。因为有人要你接受他的结论，所以你才去寻找那些影响你接受这个结论与否的词或短语。这样说来，你应该在理由和结论中寻找这些词或短语。对于那些并未包括在基本论证结构内的关键词或短语，你就可以把它们"从淘金盘里扔出去"。

of the argument we are examining that we need further clarification before we can judge the adequacy of the reasoning.

When any of us is ambiguous, we have not necessarily done something either unfair or improper. In fact, many documents, like American constitutions, are intentionally left ambiguous so that the document can evolve as different meanings of key terms, like "liberty" and "bear arms," become practical necessities. Indeed, because we rely on words to get our points across when we communicate, there is no way to avoid ambiguity. But what can and should be avoided is ambiguity in an argument. When someone is trying to persuade us to believe or do something, that person has a responsibility to clarify any potential ambiguity before we consider the worth of the reasoning.

LOCATING KEY TERMS AND PHRASES

The first step in determining which terms or phrases are ambiguous is to use the stated issue as a clue for possible key terms. Key terms or phrases will be those terms that may have more than one plausible meaning within the context of the issue; that is, terms that you know must be clarified before you can decide to agree or disagree with the communicator. To illustrate the potential benefit of checking the meaning of terminology in the stated issue, let's examine two issues:

1. Does a high income produce happiness?
2. Do reality shows create a misleading picture of how we live?

 Attention: *Ambiguity refers to the existence of multiple possible meanings for a word or phrase.*

Each of these stated issues contains phrases that writers or speakers will have to make clear before you will be able to evaluate their response to the issue. Each of the following phrases is potentially ambiguous: "high income," "happiness," and "misleading picture." Thus, when you read an essay responding to these issues, you have to pay close attention to how the author has defined these terms.

The next step in determining which terms or phrases are ambiguous is to identify what words or phrases seem crucial in determining how well the author's reasons support his conclusion; that is, to identify the *key* terms in the reasoning structure. Once you locate these terms, you can determine whether their meaning is ambiguous.

When searching for key terms and phrases, you should keep in mind why you are looking. Someone wants you to accept a conclusion. Therefore, you are looking for only those terms or phrases that will affect whether you accept the conclusion. *So, look for them in the reasons and conclusion.* Terms and phrases not included in the basic reasoning structure can thus be "dumped from your pan."

在寻找关键词或短语时还有个好帮手，那就是牢记下面这个原则：一个词或短语越抽象，人们就越有可能对其做出多重解读。

为了避免在使用"抽象"这个术语时意思不明确，我们这样来给它下定义：一个词所指代的对象离特定、具体的事例越遥远，它也就越抽象。因此，像"平等""责任""色情"和"侵犯"这些词，就要比"同样有机会获得生活必需品""直接引发某个事件""男女生殖器图片"和"故意伤害他人身体"这些短语抽象得多。后面这些短语提供了更加具体的图像，所以表意更明确。

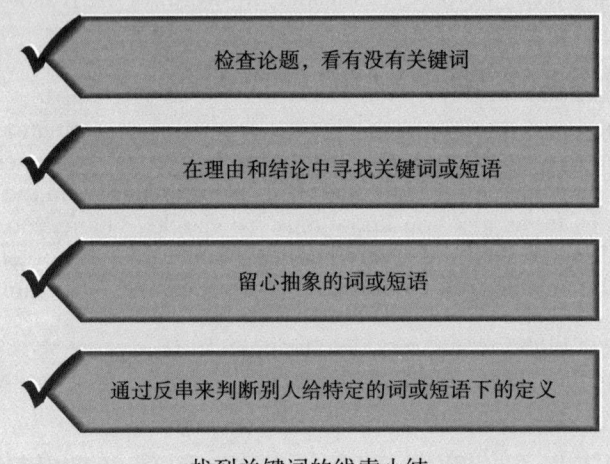

找到关键词的线索小结

你还可以通过反串来找出潜在的重要又有歧义的词或短语。问问自己，如果你采取和写作者相反的立场，那么你会不会选择用不同的方式来定义某些词或短语？如果是这样，你就找到了一处可能存在的歧义。例如，对于"对动物残忍"这个短语，把看宠物秀作为娱乐的人给出的定义肯定和把看宠物秀视为剥削动物的人给出的定义大相径庭。

检查有没有歧义

现在，你知道在哪儿找意思不明确的词或短语了。接下来要做的就是重点关注每个词或短语，问问自己："我知不知道它的意思？"要回答这个重要问题，你需要克服几个主要障碍。

第一个障碍是你自认为理解的和写作者想表达的是同一个意思。因此，在开始寻找之前，你就要避免这种和写作者心心相印的想法。你要养成不断提问的习惯，不停地问"你这样说是什么意思"而不是"我就知道你是这个意思"。第二个障碍是认为词语只存在一个明显的定义。很多词语都不是这样。因此，别忘了问自己："这些词语或短语是否可能有不

Another useful guide for searching for key terms and phrases is to keep in mind the following rule: *The more abstract a word or phrase, the more likely it is to be susceptible to multiple interpretations.*

To avoid being unclear in our use of the term *abstract*, we define it here in the following way: A term becomes more and more abstract as it refers less and less to particular, specific instances. Thus, the words *equality, responsibility, pornography,* and *aggression* are much more abstract than are the phrases "having equal access to necessities of life," "directly causing an event," "pictures of male and female genitals," and "doing deliberate physical harm to another person." These phrases provide a much more concrete picture and are therefore less ambiguous.

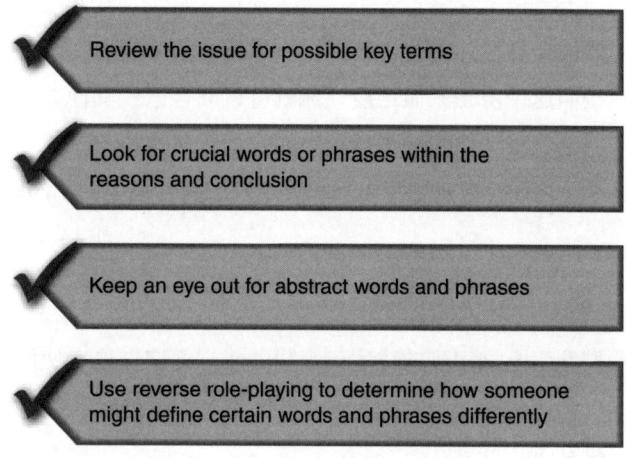

Summary of Clues for Locating Key Terms

You can also locate potential important ambiguous phrases by *reverse role-playing*. Ask yourself, if you were to *adopt a position contrary to the author's*, would you choose to define certain terms or phrases differently? If so, you have identified a possible ambiguity. For example, someone who sees dog shows as engaging entertainment is likely to define "cruelty to animals" quite differently from someone who sees them as exploitative of animals.

CHECKING FOR AMBIGUITY

You now know where to look for ambiguous terms or phrases. The next step is to focus on each term or phrase and ask yourself, "Do I understand its meaning?" To answer this very important question, you will need to overcome several major obstacles.

One obstacle is assuming that you and the author mean the same thing. Thus, you need to begin your search by avoiding mind reading. You need to get into the habit of asking, "What do you mean by that?" instead of, "I know just what you mean." A second obstacle is assuming that terms have a single,

同的意思?"

做一个如下的测验,你就可以肯定自己已经找到了一个意思不明确的重要词语。如果对于一个词语,你能找到两种或两种以上不同的含义,并且每种含义放到这个论证的上下文语境里都能说得通,而采用不同的含义会使这个论证中理由支撑结论的效力大受影响,那么你就找到了一处重要的歧义。因此,要判定你是否找到了一处重要的歧义,最好的检测方法就是将这个词的不同含义代入论证的结构中,看看改变这个词的意思是否会对论证中理由支撑结论的效力产生重大影响。

使用这个批判性问题

前面一段需要你全神贯注地加以细读。它详细说明了使用这个关于歧义的批判性问题的具体步骤。在按照这个步骤来做之后,你就可以向自己或其他人说明,为什么这个论证过程还需进一步完善。虽然你也想相信别人说的话,但作为一个批判性思维者,你就是不肯轻易同意他的论证过程,直到他将影响论证过程的那些意思不明确的地方解释清楚才行。

判定歧义

现在就让我们利用上文提到的线索,来帮助我们确定立论者使用的哪些关键词意思不清楚。请记住,在做这个练习的时候,你要不断地问:"写作者这样说是什么意思?"尤其要注意写作者对抽象词语的使用。

我们将从简单的论证结构——广告开始。

> 大家牌安眠药:只需 30 分钟就可见效。
> 论题:你应该购买什么样的安眠药?
> (暗含)结论:购买大家牌安眠药。
> 理由:只需 30 分钟就可见效。

短语"购买大家牌安眠药"和"30 分钟"的意思好像十分具体和清楚,但是,"见效"这个词怎么样呢?这个词的意思清楚吗?我们认为并不清楚。何以见得?让我们一起来做个测验。"见效"这个词可不可能包含一个以上的意思?当然可能。它可能意味着这种安眠药让你昏昏欲睡,也可能意味着这种安眠药让你一下子昏睡过去,直到第二天早上也醒不来。或者它还可能有许多其他的意思。如果说这种药很快"见效",意思就是它的效果恰如你心中所想的那样不多不少,那你是不是会急不可耐地听从这则广告的建议呢?因此,这里的歧义很重要,因为它影响到你被这则广告说服的程度。

obvious definition. Many terms do not. Thus, always ask, "Could any of the words or phrases have a different meaning?"

You can be certain you have identified an especially important unclear term by performing the following test. If you can express two or more alternative meanings for a term, each of which makes sense in the context of the argument, and if the extent to which a reason would support a conclusion is affected by which meaning is assumed, then you have located a significant ambiguity. Thus, a good test for determining whether you have identified an important ambiguity is to *substitute* the alternative meanings into the reasoning structure and see whether changing the meaning *makes a difference* in how well a reason supports the conclusion.

USING THIS CRITICAL QUESTION

The preceding paragraph deserves your full attention. It is spelling out a procedure for putting this critical question about ambiguity to work. Once you have followed the procedure, you can demonstrate to yourself or anyone else why the reasoning needs more work. Try as you might to want to believe what is being said, you just cannot, as a critical thinker, agree with the reasoning until the ambiguity that affects the reasoning is repaired.

DETERMINING AMBIGUITY

Let's now apply the above-mentioned hints to help us determine which key terms a communicator has left unclear. *Remember*: As we do this exercise, keep asking, "What does the author mean by that?" and pay particular attention to abstract terms.

We will start with a simple reasoning structure: an advertisement.

> OurBrand Sleep Aid: Works great in just 30 min.
>
> Issue: What sleep aid should you buy?
>
> Conclusion (implied): *Buy OurBrand Sleep Aid.*
>
> Reason: Works great in 30 min.

The phrases "Buy OurBrand Sleep Aid" and "in 30 min." seem quite concrete and self-evident. But, how about "works great?" Is the meaning obvious? We think not. How do we know? Let's perform a test together. Could "works great" have more than one meaning? Yes. It could mean the pill makes you drowsy. It could mean the pill completely knocks you out such that you will have difficulty waking up the next morning. Or it could have many other meanings. Isn't it true that you would be more eager to follow the advice of the advertisement if the pill worked great, meaning it works precisely as you want it to work? Thus, the ambiguity is significant because it affects the degree to which you might be persuaded by the advertisement.

广告里常常充满了意思模棱两可的词语。广告商有意使用这些歧义来说服你，让你相信他们的产品比所有竞争对手的产品都要棒。下面的广告词样本里的广告词都有意使用了一些意思不明确的词语。黑体部分的词或短语都包含了其他可以解释得通的含义，看看你能否找出这些含义。

疼痛消是超强的止痛药。
这本书千呼万唤始出来，它将告诉你如何找到好朋友并维持友谊。

在每个例子中，广告商都希望你用最吸引人的意思来理解含义模糊的词语。批判性思维能为你保驾护航，让你不贸然做出事后追悔莫及的购物决定。

下面，我们来看一个复杂一些的例子，其中同样使用了含义模糊的词语。请记住一开始就要找准论题、结论和理由。要控制自己，不要一遇到含义模糊的词，马上就全部记下来。对一名批判性思维者来说，只有在论证过程中出现的含义模糊的词才是重要的。

我们绝对有必要对晒肤施加限制。晒肤会构成实实在在的健康威胁，带来很多严重后果。研究表明，晒肤的人罹患皮肤疾病的危险会大大增加，这都是暴晒的结果。

我们检查一下写作者的论证，看看有没有什么词或短语会影响我们接受这一结论的意愿。首先，我们检查一下论题，看看论题中有没有什么我们希望写作者进一步解释清楚的术语。毫无疑问，只有写作者明确指出他所说的"晒肤"到底是什么意思，我们才能决定是否同意他的结论。他说的是户外晒肤，还是人工晒肤？我们需要检查一下他在论证中所下的定义到底有多清楚。

接下来，我们列举出结论和理由中包含的所有关键词和短语："健康威胁""严重后果""研究表明""晒肤的人危险大大增加""皮肤疾病"和"我们有必要对晒肤施加限制"。我们进一步查看这些词，判断它们是否可能还有别的意思，这些意思可能对我们如何评价其论证过程产生重要影响。

第一，写作者的结论不明确。"对晒肤施加限制"到底要表达什么意思？是说要禁止人们使用人工晒肤器具，还是说要限制人们晒肤的时间长度？在你决定是否同意写作者或发言者的观点之前，首先要判断他到底想要我们相信什么。

第二，他说"晒肤的人罹患皮肤疾病的危险大大增加"，正如我们提到的，对于写作者所谓的"晒肤的人"到底是什么意思，我们还不太确定。这里，他所说的"皮肤疾病"又是什么意思？他可能指的是由太阳暴晒引发的任何形式的疼痛感，也可能指的是像皮肤癌那样严重的疾病。如果他要说服我们相信晒肤带来的危险以及他提出的限制晒肤的结论，

Advertising is often full of ambiguity. Advertisers intentionally engage in ambiguity to persuade you that their products are superior to those of their competitors. Here are some sample advertising claims that are ambiguous. See if you can identify alternative, plausible meanings for the italicized words or phrases.

No-Pain is the *extra-strength* pain reliever.

Here is a book that shows at last how to find and keep a *good friend*.

In each case, the advertiser hoped that you would assign the most attractive meaning to the ambiguous words. Critical reading can sometimes protect you from making purchasing decisions that you would later regret.

Let's now look at a more complicated example of ambiguity. Remember to begin by identifying the issue, conclusion, and reasons. Resist the temptation to make note of the unclear meaning of any and all words. Only the ambiguity **in the reasoning** is crucial to critical thinkers.

> We absolutely must put limits on tanning. Tanning is a substantial health risk with severe consequences. Studies have shown that those who tan are at a higher risk of skin diseases as a result of tanning.

Let's examine the reasoning for any words or phrases that would affect our willingness to accept it.

First, let's inspect the issue for terms we will want the author to make clear. Certainly, we would not be able to agree or disagree with this author's conclusion until she has indicated what she means by *tanning*. Does she mean tanning outdoors or artificial tanning? Thus, we will want to check how clearly she has defined it in her reasoning.

Next, let's list all key terms and phrases in the conclusion and reasons: "health risk," "severe consequences," "studies have shown," "those who tan are at a higher risk," "skin diseases," and "we should put limits on tanning." Let's take a close look at a few of these to determine whether they could have different meanings that might make a difference in how we would react to the reasoning.

First, her conclusion is ambiguous. Exactly what does it mean to "put limits on tanning"? Does it mean to prevent people from using artificial tanning devices, or might it mean putting a limit on the amount of time spent tanning? Before you could decide whether to agree with the speaker or writer, you would first have to decide what she wants us to believe.

Next, she argues that "those who tan are at a higher risk of skin diseases." We have already talked about how we are not sure what she means by "those who tan," but what does she mean by "skin diseases"? She could mean any number of irritations that can occur from sun exposure, or she could be talking about something as severe as skin cancer. It is significant to know which of these she is addressing if she wants to convince you of the

那么搞清楚他到底指的是哪一类的皮肤疾病就很重要。我们可以试着在脑海里想一想这些短语代表的现实画面。如果想象不出来，这就说明这些短语的意思不明确。如果不同的画面会导致你对他的理由做出不同的评价，那么你就找到了一处重要的歧义。

现在，你可以检查一下我们列举的其他短语。看看它们是否需要得到进一步的解释说明。你会了解到，如果你没有要求写作者将这些含义模糊的词语解释清楚，就贸然接受他的论证，那你根本就不理解你选择相信的到底是什么东西。

上下文和歧义

写作者和发言者很少会给关键词下定义。因此，想理解一个模糊不清的陈述到底是什么意思，通常唯一可以依靠的线索就是这些词语的上下文语境。这里所说的语境，意思是指写作者或发言者的背景、这一词语在某一争论中的习惯用法，以及位于歧义词前后的其他词语和陈述，这些因素合在一起为潜在的关键词或短语的意思提供了线索。

如果你想搞清楚一篇文章中"人权"这个词的意思，你就要先问自己："人权到底是些什么权利？"如果你检查这个词出现的语境之后，发现作者是挪威政府的一位主要领导，那你基本可以确定他心目中所谓的"人权"是指就业权、免费医疗权和房屋居住权。而一个美国参议员所说的"人权"，意思可能完全不一样。他心目中的"人权"可能是指言论自由、宗教信仰、旅行以及和平集会的权利。注意这两个版本的"人权"并不需要彼此一致。一个国家可能在保障一种形式的"人权"的同时侵犯另一种形式的"人权"。你必须设法通过检查语境来弄清楚这些词的确切含义。

使用这个批判性问题

聚焦歧义现象的批判性问题为你奠定了不偏不倚的基础，让你可以理直气壮地不同意某些论证。如果你与想要说服你的人对论证过程中用到的关键词的意思在理解上存在分歧，那么你首先要解决这些分歧，然后才能接受他提供的论证。

要通过仔细检查语境来判定关键词或短语的意思。在这样做之后，如果它们的意思还是难以确定，那你就找到了一处重要的歧义。如果检查语境后你清楚了词语的意思但并不认同，那你就要对包含这个词或短语的论证心存警惕。

歧义、定义和字典

前面的讨论让我们明白，要找到并解释清楚歧义词，我们必须要知道这些词可能包含

dangers of tanning and her conclusion to limit it. Try to create a mental picture of what these phrases represent. If you can't, the phrases are ambiguous. If different images would cause you to react to the reasons differently, you have identified an important ambiguity.

Now, check the other phrases we listed earlier. Do they not also need to be clarified? You can see that if you accept this writer's argument without requiring her to clarify these ambiguous phrases, you will not have understood what you agreed to believe.

CONTEXT AND AMBIGUITY

Writers and speakers only rarely define their key terms. Thus, typically your only guide to the meaning of an ambiguous statement is the context in which the words are used. By *context*, we mean the writer's or speaker's background, traditional uses of the term within the particular controversy, and the words and statements preceding and following the possible ambiguity. All three elements provide clues to the meaning of a potential key term or phrase.

If you were to see the term *human rights* in an essay, you should immediately ask yourself, "What rights are those?" If you examine the context and find that the writer is a leading member of the Norwegian government, it is a good bet that the human rights he has in mind are the rights to be employed, receive free health care, and obtain adequate housing. An American senator might mean something very different by human rights. She could have in mind freedoms of speech, religion, travel, and peaceful assembly. Notice that the two versions of human rights are not necessarily consistent. A country could guarantee one form of human rights and at the same time violate the other. You must try to clarify such terms by examining their context.

USING THIS CRITICAL QUESTION

The critical question focusing on ambiguity provides you with a fair-minded basis for disagreeing with the reasoning. If you and the person trying to persuade you are using different meanings for key terms in the reasoning, you would have to work out those disagreements first before you could accept the reasoning being offered to you.

Examine the context carefully to determine the meaning of key terms and phrases. If the meaning remains uncertain, you have located an important ambiguity. If the meaning is clear and you disagree with it, then you should be wary of any reasoning that involves that term or phrase.

AMBIGUITY, DEFINITIONS, AND THE DICTIONARY

It should be obvious from the preceding discussion that to locate and clarify ambiguity, you must be aware of the possible meanings of words. Meanings usually come in one of three forms: synonyms, examples, and what we will

的多重意义。意义通常表现为以下三种方式的一种：同义替换、举例说明以及我们所谓的"具体标准定义"。举例来说，对"焦虑"这个词，我们至少可以找到三种不同的定义方法：

1. 焦虑就是感到紧张不安（同义替换）。
2. 焦虑就是候选人打开电视观看选举结果公布时的心情（举例说明）。
3. 焦虑是一种主观上的不适感，同时伴随着自主神经系统越来越强烈的感受（具体标准定义）。

在对大多数有争议的论题进行批判性评价时，同义替换和举例说明这两种定义方式都不够充分。它们并不能明确告诉你哪些具体特征能对清楚理解一个词的意思起到决定性作用。有用的定义则会指出具体的使用标准——越具体越好。

到哪儿去找你需要的定义呢？一个很明显又很重要的来源就是在线字典。但是，字典上的定义常常包含了同义替换、举例说明或是使用标准的不完全说明。在具体文章中，这些定义往往难以确切界定词语的使用方法。在这种情况下，你就需要从文章的上下文语境中找出这个词的潜在含义，或者从你对文章主题的其他了解里发掘出它的意思。

让我们来看看字典提供的定义的一些不充分之处。请阅读下面这段文章：

这所大学的教育质量并没有出现滑坡。我在访谈中发现，绝大多数学生和老师都认为他们在这儿根本没有看到什么教育质量滑坡的现象。

弄清上面这段文字中"教育质量"这个短语的意思非常重要。如果你在字典中查"质量"这个词，你会发现它有很多种意思。考虑到这个词出现的语境，它在这里最恰当的意思就是"杰出度"和"优越性"。"杰出度"和"优越性"是"质量"的同义词，而且它们同样抽象。你仍需要准确地了解"杰出度"和"优越性"到底是什么意思。你怎么知道教育是"质量高"还是"杰出度高"呢？理想的情况是，你想让写作者准确地告诉你，他在使用"教育质量"这个词的时候，到底指的是一种什么样的表现。你能不能想出这个短语的一些不同的定义方式呢？下面我们就列出了"教育质量"这个短语可能存在的几种定义：

学生的平均成绩。
学生批判性地思考问题的能力。
有博士学位的教授人数。
考试过关通常要付出的工作量。

每个定义都指向一种衡量教育质量的不同方法，每个定义都有不同的具体标准。每个定义也都提供了这一短语可能存在的具体使用方法。注意这些定义都会影响你同意写作者

call "definition by specific criteria." For example, one could offer at least three different definitions of *anxiety*:

1. Anxiety is feeling nervous (*synonym*).
2. Anxiety is what the candidate experienced when he turned on the television to watch the election returns (*example*).
3. Anxiety is a subjective feeling of discomfort accompanied by increased sensitivity of the autonomic nervous system (*specific criteria*).

For critical evaluation of most controversial issues, synonyms and examples are inadequate. They fail to tell you the specific properties that are crucial for an unambiguous understanding of the term. Useful definitions are those that specify criteria for usage—the more specific, the better.

Where do you go for your definitions? One obvious and very important source is your online dictionary. However, dictionary definitions frequently consist of synonyms, examples, or incomplete specifications of criteria for usage. These definitions often do not adequately define the use of a term in a particular essay. In such cases, you must discover possible meanings from the context of the passage, or from what else you know about the topic.

Let's take a closer look at some of the inadequacies of a dictionary definition. Examine the following paragraph.

> The quality of education at this university is not declining. In my interviews, I found that an overwhelming majority of the students and instructors responded that they saw no decline in the quality of education here.

It is clearly important to know what is meant by "quality of education" in the given paragraph. If you look up the word *quality* in the dictionary, you will find many meanings, the most appropriate, given this context, being *excellence* or *superiority*. Excellence and *superiority* are synonyms for *quality*—and they are equally abstract. You still need to know precisely what is meant by *excellence* or *superiority*. How do you know whether education is high in quality or excellence? Ideally, you would want the writer to tell you precisely what *behaviors* he is referring to when he uses the phrase "quality of education." Can you think of some different ways that the phrase might be defined? The following list presents some possible definitions of *quality of education*:

> average grade-point of students
>
> ability of students to think critically
>
> number of professors who have doctoral degrees
>
> amount of work usually required to pass an exam

Each of these definitions suggests a different way to measure quality; each specifies a different criterion. Each provides a concrete way in which the term could be used. Note also that each of these definitions will affect the

的论证的程度。例如，如果你认为这里的"质量"应该指学生批判性地思考问题的能力，而大多数受访学生都把它定义为考试过关要付出的工作量，那么这里的理由就不一定能够支撑结论。考试过关也许根本就不需要具备批判性思考问题的能力。

可见，在很多论证中，你并不能从字典里找到合适的定义，上下文语境也不能让意思变得清晰起来。此时，有一种方法可以帮你发现词语可能存在的歧义，那就是尽量在脑海中想象这个词所代表的具体图像。如果你想象不出来，那你很可能就找到了一处重要的歧义。我们可以用下面这个例子来测试一下。

> 我们公司一直都有很多能干的员工。如果你加入我们的员工队伍，那你立刻就能享受我们刚谈到的那个工资待遇，当然，还有好多额外福利。我希望你在做决定的时候能考虑一下所有这些因素。

上面这个论证显然是想劝说某人到某个单位去就职，其理由就是工资和"额外福利"。你能不能在脑海里想象出一幅清晰的"额外福利"的画面？我们每个人都有这样那样的想法，但是，两个人的想法完全相同的概率实在是微乎其微，实际上，不同的想法很有可能千差万别。"额外福利"是指医疗保险，还是指一间拐角的新办公室？我们要评价这个论证，就得知道写作者所指的"额外福利"的具体含义。这样，我们就找到了一处重要的歧义。

澄清歧义不只是你的责任

在你设法找出并澄清歧义以后，如果你对某些关键观点的准确含义还是拿不准，那又该怎么办呢？接下来该采取什么合理步骤？对于任何一个理由，如果其中包含了歧义，让人无法判断能否接受，那我们建议你干脆忽略它。作为一个积极主动的学习者，你有责任提出各种问题来澄清歧义。但是，你的责任到此为止。写作者和发言者才是努力要说服你接受某些观点的人。身为说客，他有责任回答你对可能存在的歧义的各种疑问。

你没有义务去回应那些表述不清的观点或选择。如果有朋友说你应该去选修某一门课，因为这门课的确"与众不同"，但他却不能告诉你不同之处在哪里，那你就没有接纳或拒绝这个建议的依据。如果他不能提供一幅清晰的论证画面，那他就没有权利强迫你相信他。

像批判性思维者那样写作和发言

你已花了几周时间研究你的论文，论文声称校园性侵案正在减少。你引用了著名研究人员的观点来印证自己的断言，对自己文章的质量感觉颇好。但是，当你从老师那里拿回

degree to which you will want to agree with the author's reasoning. For example, if you believe that "quality" should refer to the ability of students to think critically, and most of the students in the interviews are defining it as how much work is required to pass an exam, the reason would not *necessarily* support the conclusion. Exams may not require the ability to think critically.

Thus, in many arguments, you will not be able to find adequate dictionary definitions, and the context may not make the meaning clear. One way to discover possible alternative meanings is to try to create a mental picture of what the words represent. If you cannot do so, then you probably have identified an important ambiguity. Let's apply such a test to the following example:

> Our company has had many competent employees. If you join our staff, you will start immediately at the rate we discussed with, of course, added benefits.
> I hope you consider all these factors in making your employment decision.

This argument is clearly an attempt to persuade someone to work at a place of employment. The reasons are the salary and "added benefits." Can you create a single clear mental picture of "added benefits"? We each have some such idea, but it is highly unlikely that the ideas are identical; indeed, they may be quite different. Do "added benefits" refer to health care insurance or a new corner office? For us to evaluate the argument, we would need to know more about the meaning the writer has for "added benefits." Thus, we have located an important ambiguity.

LIMITS OF YOUR RESPONSIBILITY TO CLARIFY AMBIGUITY

After you have attempted to identify and clarify ambiguity, what can you do if you are still uncertain about the meaning of certain key ideas? What is a reasonable next step? We suggest you ignore any reason containing ambiguity that makes it impossible to judge the acceptability of the reason. It is your responsibility as an active learner to ask questions that clarify ambiguity. However, your responsibility stops at that point. It is the writer or speaker who is trying to convince you of something. Her role as a persuader requires her to respond to your concerns about possible ambiguity.

You are not required to react to unclear ideas or options. If a friend tells you that you should enroll in a class because it "really is different," but cannot tell you how it is different, then you have no basis for agreeing or disagreeing with the advice. No one has the right to be believed if he cannot provide you with a clear picture of his reasoning.

WRITING AND SPEAKING AS CRITICAL THINKERS

You've spent weeks researching your essay that argues that rape on campus is decreasing. You have citations from noted researchers backing up your claims, and you are feeling really good about the quality of your essay. But, when you

批改后的文章时,你只得到了一个"D"等,而且老师的评语这样写道:"有充分的证据表明你的结论是错误的。请重做!"你会不会大哭一场,把论文撕个粉碎,然后重新开始?或者你会不会要求老师给你机会来澄清,向其展示你所做的研究和找到的支撑?如果你和老师讨论你的论文成绩,也许你会发现,你和老师对"性侵"的定义是不尽相同的。你可能会发现,你聚焦的是"约会性侵",而你的老师指的是各种类型的性侵或者只是陌生人的性侵。其实你只要在论文中仔细定义和澄清你的术语,就可以免除论文获得"D"等成绩带来的压力和烦恼。

在上一章节的"像批判性思维者那样写作和发言"部分,我们讨论了清晰地组织论点的重要性。在这一部分,我们将关注写作过程中词和短语表意清晰的重要性。如果我们语焉不详,词不达意,那么没有人可以有效地回应我们所写或所说的内容。我们的意图必须通顺流畅、准确无误地传递给我们所面对的受众。

作为一名写作者或发言者,我们常常认为自己要表达的内容已经清楚明晰,但很有可能受众接收到的和我们想表达的意思大相径庭。词语本身有丰富的意义,我们说出一个词是为了引起想象的共鸣。

那么,我们的工作就是把想要表达的意思传达给受众。在上面的例子中,在你和老师讨论这个话题时,你才意识到其实你想要讨论的是"约会性侵"。在确定论证之前,我们得把用到的词的意义想清楚,这样不仅可以让读者和听众心领神会,也可以让我们自己胸有定见。

为了降低模棱两可的词或短语影响到我们的正常交流的可能性,我们可以做以下几点:

1. 在论题和结论中找到关键词或短语,并定义和/或解释它们。
2. 在理由中找到关键词或短语,并定义和/或解释它们。
3. 识别核心论点中的抽象词,一个词越抽象,我们就越有必要澄清它。
4. 问一问:"那些与我持不同意见的人,可能会如何用不同方式来定义这些术语以支持他们自己的论点?"

让我们从第一步开始,看看论证中的论题,找出可能有多重含义、有歧义的词。
看看下面的论题:

1. 自动驾驶汽车会侵犯人的隐私吗?
2. 约会性侵是否随着"必须应答"政策的出现而减少?
3. 拥有更多的选择会提高人们的幸福感吗?

get it back from your teacher, you get a D and the comment, "There is ample evidence that your conclusion is wrong. Please redo." Do you burst into tears, rip up the essay, and start over? Or do you ask her for clarification, showing the teacher your research and support? Perhaps what you discover, if you do discuss your essay grade with your teacher, is that you and your teacher are using two different definitions of rape. You might find that while you focused on "date rape," your teacher meant all types of rape or only stranger rape. You would have saved yourself the stress and aggravation of the D grade if you had carefully defined and clarified your terms in your essay.

In our last writing and speaking section, we discussed the importance of clarity about how we structure our arguments; in this section, we will focus on the importance of clarity when we use words and collections of words. No one can respond effectively to what we write or say if we are not both clear and precise. Our intended meaning must be transferred smoothly and accurately to the audience we are addressing.

As a writer or speaker, we often think what we have to say is crystal clear, but it is highly possible that our audience formed an image of what we said that was quite different from what we intended. Words are rich in meaning; to say a word is to strike a chord on the imagination.

Our job, then, is to move our intended meaning of words into the minds of our audience. In the example above, it may have become clear to you only as you were discussing the topic with your teacher that you did indeed intend to discuss specifically date rape. Thinking through our terms before we finalize our arguments can bring clarity not only to readers and listeners but also to ourselves.

To decrease the likelihood that any ambiguous words or phrases are detracting from our communication, we can do the following:

1. Locate key terms or phrases in our issue and conclusion and define and/or explain them.
2. Locate key terms or phrases in our reasons and define and/or explain them.
3. Identify abstract words in our core argument; the more abstract the word, the more likely we will need to clarify it.
4. Ask, "How might someone who disagrees with me define these terms differently in order to support his own argument?"

Let's start with the first step by looking at the issue and conclusion in an argument and identifying any words that could have multiple, potentially ambiguous meanings.

Look at the following issues (questions):

1. Do self-driving cars violate people's privacy?
2. Is date rape decreasing with the advent of "must say yes" policies?
3. Will having more choices increase people's happiness?

每个问题都包含模棱两可的词或短语，在我们确定读者或听众能理解我们的论证之前，必须澄清这些词或短语的含义。花一分钟的时间，看看你是否能识别出那些模棱两可的词或短语。

这些模棱两可的词包括"隐私""约会性侵""必须应答""更多"和"幸福感"。为了确保我们的论证能够得到理解，并且关键词的模糊性不会成为批评的基础而困扰我们（就像我们在本部分一开始举的例子），我们必须对这些词语进行定义，因为它们是我们要得到的结论的核心。

我们看看第一个例子：自动驾驶汽车会侵犯人的隐私吗？我们必须问："隐私是什么？"1923 年，美国最高法院大法官路易斯·布兰代斯将隐私定义为"我们独处的权利"。有些人认为这个定义太宽泛了，他们可能会把"隐私"细分为两类：身体隐私和信息隐私。其他人有可能会说，隐私实际上是指美国宪法第四修正案规定的诸多权利：第四修正案确保美国公民的正当权利，防止政府及其代理人无合理理由地侵犯公民的私人财产。如果我们做一些研究，就会发现这只是我们定义隐私的可能方式中的三种。作为写作者或发言者，在构建论点之前，我们需要确定在结论中我们到底想让"隐私"意味着什么。我们不能假设读者或听众能读懂我们的心思，或者假设定义已是显而易见的了。

在结论中确定了关键词的定义之后，我们就需要在理由中识别出模棱两可或抽象的词。例如，如果我陈述自动驾驶汽车因为分享了个人信息而侵犯了隐私，我们将会注意到理由中的"个人信息"这个词是含糊不清的。想想我们定义个人信息的各种不同方式：姓名、地址、车牌号码、男朋友的名字、出生日期、驾驶速度、刹车频率、最喜欢的颜色、手机号码……这个列表实际上是无限长的。而且，我们真的想争论说每一种个人信息都应该受到保护吗？不言而喻，这是一个很难为之辩护的结论，考虑到我们经常不经意地就会说出自己的姓名、地址甚至生日。相反，我们必须决定哪些个人信息应该受到保护，并向读者或听众明明白白地阐明这一点。

你现在可能在想，如果整篇文章都充斥着定义，那么我们就没有空间写其他东西了！并非如此，在澄清我们的语言时，我们有其他选择，并不只有细致入微地（也许是单调乏味地）对所有的关键词都进行定义这一种方式。一种常见的方法就是利用论证本身来明确我们要表达的意思。例如，如果我们认为幸福感会随着选择的增加而降低，我们可能会构建如下论点：

> 研究表明，幸福感会随着人们选择的增加而降低。在实验中，当受试者面对 1 个、5 个或 10 个选择时，实验人员对其脑内啡的释放量进行了测量，结果表明，选择越多，脑内啡的释放量就越少。

Each of these questions includes ambiguous words or phrases that must be clarified before we can be sure our reader or listener will understand our argument. Take a minute and see if you can identify the ambiguous words or phrases.

The ambiguous phrases include "privacy," "date rape," "'must say yes' policies," "more," and "happiness." To make sure our arguments are understood and that the ambiguity of key terms does not come back to haunt us by becoming the basis for criticism (as in our beginning example), we must define each of these terms because they are central to our conclusion.

Let's take the first example: Do self-driving cars violate people's privacy? We must ask, "What is privacy?" In 1923, Supreme Court Justice Louis Brandeis defined privacy as "our right to be left alone." Some people argue this definition is way too broad, and they might break down "privacy" into two categories: physical privacy and information privacy. Others might say privacy is actually referring to our Fourth Amendment rights: the Fourth Amendment guarantees our right to keep the government and its agents from violating our private property without probable cause. And, if we do some research, we'll see that these are only three possible ways out of many that we might define privacy. As a writer or speaker, we need to determine just what we want privacy to mean in our conclusion before we construct our argument. We cannot assume that our reader or listener can read our mind or that definitions are obvious.

Once we've determined the definition of the key terms in our conclusion, we need to identify any ambiguous or abstract terms in our reasons. For example, if I state that self-driving cars violate privacy because they share personal information, we will notice that the phrase "personal information" in our reason is particularly ambiguous. Think of all the different ways we might define personal information: name, address, license plate number, boyfriend's name, birthdate, driving speed, braking frequency, favorite color, cell phone number... The list is virtually infinite. And, do we really want to argue that every piece of our personal information should be protected? Obviously, that would be a very difficult conclusion to defend given how often we freely give out our name, address, and even birthdate. Instead, we've got to decide just what personal information is worth protecting and clarify that explicitly for our reader or listener.

You may be thinking by now that our entire essay will be filled with definitions, and we won't have space for anything else! However, in clarifying our language, we do have other options than only meticulously (and perhaps tediously) defining all our key terms. One common approach is to use the argument itself to make clear what we mean. For example, if we are arguing that happiness decreases with increased choices, we might construct the following argument:

> Happiness has been shown to decrease as we increase people's choices. In experiments that measure the release of endorphins when subjects are presented with one, five, or ten choices, results indicated that as the choices increased the released endorphins decreased.

我们可以看到，这个论证在其论证结构中定义了被我们所说的"幸福感"的含义。但是，我们必须注意，这个特定的定义是否在整篇文章中一以贯之。我们不能在另一个理由中把自述的快乐作为幸福感的衡量标准而大谈特谈。我们在"幸福感"这个术语的使用上的这种转变会给读者带来困惑，而且给读者质疑结论留足了理由。

另外，请记住，我们只需要定义那些对我们的论证具有核心意义的词语，这些词语可能是模棱两可或是抽象的。我们不需要定义我们使用的每一个词或短语。

在完成了对模糊性的搜索之后，我们可以进而采用其他一些辅助手段来确定我们仍需要澄清的关键词或短语。首先，我们可以请其他人来阅读我们所写的内容，最好是请那些对主题知之甚少，并与我们写作或发言的受众有相似之处的人。让他指出他可能不理解的词。陌生的读者通常是识别模棱两可的关键词的最佳人选，因为他们不熟悉我们的论证。

最后，人们很容易认为，定义和解释词语的过程会令人感到无聊，甚至枯燥乏味：这是一项查阅和抄写字典的乏味任务。请试着把这样的练习看作厘清自己思路的一个令人兴奋的机会。对于结论而言，文字犹如砥石，思想好比锋刃，两相砥砺将使思想更加锋利，长相砥砺，会生发出更加深思熟虑的论证。

思维体操

⚠ **批判性问题**：什么词或短语意思不明确？

请在下面的文章中找出意思不明确的例子。解释一下为什么这些例子影响到论证的效果。

第一篇

学校的着装规定主要是用来限制学生衣着不当的，目的是营造一种专心致志的学习氛围。如果一个学生衣着不当地来上课，就可能分散其他同学的注意力。规定学生在校期间的着装并不是限制学生的表达自由。不像要求学生统一着装，着装规定仍然允许学生自由地选择自己的服装，只要不被视作不当着装即可。

第二篇

平板电脑应该取代从幼儿园到12年级教育阶段的教科书。支持使用平板电脑的人声称，平板电脑性价比更高，使用起来更省时，并且占用的空间更小。例如，平板电脑比教

As we can see, this argument defines what we mean by *happiness* in its argument structure. But, we must be careful that this specific definition holds up throughout the full essay. We can't start talking about self-reported joy as a measure of happiness in another reason. Such a switch in how we use the term *happiness* would cause confusion in the reader *and* give the reader reason to challenge the conclusion.

Also, keep in mind that we need to define only those terms central to our argument *and* which are ambiguous or abstract. We do not need to define every term or phrase that we use.

After we've completed our search for ambiguity, we can turn to a couple of other possible aids in identifying key terms or phrases that we still might need to clarify. First, we can get someone else to read what we've written, preferably someone who knows little about the topic and/or resembles the audience for which we are writing or speaking. Ask him to search for terms that he might not understand. Outside readers are often a good way to identify key terms that are ambiguous because they do not have any familiarity with our argument.

Finally, it's easy to think that the process of defining and explaining terms can be boring and mind numbing: a tedious task of dictionary reading and copying. Try to see such exercises as an exciting opportunity to clarify your own thinking. Identifying what words mean within the context of your conclusion often leads to a sharpening of your own thinking and, in the long run, more reflective arguments.

PRACTICE EXERCISES

 Critical Question: **What words or phrases are ambiguous?**

In the following passages, identify examples of ambiguity. Try to explain why the examples harm the reasoning.

Passage 1

School dress codes are limits put on inappropriate clothing to help keep the learning environment focused. It can be quite a distraction for students if a classmate wears inappropriate clothing. The use of a dress code during school is not preventing freedom of expression. Unlike required uniform dress codes, the dress code still allows for students to choose what they wear as long as it is not deemed inappropriate.

Passage 2

Electronic tablets should replace textbooks in K–12 schools. Proponents of tablet use assert that tablets are more cost-effective, are more time-efficient, and take up less space. For example, tablets cost less than

科书便宜，而且能通过减少印刷和纸张的使用量，对环境产生积极的影响。此外，一项在美国加利福尼亚州完成的研究表明，使用交互式平板电脑的学生在标准化测试中的得分明显高于使用纸质印刷教科书学习的学生。

第三篇

美国政府在医疗保健事业上花费的资金数目巨大。政府可以继续为有需要的人提供医疗服务，但不应该为那些不值得获此医疗服务的人提供医疗服务。削减医疗支出将促使降低税赋，并赋予公民更强的消费能力。在美国目前的经济状况下，从长远来看，任何惠及居民钱袋子的事情都是有利于美国的。

给个提示

对于第一篇练习文章，我们的参考答案将与你深度分享"自问自答"模式的批判性思维全过程，在本章以及第2章和第3章中，我们已经详细描述过这个过程。

第一篇

- 如果这篇文章的什么地方存在重要的歧义，本书告诉我可以在论题、结论或理由中寻找。第一步就是在论证中找出这几个部分。这篇文章既没有明确说出论题也没有明确说出结论，同时也没有给出明确的指示词。我只有利用其他工具来找出论题和结论了。要找论题，本书告诉我要问一问："写作者是在对什么进行评价？"着装规定，我想是吧。着装规定是不是个好主意。好了，这样我就可以把这个意思重新整理成一个问题："学校该不该有着装规定？"这篇文章所有的句子都试图说服我相信，学校应该有一套着装规定。因此结论一定是："是的，学校应该有一套着装规定。"

- 同样，文章里也没有指示词来帮我找到理由，因此我只能试试别的方法。要找到理由，我就得先把自己放到写作者的位置上，然后问："为什么学校要有一套着装规定？"我可以从这篇文章里推测出两个理由：首先，奇装异服容易分散学生的注意力；其次，着装规定并不侵犯学生的表达自由。

- 既然我已经将论证分解到最基本的元素，现在我就可以着手找出重要的歧义了。我要首先找出论题、结论和理由中的关键词和短语，因为这些词和短语对论证过程起着决定性作用。它们在这个语境里可能有不止一种说得通的含义。例如，它们可能是抽象词汇或是被附加了感情色彩的语言。"不当着装"无疑是这个论证中的一个重

textbooks, and also positively impact the environment by lowering the amount of printing and paper usage. In addition, one study completed in California shows that students who used interactive tablets scored significantly higher on standardized tests than students who studied with print textbooks.

Passage 3

The government is spending far too much money on health care. The government could continue to fund health care for those who need it, but should not fund health care for people who do not deserve it. Cutting health care expenditures would lead to lower taxes and give citizens more spending power. With the current state of the American economy, anything that helps the citizen's pocketbook would help America in the long run.

Sample Responses

For the first practice passage, our sample response shares with you an in-depth "thinking aloud" model of the critical-thinking process we have been describing in this chapter and in Chapters 2 and 3.

Passage 1

- *If this passage has any significant ambiguity, the text said that I'll find it in the issue, conclusion, or reasons. So my first step will be to find those parts of the argument. Neither the issue nor the conclusion is explicitly stated in this passage. No indicator words are present. I'll have to try other tools to identify the issue and conclusion. To find the issue, the text suggests that I ask, "What is the author reacting to?" Dress codes, I guess. Whether they are a good idea. Okay, so I'll word that idea as a question: "Should schools have a dress code?" All the sentences in this passage are trying to convince me that we should have a dress code, so the conclusion must be, "Yes, schools should have a dress code."*

- *Again, there are no indicator words to help me find the reasons. So I'll try something else. To find the reasons, I need to put myself in the author's shoes and ask, "Why should schools have a dress code?" I can deduce two reasons from the passage: First, inappropriate clothing distracts from learning, and second, dress codes do not violate freedom of expression.*

- *Now that I have broken the argument down into its most basic elements, I can start the process of finding significant ambiguity. I'll start by identifying the key words or phrases in the issue, conclusion, and reasons because these words and phrases are crucial to the argument. They may have more than one plausible meaning within the context. For instance, they*

要元素，而写作者根本没有告诉我什么样的服装才属于不当着装。我怀疑这个词是不是还有别的可以说得通的意思。

- "不当着装"，据我所知，是指上面印有伤害人或侮辱人的字句的衣服。如果是这样的话，我也会禁止不当着装出现在学校里！印有取笑他人的文字的T恤衫自然属于不当着装，我对这一点非常清楚。当然了，本书说过，也许我觉得一个术语的定义显而易见，然而其实根本不是这样。所以，我应该不断追问：这个短语会不会有不同的意思呢？

- 本书建议的一个线索是要留心观察抽象词汇的意思，如"淫秽"和"责任"等。这些词是抽象词汇，而且意思模糊不清，因为它们没有为我们提供明确的定义，或者没有一套判定的标准。本文中提到的"不当着装"同样没有一个明确的定义，没有一套判定的标准。写作者从没有提到不当着装的意思是上面印有伤害人的文字的T恤。这个意思是我假设的，因为在我看来这些T恤就属于不当着装。写作者也没有说不当着装的意思是女生裙子短到一定地步，或是男生裤子短得能让人见到底裤。这个术语的意思看起来比我一开始想的要模糊一点。

- 在认定之前，我想试一试反串这个建议。反对这个结论的人会怎么定义"不当着装"这一短语？持反对立场的人也许会反驳说，着装规定确实限制了表达自由。学生穿衣服是希望表达哪些东西呢？在T恤上面经常可以看到政治信息。我就看到过有的少年穿的T恤衫上面印了反战口号或支持他们喜爱的总统候选人的口号。学生有权表达自己对重要问题的关心，反对着装规定的人可能是在担心学生的这种权利遭到剥夺。

- 现在我被困住了。如果写作者是在讨论印有伤害人的信息的T恤，那我同意他的观点。就让我们禁止好了。但是，如果写作者讨论的是限制学生表达政治观点的权利，那么我强烈反对。我对这个论题无法做出判断，除非消除掉这个歧义才行。

第二篇

论题：平板电脑应不应该取代从幼儿园到12年级教育阶段的教科书？

结论：应该这样做。

理由：1. 它们的价格更便宜。

2. 它们对环境更有利，因为用的纸张会更少。

3. 如果学生使用平板电脑而不是纸质教材，他们在标准化测试中的分数会更高。

could be abstract terms or loaded language. "Inappropriate clothing" is definitely an important element of the argument. And the author never tells me what qualifies as inappropriate. I wonder if there are other possible meanings for the term.

- *"Inappropriate clothing," as far as I'm concerned, is clothing with hurtful or insulting text. I'd prohibit that from schools too! T-shirts that make fun of people are definitely inappropriate. It's pretty clear to me. Of course, the text said that I might think the definition of a term is obvious, even if it's not. So I should keep questioning. Could this phrase have a different meaning?*

- *One of the clues that the text suggested was to pay attention to abstract words like* obscenity *and* responsibility. *These words are abstract—and also ambiguous—because they don't have a specific definition or set of criteria for us.* Inappropriate *similarly does not have a specific definition or set of criteria in this passage. The author never says that inappropriate means hurtful text on T-shirts. I just assumed that meaning because I think those T-shirts are inappropriate. The author also doesn't say that inappropriate means skirts of a certain length or wearing pants so low that one can see a guy's boxers. The term is starting to seem a little less obvious than I originally thought.*

- *Before I can be sure, I want to try the reverse role-playing suggestion. How would an opponent of this conclusion define the term "inappropriate clothing"? Opponents of this argument would probably argue that dress codes DO prohibit freedom of expression. What might students want to express with their clothing? Political messages are often seen on T-shirts. I've seen teenagers wearing T-shirts with antiwar slogans or slogans supporting their favorite presidential candidate. An opponent of dress codes probably would fear that students would be denied the right to voice their opinions about important issues.*

- *Wow. Now I'm stuck. If the author is talking about messages on T-shirts that hurt people, I agree. Let's prohibit them. But if the author's talking about limiting students' ability to voice their political opinions, I strongly disagree. I can't come to a decision about this issue until the ambiguity is resolved.*

Passage 2

ISSUE: *Should electronic tablets replace books in K–12 classrooms?*

CONCLUSION: *They should.*

REASONS: 1. *They are cheaper.*

2. *They are more environmentally friendly because they use less paper.*

3. *Standardized test scores are higher when students use tablets rather than texts.*

哪些词或短语有其他的含义，其含义变化程度之大，足以影响我们想要跑到学校董事会上，要求用平板电脑来取代课本的程度？首先，这些平板电脑据说能减少花销的费用。提倡者指的是一开始的费用还是长期的费用？换句话说，平板电脑的寿命是不是和教材一样长？其次，提倡者保证说平板电脑使用的纸张更少。这个向我们推销平板电脑的人的意思是不是课本是由纸张印刷出来的，而平板电脑则不是这样？这个断言自然是真实的。又或许"更少的纸张"是不是说即使算上在学校里因为使用平板电脑而带来的额外的印刷量，使用平板电脑导致的纸张用量仍然更少？如果后者是倡导者要表达的意思，那么论证就更有说服力。

What words or phrases have alternative meanings that could change the extent to which we should want to rush to our school board meetings to demand that students have tablets rather than texts? To start, the texts are said to "cost less." Is the advocate referring to the initial cost or to cost over time? In other words, do the tablets last as long as the textbooks? Next, we are assured that the tablets "use less paper." Does the person pitching the tablets to us mean simply that texts are made of paper and tablets are not? That claim is certainly true. Or does "less" mean that even counting the extra printing that results from using tablets in schools, tablets result in less use of paper? If the latter meaning is intended, then the "reduced cost argument" is much stronger.

第 5 章

价值观假设和描述性假设是什么

| 学 习 目 标 |

1. 解释找出隐含在论证中的假设的重要性。
2. 识别论证中的价值观假设。
3. 区别价值观假设与描述性假设之间的异同。
4. 理解典型的价值观冲突。
5. 运用线索找出描述性结论。

　　任何一个想说服你相信某个立场的人，都会尽量拿出与其立场相一致的理由。换句话说，理由和结论相辅相成，构成一个完整的说法。因此，乍一看，几乎每个论证都显得"言之有理"，其外表结构看起来都显得完美无缺。但是表面的、明说出来的理由并不是唯一用来证明或支撑其结论的观点。有些内在的、没有说出来的想法提供了一个不可见的结构，使可见的结构说得通。因此，就理解和评价论证而言，本章关于假设的讨论可能是本书最有影响力的部分。

　　让我们思考下面这个论证，看看假设的隐蔽结构到底有多重要。

CHAPTER 5

What Are the Value and Descriptive Assumptions?

LEARNING OBJECTIVES

1. Explain the importance of locating the assumptions implicit in an argument.
2. Identify value assumptions in an argument.
3. Distinguish between value and descriptive assumptions.
4. Develop an appreciation of typical value conflict.
5. Apply the clues for locating descriptive conclusions.

Anyone trying to convince you to believe a particular position will make an attempt to present reasons consistent with that position. In other words, the reasons and conclusion fit to form a story. Hence, at first glance, almost every argument appears to "make sense." The visible structure looks good. But the visible, stated reasons are not the only ideas that serve to prove or support the conclusion. Hidden or unstated beliefs provide an invisible structure that permits the visible structure to make sense. Consequently, this chapter about assumptions may well be the most powerful part of the text in terms of understanding and evaluating an argument.

Let's examine the importance of the hidden structure of assumptions by considering the following argument.

> 地方执法机构应采取更多措施来让公众场合醉酒者承担严重后果。很显然，人们不会积极主动地遵纪守法，因此城市警察必须采取行动。如果不强制执行法律，我们怎么能期待会有改变发生？

乍一看，这个论证的理由支撑了它的结论。如果城市期待市民在行为上有所改变，城市的执法机构就必须强制推行这种改变。

但是也有这种可能：给出的理由确实有道理，但并不足以支撑其结论。假如你认为制止公共场合醉酒的行为是个人的责任，而不是政府的集体责任呢？这样的话，从你的角度看来，上述理由就不再能证实其结论了。只有你认同写作者以为理所当然而没有明说出来的那些特定信念，这个论证对你而言才是令人信服的。在这个例子里，写作者认为理所当然的一个观点就是一种价值观（集体责任）要比另一种价值观（个人责任）更为可取。

在所有的论证中，都有一些写作者认为理所当然的特定信念。通常情况下写作者不会将这些信念明说出来。就像你必须挖掘词语所指的含意一样，你也必须通过阅读字里行间的内容来找到假设。这些信念是论证结构中无形的重要纽带，是将整个论证连接在一起的黏合剂。它们回答了一个非常重要的问题："必须持有什么样的观点才能将理由和结论从逻辑上联系起来？"这些纽带的必要性很明显。没有了这些纽带，在成千上万不同的观点中，人们又怎么能判断哪些才有资格充当理由？只有获得了这些纽带以后，你才能真正理解一个论证。

本章对于你成为一个批判性思维者特别有帮助，因为它帮助你关注整个论证的方方面面，而不仅仅关注那些比较吸引人的特征。提供论证的人可能想要对你隐藏一部分内容，而你的思维却在抓紧时机，补足论证的这些组成部分。

我们再用一种方式来看看假设的重要性，请你想一想为什么你要努力掌握本书介绍的这些技能和态度。有各种各样的理由证明你完全可以不用学习批判性思维。独立认真的思考需要我们付出更多的精力，远不如抛一块硬币做决定，或者问问身边踌躇满志的专家该怎么办那样轻松。但是本书鼓励你学习批判性思维。我们在告诉你，批判性思维对你非常有好处。

我们的建议都是基于一些潜在的观点，如果你不认同这些观点，那么你完全可以不理会这些建议。批判性思维者相信，自己做主、遇事好奇、通情达理等价值观是人类最重要的目标。批判性思维的最终结果就是要求一个人虚怀若谷地接纳各种观点，理性而又有理有据地评判这些观点，然后在理性判断的基础上决定接受哪些观点或采取哪些行动。我们相信你赞同这种对人生的刻画，并因此想成为一个批判性思维者。

Local law enforcement needs to do more to impose consequences for public intoxication. Obviously, people are not taking enough initiative on their own to follow the laws; therefore, city police have to do something. How can we expect change without active enforcement?

The reason—at first glance—supports the conclusion. If the city expects change in the behavior of its citizens, it follows that the city's law enforcement should have to enforce that change.

But it is also possible that the reason given can be true and yet not necessarily support the conclusion. What if you believe that it is the individual's responsibility—not the collective responsibility of the government—to curb the extent of public intoxication? If so, from your perspective, the reason no longer supports the conclusion. This reasoning is convincing to you only if you agree with certain unstated beliefs that the writer has taken for granted. In this case, one belief taken for granted is that one value, collective responsibility, is more desirable than another value, individual responsibility.

In all arguments, there will be certain beliefs taken for granted by the writer. Typically, these ideas will not be stated. Just as you had to dig to find the intended meaning of words, you will have to find assumptions by reading between the lines. These beliefs are important invisible links in the reasoning structure, the glue that holds the entire argument together. Such beliefs answer the very important question of, "What idea is necessary to logically connect a reason to a conclusion?" The necessity of such links should seem obvious. Without such links, how could one decide which of thousands of ideas qualify as reasons? Until you supply these links, you cannot truly understand the argument.

This chapter can be particularly useful to you as a critical thinker because it prepares you to look at the full argument, not just its more attractive features. Your mind is taking its time, creating the components of the argument that the person offering it probably wishes to hide from you.

As another illustration of the importance of assumptions, consider why you should work hard to master the skills and attitudes contained in this text. There are all kinds of reasons why you should not learn critical thinking. Careful thought is much more demanding of our energies than another decision-making approach like flipping a coin or asking the nearest self-confident expert what you should think and do. But this text is encouraging you to learn critical thinking. We are telling you that critical thinking is advantageous for you.

Our advice is based on some invisible beliefs, and if you do not share those beliefs, our advice should not be followed. Critical thinkers believe that such values as autonomy, curiosity, and reasonableness are among the most important of human objectives. The end-product of critical thinking is someone who is open to multiple points of view, assesses those perspectives with reasons and evidence, and then uses that assessment to make decisions about what to believe and what actions to take. We trust that you favor that portrayal of life and, consequently, that you will want to be a critical thinker.

当你努力理解一个人的时候，你面临的任务在很多方面都像没有亲眼看到魔术师的表演诀窍，就自己动手去复制那个魔术。你眼看着手帕被放进了帽子里，出来的却是一只兔子，而你压根就不知道魔术师暗地里玩的是什么把戏。要理解这个魔术，你就得搞清楚魔术师暗地里的那些把戏。同样，在论证中，你也要找到那些暗藏的把戏。实际上，这些把戏就是没有明说出来的观点或信念。我们把这些没有明说出来的想法称为假设。要全面理解一个论证，你就得找出这些假设。

假设具有下面这些特征：

1. 隐藏或没有明说出来（大多数情况下如此）；
2. 论证者认为是理所当然的；
3. 对决定结论有较大的影响；
4. 可能具有欺骗性。

⚠ **批判性问题**：假设是什么？

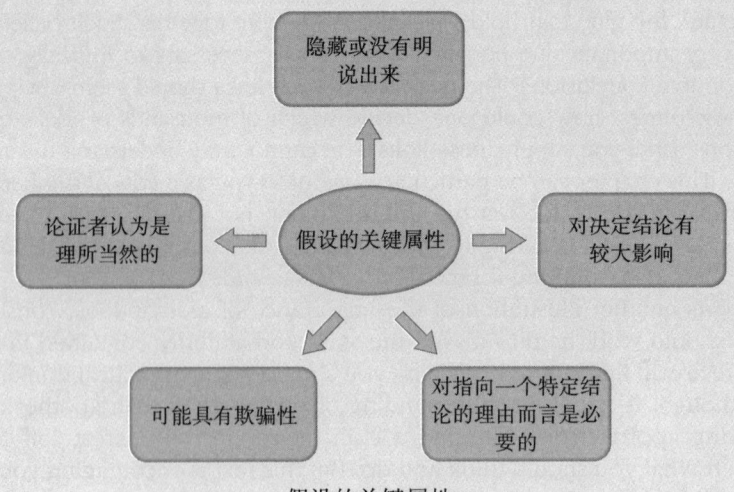

假设的关键属性

到哪儿去找假设

在寻找假设的时候，你应该在什么地方找？怎么找？每本书、每场讨论、每篇文章里都有无数的假设，但你需要关心的只有很小的一部分。你一定记得，一个论证的表面结构由理由和结论两部分组成。你需要关心的只有影响到论证结构的质量的假设。因此，你可以限定寻找假设的范围，只在你已经学会怎样找到的论证结构中去寻找这些假设。

When trying to understand someone, your task is similar in many ways to having to reproduce a magic trick without having seen how the magician did the trick. You see the handkerchief go into the hat and the rabbit come out, but you are not aware of the magician's hidden maneuvers. To understand the trick, you must discover these maneuvers. Likewise, in arguments, you must discover the hidden maneuvers, which, in actuality, are unstated ideas or beliefs. We shall refer to these unstated ideas as assumptions. To fully understand an argument, you must identify the assumptions.

Assumptions are

1. hidden or unstated (in most cases);
2. taken for granted;
3. influential in determining the conclusion; and
4. potentially deceptive.

Critical Question: **What are the assumptions?**

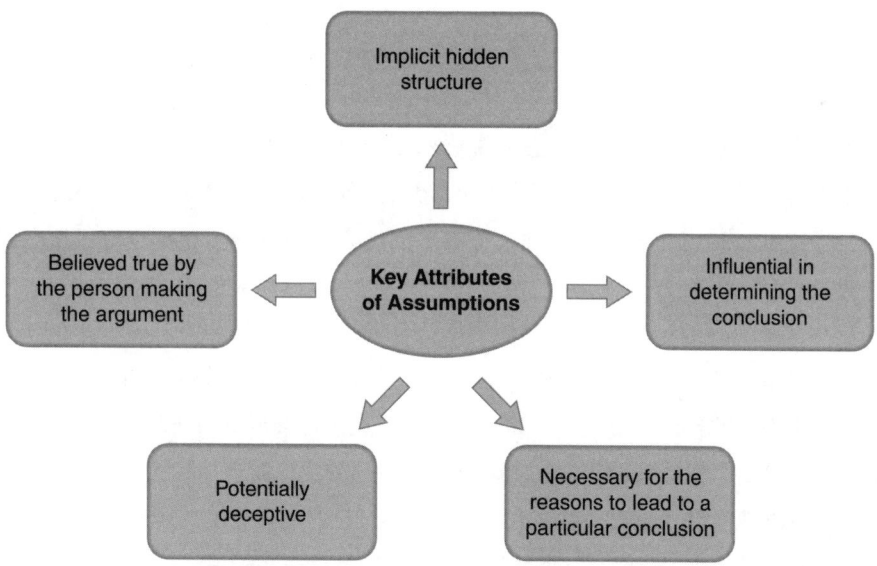

Key Attributes of Assumptions

GENERAL GUIDE FOR IDENTIFYING ASSUMPTIONS

When you seek assumptions, where and how should you look? Numerous assumptions exist in any book, discussion, or article, but you need to be concerned about relatively few. As you remember, the visible structure of an argument consists of warrants and conclusions. But you are interested only in assumptions that affect the quality of this structure. You can restrict your search for assumptions, therefore, to the structure you have already learned to identify.

具体来说，你可以在两个地方寻找假设。其一，在理由和结论之间，寻找理由要证明结论所必不可少的假设（连接假设）；其二，在理由中，寻找理由成立所必不可少的假设。

我们首先向你介绍价值观假设，然后介绍描述性假设。两者在形成论证的过程中都极为重要。

⚠ **注意**：在理由和结论之间寻找价值观假设和描述性假设。

注意，理由和结论部分也是我们寻找重要歧义的地方。在这里，我们又一次表现出对文章或发言中理由和结论的充分尊重和重视。

⚠ **注意**：所谓假设，就是一个通常没有明说出来的信念，写作者认为其是理所当然的，并用来支持外显的论证。

价值观冲突和价值观假设

为什么有些很讲道理的人会怒斥说堕胎就是谋害生命，而另一些同样讲道理的人却把堕胎看作人的一项基本权利？你有没有想过为什么每一任美国总统，不管其政治信仰如何，最终都会和媒体大打口水仗，喋喋不休地争论该不该公布那些他不愿公之于众的政府信息？

存在这些不同结论的一个最重要的原因，就是价值观的冲突，或者说由不同参照系衍生出来的不同价值观。对于道德论证或者说规定性论证，一个人的价值观会影响他提供的理由，进而影响到他的结论。例如，有一所重点大学最近宣布要裁掉100名教职工。学生的反应立刻变得很负面，闹出的动静非常大，这种反应源自他们关注的价值观，如公平（对被裁的教职工）和可信度（学校承诺过要保持一定的班级规模）；但是大学行政机构所依赖的价值观却是高效和节俭，这是他们采用的削减经费策略的基础。

实际上，只有把这些价值观假设添加到论证过程中，各种理由才能从逻辑上支持结论。下面这个例子展示了价值观假设在一个规定性论证中所起的作用。

> 我们不应该让消遣性毒品合法化。这些药物引发了太多的街头暴力和其他犯罪行为。

注意，这里的理由只有在假设大家都认为公共安全比个人责任更重要的情况下才能从

In particular, there are two places to look for assumptions. Look for assumptions needed for the reason(s) to support the conclusions (linkage assumptions) and look for ones necessary for a reason to be true.

We first introduce you to value assumptions and then to descriptive assumptions. Both are extremely influential in shaping arguments.

 Attention: Look for both value and descriptive assumptions in the movement from reasons to the conclusion.

Note that reasons and the conclusion are also the places where we search for significant ambiguity. Once again, we are showing great respect for the importance of the reasons and the conclusion in a speech or an essay.

 Attention: An assumption is a belief, usually unstated, that is taken for granted and supports the explicit reasoning.

VALUE CONFLICTS AND ASSUMPTIONS

Why is it that some very reasonable people shout that abortion is murder, while other equally reasonable observers see abortion as an essential human right? Have you ever wondered why every U.S. president, regardless of his political beliefs, eventually gets involved in a dispute with the press over publication of government information that he would prefer not to share?

One extremely important reason for these different conclusions is the existence of value conflicts, or the differing values that stem from different frames of reference. For ethical or prescriptive arguments, an individual's values influence the reasons he provides and, consequently, his conclusion. For example, a major university recently announced the firing of 100 faculty members. Student reaction was immediately loud and negative, fueled by their concerns for the values of fairness (to fired faculty) and reliability ("We were promised particular sizes of classes"), but the university administration relied on the values of efficiency and frugality as the basis for its cost-cutting strategies.

In fact, reasons will logically support the conclusion only when the value assumption is added to the reasoning. The argument that follows illustrates the role of a value assumption in a prescriptive argument.

> We should not legalize recreational drugs. Such drugs cause too much street violence and other crimes.

Note that the reason logically supports the conclusion only if one takes for granted the idea that it is more important to value public safety than it is to value individual responsibility. Value assumptions are very important assumptions for

逻辑上支持结论。价值观假设对于这样的论证非常重要，因为它在幕后指导着论证的过程。和你交流的人可能意识到了这些假设，也可能根本意识不到。你应该养成习惯，一定要找出理由赖以成立的那些价值观假设。

我们这里所谓的价值观假设，是指一种想当然的看法，认为某些相互对立的价值观中的一个比另一个更重要。当写作者对社会论争采取了一种立场，他们通常都会选择一种价值观而排斥另一种价值观，因为他们有自己的价值倾向或者说价值偏好，这些倾向就是他们的价值观假设。要找出这些价值倾向，你需要较好地把握价值观的含义。因此，现在是回顾一下第 1 章里介绍过的价值观的好时机。

从价值观到价值观假设

要找出价值观假设，我们不能仅仅简单罗列一个个价值观。你的很多价值观别人也一样认同。例如，几乎每个人都说办事灵活、愿意合作和待人诚实这些价值观较为可取，不是吗？

再看一次价值观的定义，你就会发现，从定义来看，大多数价值观都在每个人罗列的清单上。因为很多价值观都是大家共同拥有的，所以价值观本身并不能做理解的有力向导。能让你对一个规定性问题做出与别人不一样回答的，是你对所持的特定价值观的相对认同度。

想一想当一对对价值观发生冲突或抵触的时候，我们对这些论争的回应，就能理解我们对特定价值观持有不一样的认同度。发现大多数人都既看重竞争又看重合作对我们并没有很大的启示意义，但当我们发现在竞争和合作发生冲突时，一些人偏向于竞争而非合作的时候，我们就会对规定性的选择有更加全面的理解。

一个人常常不会明说对特定价值观的偏好，但是这个价值偏好一定会对他的结论产生重大影响，同时也会影响他为这一结论辩护的方式。这些没有明说的对于价值倾向的主张所起的就是价值观假设的作用。有些人称这些假设为价值判断。认识到人们对产生冲突的价值观或多种价值观的相对支持程度，你就既能进一步理解阅读的材料，又能为最终评价这些规定性论证打下坚实的基础。

⚠ **注意**：所谓价值观假设，就是在特定情形下没有明说出来的偏好一种价值观超过另一种的倾向。我们把价值倾向和价值偏好作为同义词使用。

当你在一个论证中发现了一个人的价值偏好时，不应该期待这个人在讨论不同的争议话题时还会持有相同的价值倾向。离开了讨论的问题，一个人也就不再持有同样的价值倾

such arguments because they are directing the reasoning from behind a screen. The person trying to communicate with you may or may not be aware of these assumptions. You should make it a habit to identify the value assumptions on which the reasons are based.

By *value assumption*, we mean a taken-for-granted belief about the relative desirability of certain competing values. When authors take a position on a social controversy, they typically prefer one value over another value—they have value priorities or preferences; these preferences are their value assumptions. To identify these priorities, you need to have a good grasp of what is meant by *values*. Consequently, this is a good time to review the introduction to values in Chapter 1.

FROM VALUES TO VALUE ASSUMPTIONS

To identify value assumptions, we must go beyond a simple listing of values. Others share many of your values. For example, wouldn't almost anyone claim that flexibility, cooperation, and honesty are desirable?

Look again at the definition, and you will immediately see that, by definition, most values will be on everyone's list. Because many values are shared, values by themselves are not a powerful guide to understanding. What leads you to answer a prescriptive question differently from someone else is the relative intensity with which you hold specific values.

That we attach different levels of intensity to specific values can be appreciated by thinking about responses to controversies when pairs of values collide or conflict. While it is not very enlightening to discover that most people value both competition and cooperation, we do gain a more complete understanding of prescriptive choices as we discover who prefers competition to cooperation when the two values conflict.

A person's preference for particular values is often unstated, but that value preference, nevertheless, will have a major impact on her conclusion and on how she chooses to defend it. These unstated assertions about value priorities function as value assumptions. Some refer to these assumptions as *value judgments*. Recognition of relative support for conflicting values or sets of values provides you with both an improved understanding of what you are reading and a basis for eventual evaluation of prescriptive arguments.

> *Attention: A value assumption is an implicit preference for one value over another in a particular context. We use value preferences and value priorities as synonyms.*

When you have found a person's value preference in a particular argument, you should not expect that same person to necessarily have the same value priority when discussing a different controversy. A person does not have the same value priorities without regard to the issue being discussed. The

向。与论争相关的语境和事实问题也会极大影响我们对某个价值偏好的忠诚度。我们持有的价值偏好只会坚持到一定的程度。比如说，有人在大多数情况下都认为自由选择比集体福祉更重要（例如在讨论是否应该穿印有国旗图案的服装这一话题时），但是当他看到可能会对集体福祉造成极大损害的行为时（例如在讨论一个人在有很多亲人丧生于种族大屠杀的集体里发表种族歧视演说的权利这一话题时），就有可能改变自己的价值偏好。换言之，价值观假设会随着情境的改变而改变，它们在一种情况下适用，可一旦规定性论题的具体条件发生改变，我们就有可能出现非常不同的价值倾向。

典型的价值观冲突

如果你能意识到典型的价值观冲突，你就能更快地认出一个写作者在得出特定结论时做出的价值观假设。我们列举了一些伦理道德论题上常见的价值观冲突，而且提供了可能出现这些价值观冲突的论争例证。在识别重要的价值观假设时，你可以把这些列举出的价值观冲突作为出发点。

在识别价值观冲突的时候，你常常发现在某一个论争中，似乎存在好几个价值观冲突，并且它们对形成结论似乎都很重要。所以在你评价一个论争的时候，请尽量找出几个价值观冲突，以此来检验一下自己的看法。

典型的价值观冲突和论争的具体例证	
1. 忠诚 – 诚实	该不该告诉父母姐姐有药瘾
2. 竞争 – 合作	你是否支持把评分制作为激励学习的机制
3. 媒体自由 – 国家安全	每周开总统新闻发布会是否明智
4. 秩序 – 言论自由	我们应不应该监禁那些有极端思想的人
5. 理性 – 冲动	打赌下注的时候该不该先查一下赔率

把立论者的背景作为寻找价值观假设的线索

我们已经说过，要找到价值观假设，一个比较好的起点就是检查一下作者的背景。尽量找出像写作者或发言者这样的人通常持有的价值偏好，越多越好。他是公司高管、工会领导、共和党官员、医生，还是公寓的一个租客？这样的人最希望保护的是什么利益？追求自身利益本身自然没什么错，但是这样的追求常会限制一个作者所能包容的价值观假设。

context and factual issues associated with a controversy also greatly influence how far we're willing to go with a particular value preference. We hold our value preferences only up to a point. Thus, for example, those who prefer freedom of choice over the welfare of the community in most situations (such as wearing clothing that displays an image of the flag) may shift that value preference when they see the possibility of too much damage to the welfare of the community (such as in the case of the right of a person to give a racist speech in a community with many relatives of Holocaust victims). In other words, value assumptions are very contextual; they apply in one setting, but we may make quite a different value priority when the specifics of the prescriptive issue change.

TYPICAL VALUE CONFLICTS

If you are aware of typical conflicts, you can more quickly recognize the assumptions being made by a writer when he reaches a particular conclusion. We have listed some of the more common value conflicts that occur in ethical issues and have provided you with examples of controversies in which these value conflicts are likely to be evident. You can use this list as a starting point when you are trying to identify important value assumptions.

As you identify value conflicts, you will often find that there are several that seem important in shaping conclusions with respect to particular controversies. When evaluating a controversy, try to find several value conflicts, as a check on yourself.

Typical Value Conflict and Sample Controversies

1. Loyalty–honesty	Should you tell your parents about your sister's drug habit?
2. Competition–cooperation	Do you support grades as a motivator to learning?
3. Freedom of press–national security	Is it wise to hold weekly presidential press conferences?
4. Order–freedom of speech	Should we imprison those with radical ideas?
5. Rationality–spontaneity	Should you check the odds before placing a bet?

THE COMMUNICATOR'S BACKGROUND AS A CLUE TO VALUE ASSUMPTIONS

We suggested earlier that a good starting point in finding assumptions is to check the background of the author. Find out as much as you can about the value preferences usually held by a person like the writer or speaker. Is she a corporate executive, a union leader, a Republican Party official, a doctor, or a

例如，一个大烟草公司的总裁就不大可能特别重视对敏感人群的同情，如果对同情弱势人群的偏好取代了对公司业绩稳定增长的关注，就有可能导致他丢掉饭碗。因此，作为一个批判性思考者，你如果想一想像他这样的人可能做出的价值观假设，就能很快发现他的价值偏好。

这里要注意一点：并不是因为一个人身为某个团体的一员，所以他就一定会认同这个团体的特定价值观假设。如果你以为一个团体中任何一员的想法都完全相同，那就大错特错了。我们都知道商人、农民、消防员在讨论具体论争时，彼此之间常常意见不一。在你调查发言者或写作者的背景以作为判断他的价值观假设的线索时，所得到的只不过是个线索而已，并且和其他线索一样，有可能会产生误导，你需要小心谨慎地对它加以利用。

把可能发生的后果作为寻找价值观假设的线索

在规定性论证中，对某个论题采取的每一种立场都会指向不同的结果或后果。每个潜在的后果都有一定的可能性会发生，每个后果都会在一定程度上让人觉得可以接受或难以接受。

某个结论对应的后果是否可以接受，主要取决于个人的价值偏好。对于一个论证来说，结论到底可不可以接受，主要取决于各种潜在的后果发生的可能性，以及这些后果受重视的程度。因此，要判断一个人的价值观假设，一个重要的方法就是注意他用来证实结论的各种理由，然后判断哪些价值倾向会导致作者认为这些理由比其他理由更可取，那些被抛弃的理由本可以从论题的反面进行论证。我们来看一个具体的例子。

> 根本就不该建核电厂，因为核电厂里那些危险的核废料会给环境带来污染。

这里提出的反对理由是建造核电厂带来的较为具体的潜在后果。作者明显将环境污染视作极为不可接受的后果。为什么这个后果在作者的思想中占有这么大的分量？防止污染有助于实现哪些更普遍的价值观？我们可以猜测，作者很可能特别重视公众健康或环境保护。如果换成另一个人，那么他在这个论证里可能会强调完全不一样的后果，例如核电厂给消费者的供电带来的积极影响。为什么？很可能因为他非常注重效率！因此，只有当一个人选择的价值观假设是公共健康或环境保护比效率更重要时，上述理由才能支持结论。

tenant in an apartment? What interests does such a person naturally wish to protect? There is certainly nothing inherently wrong with pursuing self-interest, but such pursuits often limit the value assumptions a particular writer will tolerate. For example, it is highly unlikely that the president of a major cigarette firm would place a high value on compassion for the vulnerable when a preference for compassion for the vulnerable rather than stability would lead to his losing his job. Consequently, you as a critical reader or listener can often quickly discover value preferences by thinking about the probable assumptions made by a person like the communicator.

A note of caution: It isn't necessarily true that because a person is a member of a group, she shares the particular value assumptions of the group. It would be a mistake to presume that every individual who belongs to a given group thinks identically. We all know that business people, farmers, and firefighters sometimes disagree among themselves when discussing particular controversies. Investigating the speaker or writer's background as a clue to her value assumptions is only a clue, and, like other clues, it can be misleading unless it is used with care.

CONSEQUENCES AS CLUES TO VALUE ASSUMPTIONS

In prescriptive arguments, each position with respect to an issue leads to different consequences or outcomes. Each of the potential consequences will have a certain likelihood of occurring, and each will also have some level of desirability or undesirability.

How desirable a consequence is for a particular conclusion will depend on personal value preferences. The desirability of the conclusions in such cases will be dictated by the probability of the potential consequences and the importance attached to them. Thus, an important means of determining an individual's value assumptions is to note the reasons given in support of a conclusion and then to determine what value priorities would lead to these reasons being judged as more desirable than reasons that might have been offered on the other side of the issue. Let's take a look at a concrete example.

> Nuclear power plants should not be built because they will pollute our environment with dangerous waste material.

The reason provided here is a rather specific potential consequence of building nuclear plants. This writer clearly sees environmental pollution as very undesirable. Why does this consequence carry so much weight in this person's thinking? What more general value does preventing pollution help achieve? We are only guessing, but probably public health or conservation is being weighted especially heavily by this person. Someone else might stress a different consequence in this argument, such as the effect on the supply of electricity to consumers. Why? Probably because he values efficiency very

可见，判定价值观假设的一个重要方法就是问一下这个问题："为什么那个人如此看重他用作理由的特定后果或结果？"

寻找价值观假设的其他方法

另一个找出价值观冲突的有用技巧就是反串。问一下这个问题："如果争辩中的这些人采取相反的立场，他们会关心些什么？"如果有人说我们根本不应该用猴子来做实验，你就应该问问自己："如果我要为用猴子做实验辩护，那么我应该关心些什么？"

最后，你应该经常看看，不同的意见是不是来源于以下这个价值观冲突：个人以特定的方式行事的权利和这种行事方式给集体福祉带来的影响之间的冲突。很多论证隐含的立场都和这个几乎无处不在的价值观冲突有关。我们能够回忆起无数个这样的例子：我们的思考活动需要我们去衡量这两个重要的价值观及其产生的影响。

例如，当我们质疑在公立学校里使用金属探测器对学生进行安检的时候，我们构建论证的过程通常是这样的：首先我们想到的是学生的隐私权，然后我们又想到，如果有学生携带武器来学校，这势必会威胁其他学生的人身安全。接着，我们尝试在这些价值观之间权衡，如在这个例子里，单个学生的隐私权是否应该比其他学生在校期间的福祉更值得被保护？这一价值冲突还涉及哪些论题？

了解其他人的价值倾向的价值

我们获取信息的大部分来源，比如说媒体、我们所在的大学和结交的朋友，都很少会告诉我们隐藏在他们观点下面的价值观假设。很多情况下他们自己可能也没有意识到这些假设。我们很少会听到这样的话："根据我的价值偏好……"真是太糟糕了！认识到别人的价值观假设及其价值偏好的根据，最大的好处就是让我们更能理解人们所持立场的来龙去脉。例如，乔纳森·海特的近著《正义之心》里说，在美国政治的大背景下，如果民主党和共和党彼此都了解一点对方的核心价值偏好，那么他们之间的分歧就能更具有建设性。他接着说自由主义者最核心的价值观就是"关怀"，特别是对受压迫者的关怀，而社会保守主义者最核心的价值观则是"权威"，尤其是保护各种可以维系道德共同体的机构和传统。海特希望双方能够更加清醒地意识到这些核心价值观，这样彼此才能更乐于考虑对方的论证。

highly! Thus, the given reason supports the conclusion if a value assumption is made that public health or conservation is more important than efficiency.

One important means of determining value assumptions, then, is to ask the question, "Why are the particular consequences or outcomes presented as reasons so desirable to the person?"

MORE HINTS FOR FINDING VALUE ASSUMPTIONS

Another useful technique for generating value conflicts is to reverse role-play. Ask the question, "What do those people who would take a different position from a stated argument care about?" When someone argues that we should not use monkeys in experimental research, you should ask yourself, "If I wanted to defend the use of monkeys, what would I be concerned about?"

Finally, you can always check to see whether the disagreement results from a value conflict concerning the rights of an individual to behave in a particular fashion and the welfare of the group affected by the behavior in question. Many arguments rest implicitly on a stance with respect to this enduring value conflict. As with other common value conflicts, we can all recall numerous instances when our thinking required us to weigh these two important values and their effects.

For example, when we wonder about the use of metal detectors in the public schools, we often begin to construct our arguments in terms of thinking about the privacy rights of the individual students and the threats to the student body if a student were to bring a weapon to school. Then, we try to balance those values against other values: Does the individual's right to privacy deserve greater protection than the welfare of the other students in the school in this instance? What other issues are involved in this value conflict?

THE VALUE OF KNOWING THE VALUE PRIORITIES OF OTHERS

Most of our sources of information such as the media, our universities, and our friends rarely announce the value assumptions underlying their opinions. In many cases they may not be conscious of them. Rarely do we hear the phrase, "according to my value preferences. . . ." That is too bad! A major advantage of becoming aware of others' value assumptions and their rationale for those preferences is the possibility of creating a greater appreciation of where people are coming from. For example, Jonathan Haidt's recent book *The Righteous Mind* suggests in the context of American politics that Republicans and Democrats could engage in more constructive disagreements by gaining appreciation of each other's core value preferences. He argues that the most central value of liberals is Care, in the specific sense of care for victims of oppression, while the most central value for social conservatives is Authority, in the sense of preservation of the institutions and traditions that sustain a moral community. Haidt's hope is that both sides being more aware of such core values will lead to more willingness to consider the other side's arguments.

使用这个批判性问题

在你发现一个价值观假设后,你该拿它怎么办呢?首先,回想一下每个批判性问题的目的:使你最终能够评价论证!你知道,有思想的人都会有不同的价值观假设,你有权质疑为什么他人会做出特定的价值观假设。因此,作为一个批判性思维者,你可以向他人指出这一点,即任何一个进行论证的人,都有必要提供一些解释,说明为什么你应该接受这个论证中暗含的特定价值观假设。

价值观与相对性

我们并不想在本章中给读者留下这样的印象:价值偏好就像冰淇淋一样,既然我选择了最爱吃的蓝莓奶油口味,你就没理由劝我说选择柠檬戚风口味更明智。冰淇淋的口味选择纯属个人喜好,正所谓萝卜白菜各有所爱——这不就行了?

相反,选择价值偏好需要进行论证。这种论证可能有凭有据、考虑周详并且细致入微,也可能草率马虎、固执己见。因此,价值偏好需要有一定的理由,这样,批判性思维者才能加以考虑。一个价值偏好需要有支持它的理由和证据,恰如所有其他的结论都需要理由和证据一样。

找出并评价描述性假设

当你找到价值观假设的时候,你心里已了然写作者或发言者希望这个世界变成什么样——什么样的目标在他看来最重要。但是你并不知道他对世界的本质及世人的本性是怎样认识的。比如说,世人一般都是懒惰的还是喜欢追逐成就的,是喜欢合作还是喜欢竞争,是由生理结构控制还是由成长环境控制,是自私自利还是无私奉献,是理性还是异想天开?他的外显论证取决于他的价值观,同时也取决于上述这般的想法。这些没有说出来的想法就属于描述性假设,它们同样是一个论证必要的隐藏构成因素。

下面这个试图说服你购买一辆轿车的论证取决于隐含的描述性假设,你能不能把它们找出来?

> 这辆车肯定能把你送到目的地,不管你的目的地在哪里。我在各种各样的地形都驾驶过这款车。

⚠ **批判性问题:这里的描述性假设是什么?**

所谓描述性假设,是指对这个世界过去、现在或未来是什么样的信念;而规定性假设或者说价值观假设,你应该还记得,是指对这个世界应该是什么样的信念。

USING THIS CRITICAL QUESTION

Once you have found a value assumption, what do you do with it? First, recall the purpose of every critical question—to move you toward the evaluation of reasoning! Because you know that thoughtful people have different value assumptions, you have the right to wonder why any single value assumption is being made. Thus, as a critical thinker, you would want to point out the need for anyone who is making an argument to offer some explanation for why you should accept the particular value assumption that is implicit in that argument.

VALUES AND RELATIVISM

We do not want to give the impression in this chapter that value preferences are like ice cream, such that when I choose blueberry cheesecake as my flavor, you have no basis for trying to persuade me that the lemon chiffon is a better choice. Ice cream is just a matter of personal preference—end of story!

However, the choice of value preferences requires reasoning. That reasoning can be informed, thoughtful, and caring. But it can also be sloppy and self-absorbed. Hence, value preferences require some justification that critical thinkers can consider. A value preference requires supporting reasons and evidence just as any other conclusion does.

IDENTIFYING AND EVALUATING DESCRIPTIVE ASSUMPTIONS

When you find value assumptions, you know pretty well what a writer or speaker wants the world to be like—what goals she thinks are most important. But you do not know what she takes for granted about the nature of the world and the people who inhabit it. For example, are people basically lazy or achievement-oriented, cooperative or competitive, controlled by their biological makeup or by their environment, self-interested or altruistic, rational or whimsical? Her visible reasoning depends on ideas like these, as well as upon her values. Such unstated ideas are descriptive assumptions, and they too are essential hidden elements of an argument.

The following argument urging you to buy a car depends on hidden assumptions. Can you find them?

> This car will get you to your destination, whatever it may be. I have driven this model of car on multiple occasions.

Critical Question: **What are the descriptive assumptions?**

Descriptive assumptions are beliefs about the way the world *was, is, or will be*; prescriptive or value assumptions, you remember, are beliefs about how the world *should be*.

描述性假设举例说明

我们一起看看上述关于汽车的论证，以便更清楚地展示我们所谓的描述性假设是什么意思。上述论证的结构如下：

> 结论：这种型号的汽车肯定能把你送到想去的地方。
>
> 理由：这种型号的汽车多年来在各种情况下性能都很不错。

论证到这儿还没有完成。我们知道，一个理由自身并不具备与结论之间的直接联系，必须通过其他特定的观点（通常是没有明说的观点）才能与结论产生联系。这些观点如果是可信的，就证明可以把这个理由作为结论的支撑。因此，一个理由能否证实结论，或者说是否与结论有关，主要取决于我们能否找到那些没有明说出来的观点，把理由和结论从逻辑上联系起来。如果这些没有明说出来的观点是描述性的，我们就称之为描述性假设。下面我们就为上述论证提供两个这样的假设。

> 假设1：年复一年，某型号汽车的质量始终如一。

上述论证本身并没有提供这样的陈述。但是，如果上述理由是可信的，如果这个假设也是可信的，那么上述理由就为证实结论提供了一定的支撑。相反，如果不是所有年份的车型都具有同样的可靠性（我们知道事实如此），那么驾驶过从前某年生产的车型就不能作为可靠的参照，用来决定买不买当前年份生产的这种车型。注意这个假设是一个关于事情是什么样的陈述，而非关于事情应该是什么样的陈述，因此它是一个描述性的连接假设。

> 假设2：将要使用新车进行的驾驶，和推荐这辆车的人以前进行的驾驶属于同样的类型。

说到驾驶汽车，如果我们不明确"驾驶"这个词的含义，那么驾驶概念的模糊性就可能给我们带来很多麻烦。如果推荐这辆车的人所说的驾驶，指的是常规地去趟杂货店，只在安静的郊区道路上行驶，一路没有山峦，那么这种经历和在科罗拉多州的山路驾驶一辆新车，还拉着一辆重吨位的拖车就完全没有可比性。因此，只有假设作者对驾驶给出了恰当的定义，这个结论才能被其理由所证实。

我们可以把这种描述性假设称为定义性假设，我们想当然地认定一个词的意思，而这个词可能有不止一层含义。因此，我们要寻找的一类非常重要的描述性假设就是定义性假设，即我们想当然地认定一个可能有多重意思的词代表其中的一个意思。

ILLUSTRATING DESCRIPTIVE ASSUMPTIONS

Let's examine our argument about the car to illustrate more clearly what we mean by a descriptive assumption. The reasoning structure is as follows:

> CONCLUSION: *This particular car will get you where you want to go.*
>
> REASON: *This model of car has functioned well on multiple occasions for years.*

The reasoning thus far is incomplete. We know that, by itself, a reason just does not have a direct link to a conclusion; the reason must be connected to the conclusion by certain other (frequently unstated) ideas. These ideas, if true, justify treating the reason as support for the conclusion. Thus, whether a reason supports, or is relevant to, a conclusion depends on whether we can locate unstated ideas that logically connect the reason to the conclusion. When such unstated ideas are descriptive, we call them descriptive assumptions. Let us present two such assumptions for the above argument.

> ASSUMPTION 1: *From year to year, a particular model of car has a consistent quality.*

First, no such statement was provided in the argument itself. However, if the reason is true and if this assumption is true, then the reason provides some support for the conclusion. But if not all model years have the same level of dependability (and we know they do not), then experience with a model in previous years cannot be a reliable guide to whether one should buy the car in the current model year. Note that this assumption is a statement about the way things *are*, not about the way things *should be*. Thus, it is a descriptive connecting assumption.

> ASSUMPTION 2: *The driving that would be done with the new car is the same kind of driving that was done by the person recommending the car.*

When we speak about "driving" a car, the ambiguity of driving can get us into trouble if we do not clarify the term. If the "driving" of the person recommending the car refers to regular trips to the grocery store on a quiet suburban street with no hills, that driving experience is not very relevant as a comparator when the new car is to be driven in Colorado, while pulling a heavy trailer. Thus, this conclusion is supported by the reason only if a certain definition of driving is assumed.

We can call this kind of descriptive assumption a *definitional assumption* because we have taken for granted one meaning of a term that could have more than one meaning. Thus, one very important kind of descriptive assumption to look for is a definitional assumption—the taking for granted of one meaning for a term that has multiple possible meanings.

一旦找出连接假设，你就回答了这个问题："认为结论可以从那个理由中推断出来有什么根据？"接下来自然要问："有没有什么接受这些假设的根据？"如果没有，那么对你来说，这个理由就不能证实其结论。如果有，那么这一理由就为结论提供了逻辑上的支撑。因此，当你找出连接假设并且有可靠的理由以相信这些假设时，你就可以说，这个论证非常严谨。

⚠ **注意**：描述性假设就是一种关于世界过去、现在和将来是怎么样的没有明说的信念。

你在寻找价值观假设的时候，其实就是在找立论者自认为理所当然的那些想法，这些想法让他的理由能支撑相应的结论。因为写作者和发言者常常意识不到自己的这些假设，所以他们自觉的信念往往和你确认为潜在假设的想法大不一样。

常见的描述性假设

描述性假设既多如牛毛，又格外重要。它们是任何一个论证都不可缺少的组成部分。它们是"先行者"或预先的假设，是提供论证的人没有明说以和我们分享的信念。它们无处不在并有着巨大的力量，但是作为读者或聆听者的你必须通过挖掘才能找到它们。同样要记住，描述性假设常常存在谬误。

找到描述性假设并借助它们评价其所支撑的论证，这是一种非常有用的技能，要培养这种技能，一个有效的方法就是让自己对一些较为常见的描述性假设变得特别敏感。这些假设在我们的思维里特别常见，一旦你学会如何寻找它们，就会开始理解它们影响我们思维的力量。而一旦你变得善于识别它们的影响力，就会更加迫不及待地让识别这些关键假设的能力成为你批判性思维的一个常规组成部分。

- 发生在人们身上的事主要都是人们自己选择的结果。这个假设是"躲在帘幕背后的大象"，决定我们应该在什么时候以及对哪些人给予责备和赞许。
- 发言者或写作者是个典型人物。当有人做出这样的假设时，他就是在根据自己的经验或品味来直白地推论。
- 这个世界是公正的。这个假设隐藏在幕后，支撑着论证的形式：某件事应该是真的就意味着它必然会是真的。我们觉得你应该能理解为什么这种形式的论证常常被称为"浪漫主义谬误"。

Once you have identified the connecting assumptions, you have answered the question, "On what basis can that conclusion be drawn from that reason?" The next natural step is to ask, "Is there any basis for accepting the assumptions?" If not, then for you, the reason fails to provide support for the conclusion. If so, then the reason provides logical support for the conclusion. Thus, you can say reasoning is sound when you have identified connecting assumptions and you have good reason to believe those assumptions.

Attention: *A descriptive assumption is an unstated belief about how the world was, is, or will become.*

When you identify assumptions, you identify ideas the communicator needs to take for granted so that the reason is supportive of the conclusion. Because writers and speakers frequently are not aware of their own assumptions, their conscious beliefs may be quite different from the ideas you identify as implicit assumptions.

COMMON DESCRIPTIVE ASSUMPTIONS

Assumptions are as numerous as they are important. They are a required component of any argument. They are priors or givens, the unannounced beliefs that the person presenting the argument does not share with us. They are present and powerful, but you as the reader or listener have to dig for them. Keep in mind as well that descriptive assumptions are often wrong.

An effective way to develop the skill of discovering and using assumptions as an aid to evaluation of the arguments they are sustaining is to sensitize you to some of the more common assumptions. These assumptions are present so frequently in our thinking that once you learn to look for them, you will start to appreciate the power they have over our thinking in general. Once you become skilled at recognizing the influence they have, you will be much more eager to make identification of key assumptions a regular part of your critical thinking.

- The events that happen to people are primarily the result of personal choices. This assumption is the elephant behind the curtain shaping when and whom we blame and give credit to.

- The speaker or writer is a typical person. When someone makes this assumption, she reasons explicitly based on her own experience or tastes.

- The world is just. This assumption is in the background, holding up reasoning of the form: That something should be true means that it will be true. We think you can understand why this type of reasoning is often called *the romantic fallacy.*

- 因为以前发生过这样的事，所以今后它还会发生。这种假设代表着对一个人甚至一个国家的过去所持的一种不加评判、过于简单化的反应。
- 我的世界就是这个宇宙的中心。这种假设让我们很难去支持为别人谋福利的法律或政策；也就是说，它抑制了我们对弱势者的共情。这个假设还让我们很难去欣赏文化多样性。

请注意，以上每个描述性假设都是有争议的，也就是说有理性的人对于这些假设的准确性持否定的态度。批判性思维者应该了解，当人们做出这些假设时，我们应当要求他们解释一下原因，然后我们才能评价完整的论证，而不只是可见的部分。

找到假设的线索

在寻找假设时，你的任务是通过填补缺失的纽带来重建论证的过程。你需要补充一些观点，让立论者的论证显得"有道理"。一旦有了一幅完整论证的画面，囊括看得见和看不见的所有因素，你就能在判断这个论证的强弱时更加游刃有余。

我们怎么才能找到这些缺失的重要环节呢？这需要我们付出努力、想象力和创造力。找到重要的假设是个艰巨的任务。在本章前面部分，我们已经提供了几条线索，帮你寻找价值观假设。这里我们再提供几条线索，以帮助你成功找到描述性假设。

不断思考结论和理由之间存在的鸿沟。你为什么寻找这些假设？因为你想让自己可以判断理由能在多大程度上支持结论。为此，一定要去寻找写作者或发言者心里认为理所当然的那些想法，这些想法将他的理由和结论联系在一起。你要不断追问："你是如何从这个理由得出这个结论的？"问他："如果理由成立，要得出这个结论，还需要什么条件成立才行？"为了回答这个问题，你会发现问一下"假设这些理由都成立，有没有可能这个结论仍然是错误的"将很有帮助。

找一找这种鸿沟，对发现价值观假设和描述性假设都会有帮助。

寻找支撑理由的那些没有明说的想法。有时候立论者在给出理由时没有提供外显的支撑，但这不一定说明理由没有道理。理由有没有道理，主要取决于立论者认为理所当然的那些想法能不能被接受。这些想法就是描述性假设。下面这个简短论证的提纲就提供了一个例子。

- Because something happened in the past, it will happen in the future. This assumption represents an uncritical and overly simplified reaction to the history of a person or even a country.
- My world is the center of the universe. This assumption makes it difficult for us to support laws or policies that primarily benefit others; that is, it inhibits empathy for the vulnerable. This assumption also makes it difficult for us to appreciate cultural diversity.

Notice that each of these assumptions is debatable, meaning that reasonable people disagree about the accuracy of the assumption. The point for critical thinkers is that when people make these assumptions, we should ask them to explain why. Then we can react to the complete argument, not just the visible part.

CLUES FOR LOCATING ASSUMPTIONS

Your task in finding assumptions is to reconstruct the reasoning by filling in the missing links. You want to provide ideas that help the communicator's reasoning "make sense." Once you have a picture of the entire argument, both the visible and the invisible elements, you will be in a much better position to determine its strengths and weaknesses.

How does one go about finding these important missing links? It requires hard work, imagination, and creativity. Finding important assumptions is a difficult task. Earlier in this chapter, we gave you several hints for finding value assumptions. Here are some clues that will make your search for descriptive assumptions successful.

Keep thinking about the gap between the conclusion and reasons. Why are you looking for assumptions in the first place? You are looking because you want to be able to judge how well the reasons support the conclusions. Thus, look for what the writer or speaker would have had to take for granted to link the reasons with the conclusion. Keep asking, "How do you get from the reason to the conclusion?" Ask, "If the reason is true, what else must be true for the conclusion to follow?" And, to help answer that question, you will find it very helpful to ask, "Supposing the reason is true, is there any way in which the conclusion nevertheless could be false?"

Searching for the gap will be helpful for finding both value and descriptive assumptions.

Look for unstated ideas that support reasons. Sometimes a reason is presented with no explicit support; yet the plausibility of the reason depends on the acceptability of ideas that have been taken for granted. These ideas are descriptive assumptions. The following outline of a brief argument illustrates such a case:

结论：高中英语班级里的所有学生至少应该看一部莎士比亚的戏剧。

理由：直接体验和感受莎士比亚的著作会大有裨益。

必须要以什么样的想法作为假设，才能让这个理由被人接受？我们必须要假设：

a. 表演必须非常到位，准确反映莎士比亚在著作中表达的内容；
b. 学生会理解这部戏剧，并能将它与莎士比亚联系起来。

a 和 b 这两个想法都必须被认为是理所当然的，才能让理由被人接受，理由因此才有可能支持得出的结论。

让自己站在写作者或发言者的立场上。如果你想象要为这个结论辩护，那么找出作者的假设通常就会容易得多。如果有可能，尽量将自己放在得出结论的这个人的位置。了解他的背景。如果一个煤炭公司的高管说，露天采矿对自然环境的美丽不会造成实质性的损害，那他很可能持有这样的信念：露天采矿对国家而言大有好处。这样他就有可能对"美丽"的定义持有一个假设，这个假设和他的论证一致，而"美丽"的其他定义则可能会引发对露天采矿的强烈谴责。

让自己站在反对者的立场上。如果你站在写作者或发言者的立场上无法找到假设，不妨换换角色。问问自己，为什么别人会不认同这个结论。什么样的论证会让别人不认同你正在评价的这个结论？如果你能扮演一个不接受这个结论的人的角色，你就更容易识别论证的外显结构中隐含的那些假设。

像批判性思维者那样写作和发言

我们都会把自己的价值观和假设带入论证中。若不仔细考量，我们常常意识不到自己的观点背后的价值观和假设。（即便仔细考量，我们仍难免有所遗漏。）

我们在形成不管是书面的还是口头的论证时，必须认识到一些因素是如何影响我们在一些问题上的立场的，这些因素包括我们是谁、如何被抚养长大、受教育经历、宗教背景、家庭和朋友圈。在我们一生大部分时间里，这些因素会引导我们做出决定、形成信念，这一过程不一定运用到批判性思维能力。可见，我们当中许多人都没有认真思考过我们所持立场背后的价值观，也没有反思过哪些价值观对我们而言更重要。更复杂的

CONCLUSION: *All high school English class students will go see at least one Shakespeare play.*

REASON: *It is beneficial to experience Shakespeare's works firsthand.*

What ideas must be taken for granted for this reason to be acceptable? We must assume the following:

a. The performance will be well done and reflective of what Shakespeare would encourage, and
b. students will understand the play and be able to relate it to Shakespeare.

Both (a) and (b) are ideas that have to be taken for granted for the reason to be acceptable and, thus, potentially supportive of the conclusion.

Identify with the writer or speaker. Locating someone's assumptions is often made easier by imagining that you were asked to defend the conclusion. If you can, crawl into the skin of a person who would reach such a conclusion. Discover his background. When an executive for a coal company argues that strip mining does not significantly harm the beauty of our natural environment, he has probably begun with a belief that strip mining is beneficial to our nation. Thus, he may assume a definition of beauty that would be consistent with his arguments, while other definitions of beauty would lead to a condemnation of strip mining.

Identify with the opposition. If you are unable to locate assumptions by taking the role of the speaker or writer, try to reverse roles. Ask yourself why anyone might disagree with the conclusion. What type of reasoning would prompt someone to disagree with the conclusion you are evaluating? If you can play the role of a person who would not accept the conclusion, you can more readily see assumptions imbedded in the explicit structure of the argument.

WRITING AND SPEAKING AS CRITICAL THINKERS

We all bring our own values and assumptions to our arguments. Often, we aren't even aware of which values and assumptions are underlying our ideas until we look closely. (And, even then, we might miss them.)

As we create our own arguments, either in writing or verbally, we have to recognize how who we are, the ways in which we've been brought up, and our education, religious background, and even family and friends are going to influence the positions we take on issues. For most of our lives, these influences have led us to make decisions and develop beliefs without necessarily using our critical-thinking skills. Thus, many of us have not given much consideration to the values that underlie the positions we hold, nor have we reflected about which values are more important and which ones less important.

是，我们其实拥有相同的价值观，区别不在于我们所秉持的价值观，而在于我们对价值观的排序。

还记得这一章开头的例子吗？有人论证说警察需要依法在公共场合禁止醉酒闹事。我们讨论了这个结论是如何将社会责任置于个人责任之上。我们中的许多人在写到或谈论这个论点时不会感到不适，因为乍一看，谁不会反对公共场合醉酒，支持警察执法呢？但是，执法真的反映了我们潜在的信念吗？如果我们仔细想想，我们会把社会责任置于个人责任之上吗？或者，我们是否认为其他某种价值观更有可取之处呢？例如，或许你会认为教育醉酒者、鼓励其承担个人责任，比让疲于奔命的警察与酩酊大醉的聚会者周旋更可取？

确定我们自己的价值偏好的第一步是退后一步，思考不同的价值观，它们可能导致对于一个问题得出不同的结论。每个问题都可能有多种答案，我们要确定哪个答案最符合我们的信念。让我们来看看之前提出的一个问题：

> 我们应该让消遣性毒品合法化吗？

如果我们要针对这个问题写一篇文章或起草一篇发言稿，我们不能只给出"是"或"否"的答案，而是要把"是"或"否"的答案说得理明据实。我们不仅要让回答与我们的价值观一致，还要让回答背后的理由也与我们的价值观一致。为了确定我们的结论，我们必须检查各种答案背后的理由。并且，我们必须记住，无论回答"是"或"否"，每个答案都可能有很多的理由。下面我们来看看之前得出的结论：

> 我们不应该使消遣性毒品合法化。

这个结论的基础是什么价值观？这一主张的重点仍是强调社会责任高于个人责任的价值观。这里的价值观假设就是政府责任优先于个人责任。（请注意，这种价值偏好类似于主张警察应执法禁止人们在公共场所醉酒。政府和个人责任之间的价值冲突在美国经常出现。）

现在，让我们来看看上面例子中给出的理由。

> 理由一：毒品导致层出不穷的街头暴力和其他犯罪。

前文中我们确定了这一理由背后的主要价值观是公共安全价值观。在做出我们是否在

To make things even more complicated, we share the same values; the difference lies not in the values we possess, but in our ranking of the values.

Remember the example at the beginning of the chapter arguing that the police really need to enforce laws against public intoxication? We discussed how that conclusion assumes a value of community responsibility over individual responsibility. Many of us would be comfortable writing or speaking that argument ourselves because at first glance, who isn't against public intoxication and for police enforcement of laws? But does the enforcement of the law really reflect our underlying beliefs? If we gave it some thought, would we prioritize community responsibility over individual responsibility? Or, do we think that some other value, perhaps a belief in educating the individual to encourage personal responsibility, for example, is more preferable than saddling busy police with drunken partiers?

The first step in identifying our own value preferences is to step back and contemplate the different values that underlie the possible different conclusions to an issue. Each issue has many possible answers. We want to determine the answer that is most consistent with our own beliefs. Let's take a look at an issue raised earlier:

Should we legalize drugs?

If we were to write an essay or draft a speech in response to this question, we wouldn't just have a yes or no answer but rather a yes or no answer with reasons. We want to be consistent with our values not only in our answer to the issue but also in the reasons that underlie that answer. To determine where our own conclusion lies, we have to examine the reason behind various answers. And, we have to remember that each answer, whether yes or no, has many possible reasons. Let's take a look at the conclusion reached earlier:

We should not legalize recreational drugs.

What values underlie this conclusion? Again, the emphasis in this claim is on the value of community responsibility over individual responsibility. The value assumption is that government responsbility is preferable over individual responsibility. (Note that this value preference is similar to the claim that the police should enforce laws against public intoxication. The value conflict between government and individual responsi bility is one that will come up often.)

Now, let's look at the reason offered in the above example:

Reason A: Such drugs cause too much street violence and other crimes.

Earlier in the chapter, we identified the main value underlying the reason as the value of public safety. But before we can decide if we want to support

自己的论证和发言中支持这个结论和理由的决定之前，让我们看看是否可能有其他的理由，并看看这些理由背后的价值观。

理由二：我们不应该让消遣性毒品合法化。如果没有违法犯罪，人们在无法受到其后果威胁的情况下，是不会做对自己最有利的事情的。

价值观：家长式作风

理由三：我们不应该让消遣性毒品合法化。吸毒会对吸毒者造成巨大的心理和生理上的伤害，对家庭也会造成心理上伤害。

价值观：健康

我们可以对最后一个例子做一点进一步说明，即生理和心理上的伤害对国家造成以百万美元计的损失（从而在我们的论证中增加另一个价值观：财政责任）。

这些理由背后都有不同的价值观。我们的工作是找出我们最偏好的价值观，或者我们是否支持所有的价值观。公共安全、家长式作风、健康以及财政责任，我们更看重哪一个？哪些在我们看来更重要？是否有一些价值观能够引起共鸣，而另一些不能？思考一下结论和支持结论的理由的多种可能性，可以让我们确定自己的价值观假设，我们可以选择最能反映我们信念的结论和理由。

这还不是全部。尽管我们意识到在某个论证中自己更偏好哪一种价值观，并颇感舒畅惬意，但是论证的语境可能会导致我们改变价值观。我们可能因为个人权利辩护而反对禁止醉酒者上街，但在关于毒品的争论中，我们可能会对这种价值观感到不安。每一个论证及其语境将决定我们对价值观进行排序的方式。所以，仅仅对我们的价值观进行排序、整理是不够的，我们需要做好准备在各种不同的论证中评估它们。

此外，或许最重要的是，识别出支撑我们结论的价值观，让我们有机会看到我们是如何看待世界的。如果我们把家长式作风看得比个人责任更重要，那我们很有可能会把这个世界看成一个人们在可以自行其是的时候，不会行对社会或对自己最有利之事的地方。这种潜在的价值观假设使我们更倾向于选择干预，由那些有能力的人去阻止个人伤害自己或伤害他人。当我们更深入地挖掘自己的价值偏好时，我们会对自己有更多的了解。

在我们确定了一个结论和支持这个结论的理由背后的价值观，并确定了自己在这个论题上的立场之后，我们还必须小心地找出描述性假设，它们是我们可能会在论证中做出的假设。请记住，描述性假设是我们对世界过去、现在和将来所持有的信念。

this conclusion and reason in our own argument or speech, let's look at other possible reasons and the values that might underlie them.

> Reason B: We should not legalize recreational drugs. Without the threat of a consequence from breaking a law, people will not do what is best for them.
>
> Value: paternalism
>
> Reason C: We should not legalize recreational drugs. Drug use causes enormous psychological and physical damage to users and psychological damage to families.
>
> Value: health

We could take the last example a little further by adding that the medical and psychological damage costs the country millions of dollars (thus adding another value, fiscal responsibility, to our argument).

All these reasons have different values underlying them. Our job is to figure out which values we prefer or if we support all of them. Which do we value more: public safety, paternalism, health, or fiscal responsibility? Which ones are more important to us? Are there some underlying values that do resonate and others that don't? Considering multiple possibilities both for the conclusion and the reasons that support the conclusion allows us to identify our own value assumptions, and we can pick the conclusion and reasons that most reflect our beliefs.

But, that's not the full picture. Even if we feel comfortable that we have a sense of which values we prefer in a certain argument, the context of the argument might cause us to change our values. Thus, we might argue for personal freedom in supporting the legalization of drugs, but we might not feel comfortable with this value in an argument for the legalization of abortion. Each argument and its context will determine how we order our values. So, it is not enough to just rank or order our values; we need to be prepared to evaluate them across a variety of different arguments.

Also, and perhaps most importantly, identifying the values that underlie our conclusions gives us the opportunity to see how we view the nature of the world. If we value paternalism over individual responsibility, chances are that we see the world as a place where people left to their own devices will not do what is best for society or for themselves. Such an underlying value assumption leads us to prefer intervention by those in a position to keep the individual from harming himself or others. As we dig deeper into our value preferences, we learn more about who we are.

Once we've identified the values underlying a conclusion and the reasons that support that conclusion, and we have determined where we stand on the issue, we must also be careful that we identify descriptive assumptions that we might be making in our arguments. Remember, a descriptive assumption is a belief we hold about how the world was, is, or will become.

为了在我们的写作或发言中识别这些假设，我们必须问自己："在把这个理由和这个结论联系起来时，我们认为什么是理所当然的？"我们还必须问："即使我们的理由可信，结论是否仍有可能不可信？"

让我们看一个例子来帮助我们理解。阅读下面的结论：

> 我们要让学生相信阅读是有趣和愉快的，这是培养阅读流畅性必要的第一步。

结论（我们应该让学生相信阅读是有趣和愉快的）是有理由支持的——因为这对于提高阅读流畅性是必要的。这里的描述性假设是流畅性为我们偏好的结果。是否还有其他更可取的结果呢？比如滋养、充实心灵？如果我们的愿望是教授阅读从而滋养、充实学生的心灵，那么我们是否还会强调我们需要说服学生阅读是有趣和愉快的？我们的结论是否会改变？也许我们的新结论是，我们应该让学生相信阅读一定是困难的，具有挑战性，因为阅读是滋养、充实我们的心灵的唯一方法。这种说法也有它自己的假设（例如，只有通过挑战，我们才能滋养、充实我们的心灵），我们想强调的是，如果我们不同意描述性假设，这个论证就无法自圆其说。因此，在我们决定在写作或发言中支持某个结论之前，我们必须弄清楚我们到底对世界做了什么样的假设。

最后，在我们自己的写作和发言中，与读者或听众分享我们潜在的价值观和描述性假设可能有所帮助。如果我们觉得提醒读者或听众，帮助他们了解我们结论背后的价值观可能有助于说服他们接受我们的结论，那么明确指出它们是非常有益处的。

因为，你遇到的话题可能非常棘手且复杂，所以对识别价值观和描述性假设的过程加以总结也许非常有用，你可以将此过程融入自己的写作或发言中：

1. 检查所提出论题的潜在结论背后的价值观假设。
2. 选择最符合我们价值观的结论。
3. 在选择了一个结论之后，我们需要确定支持该结论的潜在理由背后的价值观假设。
4. 选择与我们所选中结论，以及与当前语境中我们所偏好的价值观假设一致的理由。
5. 如果无法找到与我们所选择的结论背后的价值观假设一致的理由，那么我们就得回到所有潜在的结论中，重新进行评判。

To identify these assumptions in our own writing or speaking, we must ask ourselves, "What are we taking for granted when we link this reason with this conclusion?" And we must also ask, "Even if our reason is valid, how is it possible that our conclusion might not be?"

Let's look at an example to help clarify this rather tricky concept. Read the following conclusion:

> Convincing students that reading is fun and enjoyable is a necessary first step to developing fluency.

The conclusion—we should convince students that reading is fun and enjoyable—is supported by the reason—because it is necessary to increase fluency. The descriptive assumption here is that fluency is our preferred outcome. What other possible outcomes might be offered as more preferable? How about an outcome of enriching the soul? If our desire is to teach reading so it enriches the soul, would we still emphasize that we need to convince students that reading is fun and enjoyable, or would our conclusion change? Perhaps our new conclusion would be, We should convince students that reading must be difficult and challenging because that is the only way we can enrich our souls. This claim has its own assumptions as well (that only through challenging can we enrich our souls, for example), but the point is that if we don't agree with the descriptive assumption, the argument falls apart. Thus, we have to understand just what assumptions we are making about the world before we decide just what conclusion we want to support in our own writing or speaking.

Finally, in our own writing and speaking, we might find it helpful to share the underlying values and descriptive assumptions with our audience. If we feel that reminding the audience of the values underlying our conclusion might help convince them to adopt our conclusion, then it can be very useful to explicitly identify them.

Because this topic can be pretty tricky and complex, it might be useful to summarize the process of identifying values and descriptive assumptions and incorporating this process into our own writing or speaking:

1. Examine value assumptions underlying possible conclusions to the issue being raised.
2. Select the conclusion that is most consistent with our values.
3. Once we've selected a conclusion, identify the value assumptions underlying the possible reasons that support that conclusion.
4. Select the reasons consistent with the conclusion we've selected and with our preferred value assumptions given the context.
5. If we are unable to find reasons that are consistent with our value assumptions for the conclusion we have selected, return to the possible conclusions and re-evaluate.

发现描述性假设的一些线索

1. 不断思考理由和结论之间的鸿沟。
2. 寻找支持理由的观点。
3. 把自己放到反对立场。
4. 对论题进一步学习了解。

思维体操

⚠ **批判性问题**：价值观假设和描述性假设是什么？

请在以下三篇文章中找出写作者的重要假设。记住首先要确定结论和理由。

第一篇

大学生外出活动和参加聚会时需要更加注意安全防范。在大学里可以参加聚会、结识新朋友、开怀畅饮而无须担心父母或长者的监视，对很多大学生来说，这一切所带来的刺激感，是他们喜欢大学生活的一个最常见的理由。可惜的是，这种对自由和刺激感的关注常常会引发各种危险情况，譬如饮酒过量、同龄人间的打架斗殴，以及性侵犯。我们最好确保大学生在享受大学生活时多注意安全防范。

第二篇

尽管美国大学生联谊会因为以大欺小和狂欢聚会而变得声名狼藉，但是加入这类组织还是有很多好处，你需要认真考虑宣誓入会的必要性。比如说，其中一个极大的好处就是让你有机会和真正的好兄弟、好姐妹建立联系，结交到一辈子的好朋友。还有个相关的好处就是可以建立人际关系网。毕业后找工作时，很多人都发现自己的前途并不是由学了多少知识决定的，而是由认识些什么人决定的。加入女生联谊会或者兄弟会，你就有机会融入专业人才的庞大关系网。还有一个好处就是可以锻炼自己担任领导角色的能力，让你今后在工作中有能力和别人竞争，因为兄弟会和女生联谊会经常组织各种活动，如晚餐会或其他聚会。最后，加入联谊会让你有很多机会在和很多人成为好友的过程中收获很多快乐。你的大学岁月应该更丰富多彩，而不仅由上课和学习组成，大学岁月应该成为你一生中的黄金岁月。

Clues for Discovering Descriptive Assumptions

1. Keep thinking about the gap between the conclusion and reasons.
2. Look for ideas that support reasons.
3. Identify with the opposition.
4. Learn more about the issues.

PRACTICE EXERCISES

 Critical Question: **What are the value and descriptive assumptions?**

For each of the three passages, locate important assumptions made by the author. Remember first to determine the conclusion and the reasons.

Passage 1

College students need to be more focused on safe practices when going out and partying. For many students, one of the more popular reasons for enjoying college is the excitement of attending parties, meeting new people, and consuming alcohol without being under the watch of a parent or authority figure. Unfortunately, this focus on freedom and excitement often leads to dangerous situations such as overconsumption of alcohol, physical fights among peers, and sexual assaults. It is best to make sure that students focus on safety when enjoying their college years.

Passage 2

College sororities and fraternities get a bad rap for their hazing and partying, but joining such organizations has many benefits that should make you seriously consider pledging them. For example, a big benefit is the bonding with true brothers and sisters that occurs, making you friends for life. A related benefit is networking. When seeking a job after graduation, most of you will find your future prospects not from what you know, but whom you know. By joining a fraternity or sorority, you gain access to a vast network of working professionals. Another major benefit is the ability to take part in leadership roles that give you the skills to compete in the working world, since fraternities and sororities are always organizing events like dinners and parties. Lastly, joining a sorority or fraternity gives you a great opportunity to have lots of fun while socializing and becoming close friends with numerous people. Your college years should be much more than going to class and studying; they should be among the best years of your life.

第三篇

在线教育已逐渐成为传统课堂教学的替代方式。在线教育当然有一些好处，但弊大于利。例如，接受在线教育的孩子错失了传统教育提供的与其他孩子进行社会互动的机会。这些互动对孩子成长的重要性不亚于他们在学校里的上课学习。如果学生不与同龄人接触，他们的性格可能会变得与社会格格不入，且孤僻乖张。接受在线教育的孩子也不会与多位任课老师建立人际关系。当学生接触到各种各样的老师时，他们在与这些老师一起学习时，会从纷至沓来的不同观点中获益。传统学校中有各种各样的教师，他们能激发学生对学习的热爱，其教学质量比昏暗的电脑屏幕上的任何程序都更高。

给个提示

在展示下列论证中的假设时，我们只列举其中的一部分假设，即那些我们认为最重要的假设。

第一篇

结论：大学生外出活动和参加聚会时需要更加注意安全防范。

理由：学生如果只关心如何享受聚会、饮酒作乐，而不关心自身安全，就常常会引发危险的情况。

这个理由强调了参加聚会和过量饮酒带来的负面后果。因此，与这个论证有关的一个价值观冲突就是自由和刺激与安全的对立。当然，有人会从相反方向来争论，说刺激和自由实际上正是上大学的最佳理由。在这篇文章中，认为安全比自由和刺激更重要的这种价值偏好，把论证的理由和结论联系在了一起。和大多数规定性的争论一样，在上述两难问题里也牵涉不止一个价值观冲突。例如，这个争论还要求我们思考理性和自发性之间孰轻孰重的问题。

第二篇

结论：大学生应该考虑加入联谊会。

Passage 3

Online schooling is starting to become a popular alternative to the traditional classroom. Online schooling certainly has some benefits, but the downsides outweigh the advantages. For example, children who use online schooling miss out on the social interactions with other children that traditional schooling brings. These interactions are just as important to a child's development as the lessons they receive in school. Students who don't experience this contact with their peers run the risk of becoming antisocial and isolated. Children enrolled in online schooling also do not experience an interpersonal connection with multiple teachers. When students have a variety of teachers, students gain from the different viewpoints they encounter while working with those teachers. Traditional schooling provides a variety of teachers who can inspire a love of learning much better than any program on a dim computer screen ever could.

Sample Responses

In presenting assumptions for the following arguments, we will list only some of the assumptions being made—those which we believe are among the most significant.

Passage 1

CONCLUSION: *College students need to be more focused on safe practices when going out and partying.*

REASON: *Focusing only on enjoying partying and consuming alcohol, instead of safety, often leads to dangerous situations.*

The reason stresses the negative consequences of partying and drinking excessively. Thus, one value conflict that relates to this argument is between freedom and excitement on the one hand and safety on the other. Of course, others would argue in the opposite direction, that excitement and freedom are in fact some of the best reasons to attend college. In this passage, the value preference for safety over freedom and excitement links the reason to the conclusion. As with most prescriptive controversies, more than one value conflict is involved in this dilemma. For example, this controversy also requires us to think about the weights we should attach to rationality over spontaneity.

Passage 2

CONCLUSION: *College students should consider joining a fraternity or sorority.*

理由：1. 学生可以和他人建立起牢固的纽带。
2. 与他人交往有助于建立人际关系网，对今后求职有帮助。
3. 联谊会活动锻炼人的领导才能。
4. 联谊会促进社交活动，能让人收获很多乐趣。

是什么将这些理由和结论联系起来？有没有可能理由都正确，但并不能支持这个结论？价值倾向是必需的纽带。这里一个假设的价值倾向就是，归属感和找乐子要比自律和优异的学业成绩更重要。还有一个有争议的描述性假设把理由和结论联系在一起：联谊会带来的好处并不能通过其他途径获取，譬如各种学校社团和组织。有没有什么想法是作者认为理所当然，但对我们接受每一个理由而言必不可少的？第三个理由只有在将来的老板把参加大学生联谊会的经历当成个人履历的一部分时才可能成立。不过，也可能有很多老板把这种经历看成一个人缺少独立性、严肃性和学习动力的标志。

REASONS: 1. *Students develop strong bonds with others.*
2. *The interaction with others promotes networking helpful for getting jobs.*
3. *Sorority and fraternity activities facilitate leadership skills.*
4. *Fraternities and sororities promote socializing and having a lot of fun.*

What links these reasons to the conclusion? Can they be true, yet *not* support the conclusion? Value priorities are a needed link. An assumed value priority that a sense of belonging and fun are more important than self-discipline and academic excellence links the reasons to the conclusion. A debatable descriptive assumption also links the reasons to the conclusion: Benefits of a sorority cannot be acquired through other choices such as on-campus clubs and organizations. Are there any ideas taken for granted that are necessary for us to accept the truth of any of the reasons? The first reason will be true only if potential employers see a history of belonging to a fraternity or sorority as a plus on one's resume. It is possible, for example, that many employers will view this background as a sign of lack of independence, seriousness, and drive.

第 6 章

6

论证中有没有谬误

| 学 习 目 标 |

1. 练习通过评判假设发现谬误。
2. 熟悉几个重要的逻辑谬误。

到目前为止,你一直在把写作者或发言者提供给你的原材料组织成一个有意义的整体结构。你已经学会了一些方法,把淘金盘里不相干的部分剔除掉,并且学会了怎样去发现那些可以把相关部分粘连在一起的"隐形黏合剂",即各种各样的假设。所有这些成绩的取得都依赖于提出批判性问题。让我们简单回顾一下这些问题:

1. 论题和结论是什么?
2. 理由是什么?
3. 哪些词或短语意思不明确?
4. 价值观假设和描述性假设是什么?

问问这些问题,一方面能让你清楚理解交流者的论证过程,另一方面也让你大致了解论证的可靠部分和薄弱环节。现在你的主要问题是:"根据所提供的理由,这个结论可不可以接受?"现在你已做好准备,要做出最核心的评价。请记住:批判性阅读和聆听的主要目标,就是判断结论是不是强有力,或者判断它的价值是大还是小。

CHAPTER

6

Are There Any Fallacies in the Reasoning?

LEARNING OBJECTIVES

1. Practice the discovery of fallacies by evaluating assumptions.
2. Become familiar with several important logical fallacies.

Thus far, you have been working at taking the raw materials a writer or speaker gives you and assembling them into a meaningful overall structure. You have learned ways to remove the irrelevant parts from your pan as well as how to discover the "invisible glue" that holds the relevant parts together—that is, the assumptions. All these things have been achieved by asking critical questions. Let's briefly review these questions:

1. What are the issue and the conclusion?
2. What are the reasons?
3. What words or phrases are ambiguous?
4. What are the value and descriptive assumptions?

Asking these questions should give you a clear understanding of the communicator's reasoning as well as a sense of where there might be strengths and weaknesses in the argument. Your major question now is, "How acceptable is the conclusion in light of the reasons provided?" You are now ready to make your central focus evaluation. *Remember*: The objective of critical reading and listening is to judge the strength or worth of conclusions.

回答上面的四个问题是评价过程的一个必要开端，接下来我们探讨的问题，需要我们对论证的价值大小或质量高低做出更加直接和明晰的判断。现在我们的任务是将破铜烂铁和真金白银区分开来。我们要分辨出最好的理由，即那些我们要特别认真对待的理由。

在评价过程的这个阶段，你的第一步就是要检查一下论证结构，判断交流者的论证是否依赖于错误的或高度存疑的假设，或交流者是否在通过逻辑上的谬误或其他带有欺骗性的论证形式来糊弄你。第 5 章的重点是找出假设，然后思考其质量高低。本章则着重探讨那些被称作"谬误"的论证过程中的"小花招"。逻辑谬误指的是某些草率的假设，这些假设屡见不鲜，为了帮助你快速记忆，我们给它们起了名字。换句话说，本章只是前一章对描述性和规定性假设的讨论的一个附加部分。

如果能找出这些小花招，你就能防止自己在不知不觉中上了别人的当。下面，让我们一起看看论证中的谬误到底是什么样的。

> 亲爱的编辑：贵报支持参议员斯彭道关于通过增税来提高国家财政收入以便改善公路状况的论辩，让我感到无比震惊。参议员先生自然喜欢增税这一套，一个自由派民主党人，自然会动不动就提议增加税收和支出，你还想从他那儿得到什么别的主意呢？

注意，这封信乍一看好像是要列出一个"理由"来反驳增税的提议，通过援引参议员所属的民主党的声名来达到目的。但是这个理由和他的结论根本就不相干。问题的关键在于增税是不是个好主意。写信的人完全忽略掉参议员的增税理由，没有提出任何具体理由来反对增税；相反，他对参议员实施人身攻击，给他扣上一顶"动不动就提议增加税收和支出的自由派民主党人"的大帽子。作者在这里犯了一个论证中的谬误，他的论证背后是一个和结论有关的荒诞不经的假设，旨在将人们的注意力从论证本身转移到他的论辩对手——参议员斯彭道身上。一个容易轻信别人并且对此类谬误毫无防备的读者很有可能中他的招，认为作者提供了一个有说服力的理由。

本章为你提供许多练习，帮助你找出这类谬误，这样你就不会轻易中别人的圈套。

⚠️ **批判性问题**：论证中有没有谬误？

⚠️ **注意**：所谓谬误，就是论证中的欺骗手段，交流者有可能利用这个欺骗手段来说服你接受他的结论。

Answering our first four questions has been a necessary beginning to the evaluation process; we now move to questions requiring us to make judgments more directly and explicitly about the worth or the quality of the reasoning. Our task now is to separate the fool's gold from the genuine gold. We want to isolate the best reasons—those that we want to treat most seriously.

Your first step at this stage of the evaluation process is to examine the reasoning structure to determine whether the communicator's reasoning has depended on false or highly doubtful assumptions or has "tricked" you through either a mistake in logic or some form of deceptive reasoning. Chapter 5 focused on finding and then thinking about the quality of assumptions. This chapter highlights those reasoning "tricks" called *fallacies*. Logical fallacies refer to names we have given to certain sloppy assumptions. These assumptions are so commonplace that for shorthand purposes we have assigned names to them. In other words, this chapter is simply an additional component of the previous chapter's discussion of descriptive and prescriptive assumptions.

Spotting such logical tricks will prevent us from being unduly influenced by them. Let's see what a fallacy in reasoning looks like.

> Dear Editor: I was shocked by your paper's support of Senator Spendall's arguments for a tax hike to increase state money available for improving highways. Of course the Senator favors such a hike. What else would you expect from a tax-and-spend liberal?

Note that the letter at first appears to be presenting a "reason" to dispute the tax hike proposal, by citing the senator's liberal reputation. But the reason is *not relevant* to the conclusion. The question is whether the tax hike is a good idea. The letter writer has ignored the senator's reasons and has provided no specific reasons against the tax hike; instead, she has personally attacked the senator by labeling him a "tax-and-spend liberal." The writer has committed a fallacy in reasoning, because her argument requires an absurd assumption to be relevant to the conclusion and shifts attention from the argument to the arguer—Senator Spendall. An unsuspecting reader not alert to this fallacy may be tricked into thinking that the writer has provided a persuasive reason.

This chapter gives you practice in identifying such fallacies so that you will not fall for such tricks.

 Critical Question: **Are there any fallacies in the reasoning?**

 Attention: *A fallacy is a reasoning "trick" that someone might use while trying to persuade you to accept a conclusion.*

用提问的方法找出论证中的谬误

论证中的谬误有很多种。你可以在很多书本或网站上找到关于这类谬误的长长清单。幸运的是，即使你不能记住所有这些谬误及其名称，你也能辨别出它们。只要你问对了问题，就能找到论证谬误，即使你叫不出它们的名字也没关系。

因此，我们采用强调自我提问的策略，而不要求你死记硬背各种谬误的名字。不过，我们相信，了解一些最常见的谬误的名字，可以让你对它们变得更敏感，也可以让你在和那些熟悉这些名称的人讨论你对错误论证的反应时，在表达上少走一些弯路。因此，我们在帮你识别那些具有欺骗性的论证过程时，会向你介绍一些谬误的名称，我们还鼓励你学会本章结束部分描述的那些最常见的谬误的名称。

我们已经在前面的"致编辑的一封信"例子里向你介绍了一种常见的谬误，我们指出这封信的作者对参议员斯彭道实施人身攻击，而不是直接反驳参议员给出的理由。这种论证方式就是所谓的"人身攻击型谬误"。拉丁文短语"ad hominem"的意思是"针对个人"。人身攻击型论证之所以属于论证谬误，是因为进行论证的人的品格、外表或态度怎么样，通常与此人所做的论证的质量毫无关系。它是在攻击"送信人"而不是在讨论送来的"信息"。

下面再举一个人身攻击型谬误的简单例子。

> 桑迪：我认为参加女生联谊会纯粹是浪费时间和金钱。
> 朱莉：你当然那样说了，反正什么联谊会都不收你。
> 桑迪：不说这个，你怎么看我用来支持自己立场的论证呢？
> 朱莉：那根本不算数。反正你就是吃不到葡萄说葡萄酸。

这是你的谬误清单中的第一个谬误。以下是其定义。

⚠ **人身攻击型谬误**：指针对个人进行人身攻击，而不是直接反驳其提供的理由。

从评价假设出发

为了说明定位错误假设作为一种识别谬误方法的力量，我们向你介绍叙述谬误。

我们知道，有些警局在逮捕非裔美国人时，会迅速使用致命的武力。现在我们看到一段视频，画面上一辆警车正在追逐一辆超速行驶的汽车。随着视频的播

A QUESTIONING APPROACH TO FINDING REASONING FALLACIES

There are numerous reasoning fallacies. You can find many lengthy lists of fallacies in numerous texts and Web sites. Fortunately, you don't need to be aware of all the fallacies and their names to be able to locate them. If you ask yourself the right questions, you will be able to find reasoning fallacies—even when you cannot name them.

Thus, we have adopted the strategy of emphasizing self-questioning strategies, rather than asking you to memorize an extensive list of possible kinds of fallacies. We believe, however, that knowing the names of the most common fallacies can sensitize you to these fallacies and also act as a language shortcut in communicating your reaction to faulty reasoning to others familiar with the names. Thus, we provide you with the names of fallacies as we identify the deceptive reasoning processes and encourage you to learn the names of the common fallacies described at the end of the chapter.

We have already introduced one common fallacy in our *letter-to-the-editor* example mentioned earlier. We noted that the writer personally attacked Senator Spendall instead of responding directly to the senator's reasons. The reasoning illustrates the *ad hominem fallacy*. The Latin phrase *ad hominem* means "against the man or against the person." An *ad hominem* argument is considered a fallacy because the character, appearance, or attitude of individuals making arguments are usually not relevant to the quality of the argument being made. It is attacking the messenger instead of addressing the message.

Here is another brief example of ad hominem reasoning.

Sandy: "I believe that joining sororities is a waste of time and money."

Julie: "Of course you would say that, you didn't get accepted by any sorority."

Sandy: "But what about the arguments I gave to support my position?"

Julie: "Those don't count. You're just a sore loser."

You can start your list of fallacy names with this one. Here is the definition:

Fallacy: *Ad Hominem*: An attack on the person, rather than directly addressing the person's reasons.

EVALUATING ASSUMPTIONS AS A STARTING POINT

To illustrate the power of locating erroneous assumptions as a method of identifying fallacies, we introduce you to the *narrative fallacy*.

> We know that some police departments are especially quick to use deadly force when the person they apprehend is an African American. Now we see a video that shows a speeding car being chased by a police car. As the video

放，我们看到逃逸的汽车停在路边。警官走过去，拔出枪，朝司机开枪。司机是一名非裔美国人。我们以前在电视上看到过这种情况。

讲故事是人类的天性。当我们看到一系列事件时，我们会倾向于把事件的点点滴滴联系起来。我们不仅想知道发生了什么，而且愿意立即编织一个故事来解释到底发生了什么。第一个出现在脑海中的"有意义"的故事通常足以满足我们对理解发生了什么的需要。我们相信"这段描述警察的文字所写的只是又一起警察践踏了非裔美国人基本人性的事件"。我们认定自己讲的故事是一个准确的解释。

但请注意，我们对这里描述的事件所知甚少。我们不知道司机是否在逃离武装抢劫。当警官走近路边那辆车时，事实并没有告诉我们他看到了什么。我们无从得知任何关于这个辖区的警察文明执法情况的信息。这里的警察是否有歧视有色人种的记录？

我们错误地假设：因为我们有一个故事，它为一系列事实提供了一个可能的解释，所以这个故事就是对这些事实相互作用的解释。

⚠ **叙述谬误**：错误地假设因为我们能讲出一个貌似可以解释一系列正在发生的事实的故事，所以我们已认识到事实和现象之间的全部联系。

在接下来的这一节中，我们将带你做一些练习，以找出其他常见的谬误。一旦你知道怎样寻找，你就有能力发现越来越多的谬误。

在评价各种假设以识别谬误时，你需要经历一个过程。为了展示这个过程，我们来检查一下下面这篇文章中论证的质量。首先我们一起来组织一下论证的结构。

  这次立法所牵涉的问题，其实并非喝酒是否有害健康，而是美国国会是否愿意让联邦通信委员会随意做出一个决定，禁止在广播和电视上播放酒类广告。如果我们允许联邦通信委员会采取这一关于禁播酒类广告的行动，那么它明年又会说糖果有害公众健康，因为糖果会导致肥胖、蛀牙和其他健康问题，我们有什么办法阻止它采取相应行动呢？接下来它拿牛奶和鸡蛋开刀又该怎么办？牛奶和鸡蛋的饱和性动物脂肪含量都非常高，无疑会增加血液中的胆固醇含量，很多心脏病专家都认为它们是引发心脏病的重要原因，那我们要不要联邦通信委员会禁止在电视上播放牛奶、鸡蛋、黄油和冰淇淋的广告？

continues, we see the fleeing car pull to the side of the road. The officer approaches, draws his gun, and shoots the driver, who is an African American. We have seen this scenario before on television.

Humans are storytellers. We are driven to connect the dots when we see a series of events. We not only want to know what happened; we are immediately willing to weave a narrative that explains what actually happened. The first "make-sense" story that comes to mind is usually enough to satisfy the need to think we understand what happened. We believe that "the police person described in this paragraph is simply one more police violator of the basic humanity of African Americans." We assume our story is an accurate explanation.

But notice how little we know about the event described here. We have no idea whether the driver was fleeing an armed robbery. The facts tell us nothing of what the officer saw when he approached the car on the side of the road. We have no information about the police culture in this jurisdiction. Does it have a track record of discrimination against people of color?

We have falsely assumed that because we have a narrative offering a possible explanation for a series of facts, that narrative *is* the explanation for the interaction of those facts.

Fallacy: Narrative: Assuming incorrectly that because we can tell a story that seems to explain the occurrence of a set of facts, we now understand the links relating the facts to one another.

The next section takes you through some exercises in discovering other common fallacies. Once you know how to look, you will be able to find most fallacies.

To demonstrate the process you should go through to evaluate assumptions and thus recognize many fallacies, we examine the quality of the reasoning in the following passage. We begin by assembling the structure.

> The question involved in this legislation is not really a question of whether alcohol consumption is or is not detrimental to health. Rather, it is a question of whether Congress is willing to have the Federal Communications Commission (FCC) make an arbitrary decision that prohibits alcohol advertising on radio and television. If we should permit the FCC to take this action in regard to alcohol, what is there to prevent it from deciding next year that candy is detrimental to public health in that it causes obesity, tooth decay, and other health problems? What about milk and eggs? Milk and eggs are high in saturated animal fat and no doubt increase the cholesterol in the bloodstream, believed by many heart specialists to be a contributing factor in heart disease. Do we want the FCC

还有，我们都知道联邦政府采取的行动，不论多么激烈，都不能也不会产生完全消除酒类消费的效果。如果人们想喝含有酒精的饮料，那么他们一定可以找到办法来满足自己。

结论：联邦通信委员会不应该禁止在广播和电视上播放酒类广告。

理由：1. 如果我们允许联邦通信委员会禁止在广播电视上播放酒类广告，那么联邦通信委员会很快就会禁止很多其他类型的广告，因为有很多种产品都有可能带来潜在的健康危险。

2. 联邦政府的行动无法起到彻底消除酒类消费的作用。

首先，我们应该注意到，这两个理由都指向限制酒类广告的非常具体的不利因素，这是挺好的开端。但是，第一条理由能否被接受，主要取决于一条隐藏的假设，即一旦我们让人出于好的考虑在一个案例中采取行动，那么他在类似案例中采取相应行动时，我们就再也无法阻止。我们并不能接受这样的假设，因为我们相信，如果这些行动没有正当的理由，美国的法律体系中还有很多步骤可以防止这样的行动。我们特别擅长采取部分行动。因此，我们判定这个理由不能成立，这样的论证方式是"滑坡谬误"的一个例证。

⚠ **滑坡谬误**：假设采取某种做法会引发一连串不可控的不利事件，而实际上有现成的程序可用来防止此类连锁事件的发生。

第二个理由的相关度也值得商榷，因为就算这个理由成立，将其和结论连接起来的假设，即禁止在广播和电视上播放酒类广告的主要目标是完全消除酒类消费，也是错误的。禁止播放酒类广告的一个更可能的目标是减少酒类消费，因此这个理由我们也无法接受。我们把这类谬误叫作"追求完美解决方案谬误"。它的形式如下：我们不应支持针对某问题的解决方案，除非它能从根本上解决这个问题。如果我们真能找到完美的解决方案，那我们自然该接受它。问题是尝试某种解决方案后还遗留一部分问题，并不意味着这个解决方案欠考虑。能提出一个部分解决问题的方案，总比束手无策要强得多，它可能让我们向彻底解决问题迈进一步。

如果坐等完美解决方案出现，我们常常发现自己固守原地，动弹不得。下面是这个谬误的又一例证：家里添一套安保系统纯粹是浪费钱。如果贼要光顾你家，他们总归想得到办法，不管你装什么系统都不管用。

to be able to prohibit the advertising of milk, eggs, butter, and ice cream on TV?

Also, we all know that no action by the federal government, however drastic, can or will be effective in eliminating alcohol consumption completely. If people want to drink alcoholic beverages, they will find some way to do so.

CONCLUSION: *The FCC should not prohibit alcohol advertising on radio and television.*

REASONS: 1. *If we permit the FCC to prohibit advertising on radio and television, the FCC will soon prohibit many kinds of advertising, because many products present potential health hazards.*

2. *No action by the federal government can or will be effective in eliminating alcohol consumption completely.*

First, we should note that both reasons refer to rather specific disadvantages of the prohibition—a good start. The acceptability of the first reason, however, depends on a hidden assumption that once we allow actions to be taken on the merits of one case, it will be impossible to stop actions on similar cases. We do not agree with this assumption because we believe that there are plenty of steps in our legal system to prevent such actions if they appear unjustified. We are quite capable of doing things partially. Thus, we judge this reason to be unacceptable. Such reasoning is an example of the *slippery slope fallacy*.

Fallacy: Slippery Slope: Making the assumption that a proposed step will set off an uncontrollable chain of undesirable events, when procedures exist to prevent such a chain of events.

The relevance of the second reason is questionable because even if this reason were true, the assumption linking the reason to the conclusion—the major goal of prohibiting alcohol advertising on radio and television is to *eliminate alcohol consumption completely*—is false. A more likely goal is to *reduce consumption*. Thus we reject this reason. We call this fallacy the *searching for perfect solutions fallacy*. It takes the following form: A solution to X does not deserve our support unless it destroys the problem entirely. If we ever find a perfect solution, then we should adopt it. But the fact that part of a problem remains after a solution is tried does not mean the solution is unwise. A partial solution may be vastly superior to no solution at all. It may move us closer to solving the problem completely.

If we waited for perfect solutions to emerge, we would often find ourselves paralyzed, unable to act. Here is another example of this fallacy: It's a

⚠ **追求完美解决方案谬误**：错误地认为如果尝试某种解决方案后还有遗留问题未解决，那么这种解决方案根本就不该采用。

发现其他常见的论证谬误

下面，我们要带你做一些练习，以帮助你发现更多常见的谬误。在你进入每个练习的时候，请试着发现谬误，找一找我们前面提到的种种提示。一旦你养成鉴别谬误的良好习惯，你就能找到大多数谬误。下面每个练习所展示的论证里都含有这样或那样的谬误。我们先指出为什么我们认为这个论证是荒谬的，然后再给出谬误的名称和定义。

练习A

是时候让大麻成为人们用来缓解慢性疼痛的一种选择了。当社会对一种毒品的药用价值达成共识时，我们就该同意使用这种毒品。现在，显然美国社会中已经达成了同意使用大麻的共识。最新的民意调查显示，有73%的人认为应该允许使用药用大麻。此外，加州艾滋病受害者治疗协会也支持将抽大麻作为艾滋病人的一种治疗选择。

分析谬误的第一步，我们先来整理一下这个论证的框架：

结论：抽大麻应该成为一种医疗选择。
理由：1. 一旦我们对某些毒品的药用价值达成共识，我们就应允许其使用，最近的调查显示，将大麻作为治疗药物的共识已经形成。
2. 加州的一个协会支持使用药用大麻。

只有我们接受"如果某个信念很流行，那么它肯定是好的"这个错误的假设，调查结果才能支持这个结论。公众往往没有充分研究过一个问题，不足以提供合理的判断。一定要警惕诉诸流行观点或公众感受的论证。我们把这种论证错误称为"诉诸公众谬误"。

⚠ **诉诸公众谬误**：试图通过引述很多人都持有这一观点，以证明某个断言有道理。错误地以为很多人支持的事就是可取的。

waste of money to add a security system to your home. If thieves want to break into your house, they will find a way to do so, regardless of any security system.

Fallacy: Searching for Perfect Solution: Falsely assuming that because part of a problem remains after a solution is tried, the solution should not be adopted.

DISCOVERING OTHER COMMON REASONING FALLACIES

We are now going to take you through some exercises to discover more common fallacies. As you encounter each exercise, try to detect the fallacy, finding hints that we listed earlier. Once you have developed good fallacy-detection habits, you will be able to find most fallacies. Each exercise presents some reasoning that includes fallacies. We indicate why we believe the reasoning is fallacious and then name and define the fallacy.

Exercise A

It's about time that we make marijuana an option for people in chronic severe pain. We approve drugs when society reaches a consensus about their value, and there is clearly now a consensus for such approval. A recent survey of public opinion reported that 73 percent thought medical marijuana should be allowed. In addition, the California Association for the Treatment of AIDS Victims supports smoking marijuana as a treatment option for AIDS patients.

As a first step in analyzing for fallacies, let's outline the argument.

CONCLUSION: *Smoking marijuana should be a medical option.*

REASONS: 1. *We approve drugs when a consensus of their medical value has been reached, and a recent survey shows a consensus approving marijuana as a medical treatment.*

2. *A California association supports medical marijuana use.*

The survey results by themselves support the conclusion *only if* we accept the assumption that when some belief is popular, then it must be good—a mistaken assumption. The public often has not sufficiently studied a problem to provide a reasoned judgment. Be wary of appeals to common opinion or to popular sentiment. We label this mistake in reasoning the *appeal to popularity fallacy*.

Fallacy: Appeal to Popularity (*Ad Populum*): An attempt to justify a claim by appealing to sentiments that large groups of people have in common; falsely assumes that anything favored by a large group is desirable.

现在，我们仔细检查一下作者的第二条理由。作者所做的假设是什么？为了证明医用大麻是可取的，他引用了可疑的权威，即加州的一个协会。一个立场并不会因为权威支持就变得光荣正确。判定这样的论证是否可信，最重要的是要看权威据以做出判断的证据。如果我们无法得知这些权威对这一论题拥有特别的知识，那么我们就将这个理由视为谬误。这种类型的谬误叫作"诉诸可疑权威谬误"。

⚠ **诉诸可疑权威谬误**：引用某一权威的话来证明结论，而该权威对这一论题并没有特别的专门知识。

下面，我们检视一下与另一论争有关的一些论证。这一论争是美国国会该不该批准联邦政府资助的儿童发展项目，为儿童修建日托护理中心。

练习 B

> 我反对政府的儿童发展项目。首先，我关心的是保护美国的儿童。他们理应受到保护，不该遭受社会规划师和自以为是的空想家的干扰。这些人往往会扰乱儿童生活的正常发展过程，将他们从母亲和家人身边夺走，使他们变成宣称会在20年后为孩子创造出无穷幸福的通用计划的试验品。儿童就该和母亲一起成长，而不是在一群临时看护或护士的帮助下长大。实际的问题是父母还该不该继续有权塑造子女的性格，或该不该将权利赋予国家，用各种工具和技术来塑造年轻人。

现在让我们先整理一下这个论证的框架。

> 结论：政府的儿童发展项目是个错误。
> 理由：1.儿童应该受到保护，免受社会规划师和自以为是的空想家的干扰，这些人会干扰儿童生活的正常秩序，将他们从家人那里夺走。
> 2.有权塑造儿童品格的应该是父母，而不是国家。

作为批判性思维者，我们应寻找和这个项目有关的具体事实，可是我们一个事实也找不到。这里的理由全是未加定义且情绪化的一般概括。这些用词通常会激起负面情绪，立论者希望读者或听众将这些负面情绪与其所攻击的立场联系起来。

Now, carefully examine the author's second reason. What assumption is being made? To prove that medical marijuana is desirable, she *appeals to questionable authorities*—a California association. A position is not good just because the authorities are for it. What is important in determining the relevance of such reasoning is the evidence that the authorities are using in making their judgment. Unless we know that these authorities have special knowledge about this issue, we must treat this reason as a fallacy. Such a fallacy is called the *appeal to questionable authority fallacy*.

Fallacy: Appeal to Questionable Authority: Supporting a conclusion by citing an authority who lacks special expertise on the issue at hand.

Now let's examine some arguments related to another controversy: Should Congress approve a federally funded child development program that would provide day-care centers for children?

Exercise B

I am against the government's child development program. First, I am interested in protecting the children of the United States. They need to be protected from social planners and self-righteous ideologues who would disrupt the normal course of children's lives and tear them from their mothers and families to make them pawns in a universal scheme designed to produce infinite happiness in 20 years. Children should grow up with their mothers, not with a series of caretakers and nurses' aides. What is at issue is whether parents shall continue to have the right to form the characters of their children, or whether the state with all its power should be given the tools and techniques for forming the young.

Let's again begin by outlining the argument.

CONCLUSION: *The government's child development program is a mistake.*

REASONS: 1. *Our children need to be protected from social planners and self-righteous ideologues, who would disrupt the normal course of children's lives and tear them from their families.*

2. *The parents, not the state, should have the right to form the characters of their children.*

As critical thinkers, we should be looking for specific facts about the program. But we find none. The reason is saturated with undefined and emotionally loaded generalities. Such terms will typically generate negative emotions that the writer or speaker hopes readers and listeners will associate with the position she is attacking.

当然，对许多事情产生情绪化的反应是没有错的。当我们看到医院遭受轰炸时，如果我们情感上一点反应都没有，那就是对人性的背叛；对成千上万女性遭受性奴役的事实无动于衷着实令人不安。同样，试图用理由和证据说服他人对某件事做出情绪化反应也是没有错的。

然而，仅仅靠情感诉求就形成一种论点，是一种逻辑上的花招。这篇文章的作者试图让我们反对政府的儿童发展项目，他对我们用了两个常见的花招。首先，他精心挑选词汇以引发我们的情感共鸣，希望我们的情绪反应会促使我们同意他的结论。当立论者激发人们的情绪反应，然后依靠这种情绪反应而非正当理由来促使人们同意他的结论时，他就犯了"诉诸感情谬误"。这类谬误的三个特别常见的地方就是广告、政治辩论和法庭辩论。

⚠ **诉诸感情谬误**：使用带有强烈感情色彩的语言来分散读者或听众的注意力，让他们忽视相关的理由和证据。常被用来加以利用的情感有：恐惧、希望、爱国主义、怜悯和同情。

其次，作者树立了一个靶子来加以攻击，让我们更容易站在他这边，而这个靶子实际上根本不存在。他有意引申反对方的立场，使其达到易于攻击的程度。这个例子里的错误假设就是作者所攻击的立场和立法中实际呈现的立场是同一立场。儿童是否真的是一些通用计划里的试验品？批判性思维者得到的教训是：如果有人攻击一个立场的某些方面，一定记住要查看一下他是否公正全面地呈现了这个立场。如果没有的话，你就找到了一个"稻草人谬误"。

稻草人不是真人，很容易被击倒。当一个人犯了稻草人谬误时，他攻击的立场就像稻草人。要检查一个立场被展现得有多公平，最好的办法就是找出反映所有立场的事实。

⚠ **稻草人谬误**：歪曲对方的观点，使其容易受到攻击，进而攻击事实上根本就不存在的观点。

现在我们进一步仔细分析第二条理由。作者说要么父母享有塑造子女性格的权利，要么国家应该被掌握主导权。让我们快速看一看布兰妮·斯皮尔斯的作品《马戏团》里出现的另一个例子："世界上只有两种人——一种人表演，另一种人观看。"

There is certainly nothing wrong with an emotional reaction to many things. When we see hospitals being bombed, it would be a betrayal of our humanity not to have an emotional reaction. Indeed to *not* react emotionally to the sexual slavery of thousands of women is indeed troubling. Similarly, there is nothing wrong with trying to persuade others with reasons and evidence to have a particular emotional reaction.

But to form an argument that relies on little but emotional appeals is a logical trick. The writer of the passage trying to get us to oppose the government's child development program plays two common tricks on us. First, she is *appealing to our emotions* with her choice of words, hoping that our emotional reactions will get us to agree with her conclusion. When communicators stimulate emotional reactions from people and then rely on that reaction, rather than warrants to create agreement with their conclusion, they commit the *appeal to emotion fallacy*. Three especially common places for finding this fallacy are advertising, political debate, and the courtroom.

Fallacy: Appeal to Emotion: The use of emotionally charged language to distract readers and listeners from relevant reasons and evidence. Common emotions appealed to are fear, hope, patriotism, pity, and sympathy.

Second, the writer has set up a position to attack, which in fact does not exist, making it much easier to get us on her side. She has extended the opposition's position to an "easy-to-attack" position. The false assumption in this case is that the position attacked is the same as the position actually presented in the legislation. Will children really be pawns in some universal scheme? The lesson for the critical thinker is, When someone attacks aspects of a position, always check to see whether she is fairly representing the position. If she is not, you have located the *straw-person fallacy*.

A straw person is not real and is easy to knock down—as is the position attacked when someone commits the straw-person fallacy. The best way to check how fairly a position is being represented is to get the facts about all positions.

Fallacy: Straw Person: Distorting our opponent's point of view so that it is easy to attack; thus we attack a point of view that does not truly exist.

Let's now look closely at the second reason. The writer states that either parents have the right to form the characters of their children or the state should be given the decisive tools. Take a quick look at another example in a statement by Britney Spears in *Circus*: "There are only two types of people in the world: the ones that entertain and the ones that observe."

要让这样的陈述成立，我们必须假设只有这两种选择。是不是这样呢？当然不是！作者创造了一种虚假的两难选择。有没有可能既让儿童发展项目存在，又让家人对孩子的成长施加重要影响呢？当争议被表现得好像只有两种可能的选择时，我们一定要特别当心，事实上总有两种以上的选择。如果一个立论者通过仅陈述两种选择来过度简化一个论题，他所犯的错误就叫作"虚假的两难选择谬误"。要找出两难选择的谬误，就要时时警惕下面这些说法：

- 不是……就是；
- 唯一的选择就是；
- 两种选择分别是；
- 因为甲不起作用，所以只有乙能

看到这些说法并不一定意味着你发现了一个谬误。有时候真的只有两种选择。这些词只不过是警告标志，见到它们的时候就要停下来想一想："在这种情况下，是不是还有其他的选择呢？"

你能不能看出下面这个对话中虚假的两难选择？

  市民：我认为，美国政府决定入侵伊拉克是一个大错误。
  政客：你为什么要恨美国呢？

⚠ **虚假的两难选择谬误**：在现实中存在两种以上的选择时，却假想只有两种选择。

我们在要为某种行为寻求解释时，常常会遇到越想越糊涂的情形。下面这段大学室友之间的简短对话就体现了这种犯糊涂的情况。

  丹：我发现查克最近行为一直有点古怪。他对别人的态度真的很粗暴，把宿舍里弄得一团糟，而且坚决不清理。你认为他到底怎么了？
  凯文：这对我来说一点都不奇怪。他就是个混账。

"解释"需要对一个行为为什么发生进行分析。解释是个耗时费力的活儿，经常考验我们脑子里有多少知识。在上述例子中，"混账"对查克的古怪行为来说是个让人难以满意的

For statements like these to be true, one must assume that there are only two choices. Are there? No! The writer has created a *false dilemma*. Isn't it possible for the child development program to exist and also for the family to have a significant influence on the child? Always be cautious when controversies are treated as if only two choices are possible; there are usually more than two. When a communicator oversimplifies an issue by stating only two choices, the error is referred to as an *either-or* or *false dilemma fallacy*. To find *either-or* fallacies, be on the alert for phrases like the following:

either . . . or

the only alternative is

the two choices are

because A has not worked, only B will

Seeing these phrases does not necessarily mean that you have located a fallacy. Sometimes there *are* only two options. These phrases are just caution signs causing you to pause and wonder: "Are there more than two options in this case?"

Can you see the false dilemma in the following interchange?

Citizen: I think that the decision by the United States to invade Iraq was a big mistake.

Politician: Why do you hate America?

Fallacy: Either-Or (or False Dilemma): Assuming only two alternatives when there are more than two.

We often encounter further confusion in our thinking when we seek explanations for behavior. A brief conversation between college roommates illustrates the confusion.

Dan: I've noticed that Chuck has been acting really weird lately. He's being really rude toward others and is making all kinds of messes in our residence hall and refusing to clean them up. What do you think is going on?

Kevin: That doesn't surprise me. He is just a jerk.

To "explain" requires an analysis of why a behavior occurred. Explaining is demanding work that often tests the boundaries of what we know. In the given example, "jerkhood" is an unsatisfactory explanation of Chuck's

解释。当别人要我们解释为什么某个特定的行为会发生时，我们常常忍不住想要隐藏自己对很多复杂因果关系的无知，干脆给那种行为贴个标签或起个名字。然后，我们错误地以为，因为我们知道那个标签或名字，所以我们知道了事情的起因。

我们之所以这样做，是因为乱扣帽子欺骗了我们，让我们相信自己找到了那个人的特点或发现那个人是什么样的人，相信是这些促使他做出相应的行为。例如，比起详细列举出导致一个人大发脾气的一套又一套复杂的内因和外因，如人际关系出了问题、父母给他的强化或者孤立无援的感觉等，我们可能更倾向于干脆说那个人脾气不好，或者说他就是喜欢与人为敌。这样的解释过度简化了问题，阻碍我们取得更深邃的理解。

下面的例子应该能提高我们对此类谬误的警惕感。

　　1. 作为对爸爸酗酒毛病的回应，当妈妈被大女儿问道"为什么爸爸行为这么古怪"时，她干脆回答说："他遇到了中年危机。"
　　2. 有个朋友老是担心其他人在背后谈论他。你问一个心理学家他为什么会这样，心理学家回答说："因为这个人有妄想症。"

两个回答问题的人都没有对所发生的行为提供让人满意的解释。比如说，爸爸基因的具体情况、工作上的压力以及夫妻间的争吵等，都有可能是老爸酗酒的合理解释。"中年危机"的解释不但不充分，而且容易误导别人。我们本以为自己知道了爸爸为什么酗酒，实际上却什么都不知道。

有时有人宣称他们发现了导致某个行为的原因，而实际上他们所做的不过是为这种问题起了个名字而已。我们要对这种"乱扣帽子谬误"保持警惕。

⚠ **乱扣帽子谬误**：错误地以为因为你给某个特定事件或行为起了个名字，所以你合理解释了这一事件。

经验告诉我们，过去的行为往往是对未来行为的最佳预测因素。但我们有时也会犯一种推理谬误，错误地认为我们过去的行为并非很好的预测因素，即使有非常可靠的理由支持这一点，我们也不管不顾。我们表达出一种乐观偏见，可以称之为"计划谬误"。这里有个例子。假设你和我要一起完成一份必须提交的报告，我们一直拖延到截止时间前的最后一刻才开始写，尽管我们都知道，过去一次又一次，我们到了截止时间还是无法完成工作。同样，很多机构常常低估完成一个项目所需的时间。计划谬误是一种一厢情愿，我们对工

behavior. When asked to explain why a certain behavior has occurred, it is frequently tempting to hide our ignorance of a complex sequence of causes by labeling or naming the behavior. Then we falsely assume that because we know the name, we know the cause.

We do so because the naming tricks us into believing we have identified something the person *has* or *is* that makes her act accordingly. For example, instead of specifying the complex set of internal and external factors that lead a person to express an angry emotion, such as problems with relationships, parental reinforcement practices, or feelings of helplessness, we say that the person *has* a bad temper or that the person *is* hostile. Such explanations oversimplify and prevent us from seeking more insightful understanding.

The following examples should heighten your alertness to this fallacy:

1. In response to dad's heavy drinking, mom is asked by her adult daughter, "Why is dad behaving so strangely?" Mom replies, "He's having a midlife crisis."
2. A friend worries constantly that other people are talking about him. You ask a psychologist why he does so. He answers, "Because he is paranoid."

Neither respondent satisfactorily explained what happened. For instance, the specifics of dad's genes, job pressures, and marital strife could have provided the basis for explaining the heavy drinking. "A midlife crisis" is not only inadequate but also misleading. We think we know why dad is drinking heavily, but we don't.

Be alert for the *explaining by naming fallacy* when people claim that they have discovered a cause for the behavior when all they have actually done is provide a name for the problem.

Fallacy: Explaining by Naming: Falsely assuming that because you have provided a name for some event or behavior, you have also adequately explained the event.

Experience tells us that past behavior is often the best predictor of future behavior. Sometimes, however, we commit a reasoning fallacy because we mistakenly assume that our past behavior is *not* a very good predictor even when there is good reason to believe that it is. We express an optimism bias, which we can label the *planning fallacy*. Here is an illustration. Suppose you or I postpone a report we are required to submit until right before it is due even though we know that time and time again we have been unable to complete our work when it was due. Also, organizations frequently underestimate the time required to complete a project. The planning fallacy is a special case

作的结果做出了过于乐观的、不切实际的预计。

> ⚠ **计划谬误**：人们或者机构倾向于低估他们完成一项工作所需的时间，尽管有很多过去的经验显示他们曾一直低估完成某件事所需要的时间。

警惕分散注意力的干扰

那些设法让人接受他的断言的人常会发现，只要他们不让听众太过仔细地审视相关的理由，他们就可以捍卫这一断言。他们使用分散注意力的策略来防止别人细细审视。在你寻找谬误的时候，会发现下面这个方法很有帮助：当立论者所使用的论证主要是为了转移你的注意力，让你不再关注最相关的理由时，你就要特别警惕。例如，人身攻击型谬误就可以通过这种方式来愚弄我们：转移我们的注意力，让我们多多关注这个人本身，从而不去关注是否有正当的理由。在这一节，我们会提供更多练习来展示其他类型的谬误，只要我们提问"作者有没有通过转移我们的注意力来欺骗我们"，就有可能发现这类谬误。

练习 C

> 政治演说：在即将到来的选举中，你迎来了为一位女性投票的良机，她代表了这个伟大国家的未来，她为实现民主长期奋斗，为捍卫国家利益不遗余力，她为追寻美国梦而当机立断、充满信心、勇往直前。这位女性充满爱心，为儿童福利出力，为环境保护奔走，为推动国家迈向和平、繁荣和自由而出谋划策。投古德哈特一票就是投真理一票，投梦想一票，投常识一票。

听起来好像古德哈特女士是个完人，是不是？但是这一演说没有提供任何具体的细节，说明这位参议员的过去记录和对重大事件的立场。取而代之的是一连串描述品行的词，它们全都倾向于和我们心中积极的情感产生联系。我们把这类词称为粉饰，因为它们都能让人产生积极的联想，而且它们都很笼统，简直可以说读者想它们是什么意思，它们就是什么意思。这种粉饰的手法会让我们赞同或接受一个结论，而根本不去检查相应的理由、证据或具体的优缺点。"粉饰谬误"就好像把乱扣帽子的方法颠倒过来，后者的重点是要我们形成负面的评价，而不去检查相应的证据。品行词的使用是政客们经常玩弄的手腕，这个手法可以分散读者或听众的注意力，让他们不去关注具体的行动或政策，这些行动和政策更容易招致反对和批评。

of wishful thinking in which we make overly optimistic, unrealistic predictions about the outcomes of projects.

Fallacy: The Planning Fallacy: The tendency for people or organizations to underestimate how long they will need to complete a task, despite numerous prior experiences of having underestimated how long something would take to finish.

LOOKING FOR DIVERSIONS

Frequently, those trying to get an audience to accept some claim find that they can defend that claim by preventing the audience from taking too close a look at the relevant reasons. They prevent the close look by diversion tactics. As you look for fallacies, you will find it helpful to be especially alert to reasoning used by the communicator that *diverts your attention* from the most relevant reasons. For example, the ad hominem fallacy can fool us by diverting our attention too much to the nature of the person and too little to the legitimate reasons. In this section, we present exercises that illustrate other fallacies that we are likely to detect if we ask the question, "Has the author tricked us by diverting our attention?"

> **Exercise C**
>
> Political speech: In the upcoming election, you have the opportunity to vote for a woman who represents the future of the United States, who has fought for democracy and defended our flag, and who has been decisive, confident, and courageous in pursuing the American Dream. This is a caring woman who has supported our children and the environment and has helped move this country toward peace, prosperity, and freedom. A vote for Goodheart is a vote for truth, vision, and common sense.

Sounds like Ms. Goodheart is a wonderful person, doesn't it? But the speech fails to provide any specifics about the senator's past record or present position on issues. Instead, it presents a series of *virtue words* that tend to be associated with deep-seated positive emotions. We call these virtue words *glittering generalities* because they have such positive associations and are so general as to mean whatever the reader wants them to mean. The *glittering generality fallacy* leads us to approve or accept a conclusion without examining relevant reasons, evidence, or specific advantages or disadvantages. The glittering generality is much like name-calling in reverse because name-calling seeks to make us form a negative judgment without examining the evidence. The use of virtue words is a popular ploy of politicians because it serves to distract the reader or listener from specific actions or policies, which can more easily trigger disagreement.

⚠ **粉饰谬误**：使用模糊、引发人们强烈情感认同的品行词，使我们倾向于同意某件事而不去细查其理由。

我们再来看看另一个很常见的转移注意力的手法。

练习 D

对于医药公司修改研究数据以让他们的止痛药对健康造成的危险显得比实际上要小一点的做法，我不明白为什么每个人都如此反感。服用那些药物的结果根本不会那样糟。毕竟有成千上万的人使用这些止痛药，并且从中获得了缓解疼痛的效果。

问题的实质是什么？在止痛药的安全性问题上，公众是不是被误导了？如果读者不够细心，他的注意力就会被转移到公众是否想使用这些药物的问题上。如果一个写作者或发言者将我们的注意力从论题上转移走，我们就可以说他故意把话题转移到和原来的主题不相干的事情上去。我们很多人都擅长利用"转移话题谬误"，正如下面这个对话所揭示的。

妈妈：你和男朋友去哪儿了？你为什么要跟我撒谎？
女儿：你总是刁难我。

如果女儿成功了，问题就变成这个妈妈是不是在刁难女儿，而不再是女儿为什么要跟妈妈撒谎。

只要你谨记真正的论题所在，同时牢记解决这一论题所需的证据的类型，那么一般而言，要找到转移话题谬误并不难。

⚠ **转移话题谬误**：插入一个不相干的话题，以将人们的注意力从原来的论题上转移开，通过将注意力从当前的论证转移到另一个论题上以赢得论证。这个例子中的谬误顺序如下：1. A 主题正被讨论；2. 将 B 主题引入，它好像和 A 主题有关，实际上两者并不相干；3. A 主题被置之不理。

这种类型的"论证"是错误的，因为仅仅改变讨论的主题，不能被当作一个反对某一断言的论证。

愚弄人的循环论证

最后要介绍的这一谬误特别具有欺骗性。有时候，你可能看到一个结论自己支持自己，只不过措辞有所改变，以愚弄那些单纯无知的人！例如，论证辍学是不可取的，因为它

Fallacy: Glittering Generality: The use of vague, emotionally appealing virtue words that dispose us to approve something without closely examining the reasons.

Let's examine another very common diversionary device.

Exercise D

I don't understand why everyone is so upset about drug companies' distorting research data in order to make their painkiller drugs seem to be less dangerous to people's health than they actually are. Taking those drugs can't be that bad. After all, there are still thousands of people using these drugs and getting pain relief from them.

What is the real issue? Is the public being misled about the safety of painkiller drugs? But if the reader is not careful, his attention will be diverted to the issue of whether the public wants to use these drugs. When a writer or speaker shifts our attention from the issue, we can say that she has drawn a *red herring* across the trail of the original issue. Many of us are adept at committing the *red herring fallacy*, as the following dialogue illustrates.

Mother: Why did you lie to me about where you were going with your boyfriend?
Daughter: You're always picking on me.

If the daughter is successful, the issue will become whether the mother is picking on her daughter, not why the daughter lied to her.

You should normally have no difficulty spotting red herrings as long as you keep the real issue in mind as well as the kind of evidence needed to resolve it.

Fallacy: Red Herring: An irrelevant topic is presented to divert attention from the original issue and help to win an argument by shifting attention away from the argument and to another issue. The fallacy sequence in this instance is as follows: (a) Topic A is being discussed; (b) Topic B is introduced as though it is relevant to topic A, but it is not; and (c) Topic A is abandoned.

This sort of "reasoning" is fallacious because merely changing the topic of discussion hardly counts as an argument against a claim.

SLEIGHT OF HAND: BEGGING THE QUESTION

Our last illustrated fallacy is a particularly deceptive one. Sometimes, a conclusion is supported by itself; only the words have been changed to fool the innocent! For example, to argue that dropping out of school is

是不好的，这实际上与不论证没什么两样。结论由同一个结论来"证明"（只是表述不同），这样的论证其实是在回避问题，而不是在回答问题，这就是"循环论证谬误"。我们来看一个不那么明显的例子。

> 阅读传统教科书比阅读电子文本的学习效果要好得多，因为以教材的形式来展现各种材料非常有利于学习。

同样，支持结论的理由只不过是用不同的措辞重申了一遍结论而已。根据定义，传统的书都是以教材的形式来阅读的。作者的论证是：这样的做法非常好，因为它非常好。而一个合理的理由肯定会指出阅读传统教材的某个具体好处，例如对所学材料的持久记忆。

只要结论本应在论证过程中被证明，结果却只能被假设出来，那么循环论证谬误就在所难免。在列出论证的结构提纲时，请检查一下理由，以确保它们并非仅仅用不同的措辞重复结论，再检查一下，确保结论不是拿来证明理由的。

⚠ **循环论证谬误**：在论证过程中假设自己的结论成立的论证。

使用这个批判性问题

当你找到一个谬误时，你也就找到了一个合理的根据以拒绝接受立论者的那部分论证。但是秉持着建设性的批判性思维精神，你还是需要考虑立论者提出的每一个不属于谬误的理由。可惜的是，一本书或一篇文章的作者并不能亲自过来与你进行深入探讨。在口头论辩中，在对方出现谬误时，你如果想要深入交谈下去，最有效的措施就是问一问犯了逻辑谬误的对方，有没有更好的理由来证实他的结论。例如，如果出现一个转移话题谬误发生，你就可以问对方是否可以回到原来的那个论题。

论证谬误小结

我们已经通过几个练习向你展示论证过程中各种可能出错的方式。我们虽然并未穷尽所有的出错方式，但是为你提供了一个良好的开端。我们把更多的谬误留到后面的章节中讨论，因为只要你专心关注后面章节中所讨论的具体问题，就极有可能自己辨别出它们。在你遇到每个新的谬误时，请确保能将它们添加到你的谬误清单中。

undesirable because it is *bad* is to not argue at all. The conclusion is "proven" by the same conclusion (in different words). Such an argument *begs the question*, rather than answering it. Let's look at an example that is a little less obvious.

> Reading traditional textbooks is superior to reading E-texts in learning effectiveness because it is highly advantageous for learning to have materials made available in a textbook format.

Again, the reason supporting the conclusion restates the conclusion in different words. By definition, traditional books are read in a textbook format. The writer is arguing that such a procedure is good because it is good. A legitimate reason would be one that points out a specific advantage to reading traditional textbooks such as greater retention of learned material.

Whenever a conclusion is *assumed* in the reasoning when it should have been proven, a *begging the question fallacy* has occurred. When you outline the structure of an argument, check the reasons to be sure that they do not simply repeat the conclusion in different words and check to see that the conclusion is not used to prove the reasons.

Fallacy: Begging the Question: An argument in which the conclusion is assumed in the reasoning.

USING THIS CRITICAL QUESTION

When you spot a fallacy, you have found a legitimate basis for rejecting that part of the communicator's argument. But in the spirit of constructive critical thinking, you want to consider any reasons offered that are not fallacies. Unfortunately, the author of a book or article is unavailable for more conversation. But for those fallacies occurring in an oral argument, your best bet for an enduring conversation is to ask the person who committed the fallacy whether there are any better reasons for the conclusion. For example, if a red herring fallacy occurs, ask the speaker if he could return to the original issue.

SUMMARY OF REASONING ERRORS

We have taken you through exercises that illustrate a number of ways in which reasoning may be faulty. We have not listed all the ways, but we have given you a good start. We have saved some additional fallacies for later chapters because you are most likely to spot them when you focus on the particular question central to that chapter. As you encounter each additional fallacy, be sure to add it to your fallacy list.

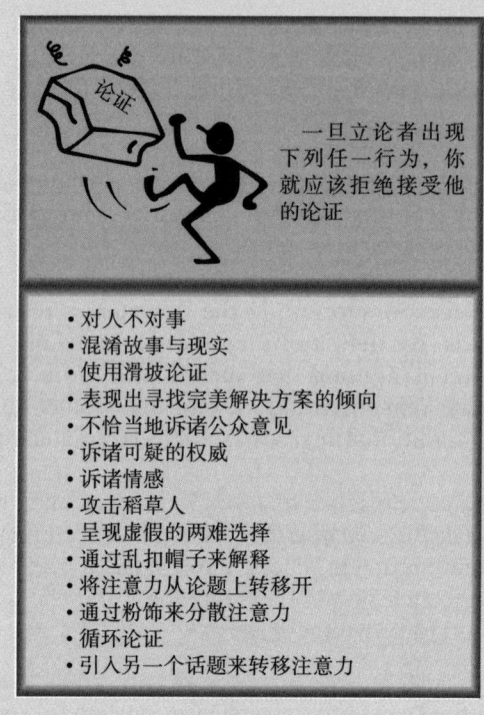

谬误总结

扩展关于谬误的知识

我们建议你参考其他教材和某些网站来扩充你对论证谬误的认识和理解。戴默的《好好讲道理：反击谬误的逻辑学训练》就是个特别好的资源，可以帮你更加熟悉论证中的谬误。

思维体操

⚠ **批判性问题：** 论证中有没有谬误？

请在以下每篇文章里找出论证中存在的谬误。

第一篇

最近，美国各州有将同性婚姻合法化的危险趋势。显而易见，同性恋议程已经渗透到这些州的立法机构，影响立法机构，使其通过立法确认同性婚姻的地位。自由主

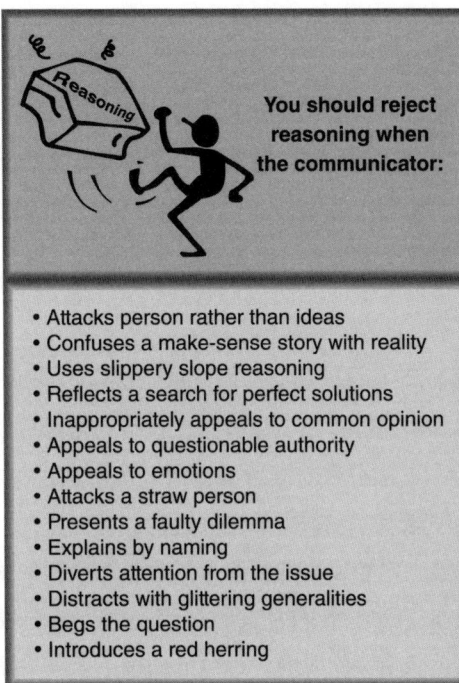

- Attacks person rather than ideas
- Confuses a make-sense story with reality
- Uses slippery slope reasoning
- Reflects a search for perfect solutions
- Inappropriately appeals to common opinion
- Appeals to questionable authority
- Appeals to emotions
- Attacks a straw person
- Presents a faulty dilemma
- Explains by naming
- Diverts attention from the issue
- Distracts with glittering generalities
- Begs the question
- Introduces a red herring

EXPANDING YOUR KNOWLEDGE OF FALLACIES

We recommend that you consult texts and some Web sites to expand your awareness and understanding of reasoning fallacies. Damer's *Attacking Faulty Reasoning* is an especially good source to help you become more familiar with reasoning fallacies.

PRACTICE EXERCISES

 Critical Question: ***Are there any fallacies in the reasoning?***

Try to identify fallacies in the reasoning in each of the three practice passages.

Passage 1

There has been a dangerous trend of states' legalizing gay marriage as of late. It's apparent that the gay agenda has infiltrated the legislative bodies of these states and has influenced the legislatures into passing laws

义媒体及其对同性婚姻的颂扬当然也起到了推波助澜的作用。这些法律破坏了美国的传统道德。

尽管有这种趋势，但一些政客仍然同意同性婚姻玷污了美国的传统和价值观。如果美国允许同性婚姻存在，那么这个国家终将允许其他非传统关系合法化，如一夫多妻或乱伦，只不过是时间问题罢了。婚姻一直是男人和女人之间的一种神圣的制度，并且应该保持这种方式。否则，美国人民的道德将不可避免地每况愈下。

第二篇

热衷于举办各种派对的大学生联谊会成员被媒体一边倒地描绘成品行不端的坏学生，聚会上偶然发生的打架斗殴或性侵犯，导致媒体对个别事件反应过度。只有傻瓜才会禁止在校园里举办学生派对。大部分针对联谊会派对的投诉都出自那些反社会、不合群的人，这些人打心眼里痛恨各种派对。我就组织过好几次联谊会派对，各种酒水随意喝，但从头到尾没出任何事。很明显，对校园里的联谊会派对施加某种形式的禁令或规范是毫无意义的行为。我见到过其他在校园之外举办的派对，不是联谊会成员举办的，其中存在性侵现象。一旦学校管理者禁止联谊会的派对，那么他们接下来就会禁止在校园里举办任何允许喝酒的社交活动。

第三篇

比尔：窝藏图谋摧毁美国的恐怖分子的国家应该被当作美国的敌人。任何不将恐怖分子主动交给美国司法体系的国家，都明白无误地站在恐怖分子的一边。这种行为意味着这些国家的领导人不愿看到这些恐怖分子受到应有的惩罚，反而更热衷于窝藏杀人犯、强奸犯、窃贼以及反民主的人士。

泰勒：如果一个人有直系亲属一直在为中情局效力，那么从他嘴里说出这样的话来，我一点也不觉得奇怪。但是在我看来，一旦你开始给那些和美国政策不一致的国家贴上敌人的标签，那么最终导致的结果就是几乎所有国家都会被当成美国的敌人，那么我们最终将成为孤家寡人，没有一个同盟。

that allow gay marriage. The liberal media and its glorification of gay marriage have certainly played a part as well. These laws destroy the traditional morals of this nation.

Despite the trend, several politicians still agree that gay marriage is an affront to America's traditions and values. If America allows gay marriage, it's only a matter of time before this nation allows other nontraditional relationships, such as polygamy or incest. Marriage has always been a sacred institution between a man and a woman and should remain that way. Otherwise, it's inevitable that the morals of the American people will fall by the wayside.

Passage 2

Fraternity members who hold college parties are unfairly depicted in the media because of overreactions to a few cases where fights or sexual assaults have occurred at a fraternity party. Only an idiot would ban fraternity parties on campus. Most of the complaints about fraternity parties come from antisocial loners, people who hate parties in the first place. I've held several fraternity parties with alcohol available and nothing has ever gone wrong. Clearly placing some kind of ban or regulation on fraternity parties on campus would be a pointless action. I have seen other parties off campus that were not held by fraternities where assaults happened. Once administrators ban fraternity parties, their next step will be to ban any sort of social event on campus where alcohol is permitted.

Passage 3

Bill: *Countries that harbor terrorists who want to destroy the United States must be considered enemies of the United States. Any country that does not relinquish terrorists to the American justice system is clearly on the side of the terrorists. This sort of action means that the leaders of these countries do not wish to see justice done to the terrorists and care more about hiding murderers, rapists, thieves, and anti-democrats.*

Taylor: *That's exactly the kind of argument that I would expect from someone who has relatives who have worked for the CIA. But it seems to me that once you start labeling countries that disagree with America on policy as enemies, then eventually almost all countries will be considered our enemies, and we will be left with no allies.*

给个提示

第一篇

结论：同性婚姻是错误的，不应该被合法化。

理由：1. 最近同性婚姻合法化是外部议程的结果。

2. 同性婚姻对美国的道德产生了负面影响。

3. 一些政客反对同性婚姻。

4. 允许同性婚姻意味着允许其他非传统的关系。

5. 婚姻一直是一男一女之间的事，而且也应该这样保持下去。

作者首先攻击了已经将同性婚姻合法化的美国部分州的立法机构，而不是这个论题本身。他声称，使同性婚姻合法化的州的数量不断增加是外部议程的结果，而不是立法机构本身的研究和论证的结果。作者认为，立法机构只是对外部议程和媒体做出反应，这种看法诟病了立法机构的性质，代表一种人身攻击。

第二个理由是诉诸情感。作者用语言让美国的"传统"道德看起来像某种读者必须保护的东西，呼吁不能让它受到同性婚姻可能带来的威胁。

第三个理由是诉诸可疑的权威。仅仅某些政客不同意同性婚姻，并不一定意味着读者也应该这样做，特别是作者没有提供任何这些政客用来支持他们立场的论据和证据。

第四个理由是，作者声称允许同性婚姻将不可避免地导致其他非传统关系的合法化，这是在使用滑坡谬误。同性婚姻的合法化并不意味着其他关系也会顺势地合法化。

最后一个理由是假设虚假的两难选择，要么美国禁止同性婚姻，要么美国公众会堕落到道德低下的境地。但是，如果同性婚姻被完全禁止，难道那些道德沦丧的人就不存在了吗？是否可能存在其他的可能性呢？是否可能让同性婚姻合法化，同时父母、家庭和学校等其他因素仍然向孩子灌输正确的道德观念？

第二篇

结论：大学校园里举办的联谊会派对不应该被禁止。

理由：1. 禁止派对的愿望来自公众对少数几例事件的恐慌反应。

Sample Responses

Passage 1

CONCLUSION: *Gay marriage is wrong and should not be legalized.*

REASONS:
1. *Recent legalization of gay marriage is the result of outside agendas.*
2. *Gay marriage negatively impacts America's morals.*
3. *Several politicians are against gay marriage.*
4. *Allowing gay marriage means allowing other nontraditional relationships.*
5. *Marriage has always been between a man and a woman and should remain so.*

The author begins by attacking the legislatures of the states that have legalized gay marriage instead of the issue itself. She claims the growing number of gay marriage states is a result of an outside agenda rather than research and reasoning on the part of the legislatures. The author's suggestion that the legislatures are simply reacting to agendas and the media undermines the character of the legislatures in question and represents an *ad hominem* attack.

The second reason represents an *appeal to emotion*. The author is using language to make the "traditional" morals of America seem like something the reader must protect against the threat gay marriage supposedly holds.

The third reason is an *appeal to questionable authority*. Just because certain politicians disagree with gay marriage does not necessarily mean the reader should as well, especially when the author does not provide any of the arguments or evidence that these politicians use to support their position.

By claiming that allowing gay marriage will inevitably lead to the legalization of other forms of nontraditional relationships, the author is using a *slippery slope* fallacy. Legalization of gay marriage does not automatically carry with it the legalization of other types of relationships.

The final reason assumes the *faulty dilemma* that either America bars gay marriage or the American public will descend to a lower moral state. But isn't it possible to have morally loose people even with gay marriage totally banned? Isn't it also a possibility that other factors, such as parents, families, and schools, can instill proper morals while gay marriage is legal?

Passage 2

CONCLUSION: *Fraternity parties on campus should not be banned.*

REASONS:
1. *Desire to ban results from hysterical public reaction to just a few cases.*

2. 大多数投诉来自反社会、不合群的人。
3. 有一个联谊会成员举办过几场派对，酒随便喝，却没有出过任何问题。
4. 禁止在校园里举办联谊会派对也解决不了问题，在其他不是由联谊会成员举办的派对上，仍存在性侵和安全问题。
5. 禁止在校园里举办联谊会派对将会演变为禁止在校园里举办其他允许喝酒的社交活动。

这篇文章一开始就出现了人身攻击和乱扣帽子的谬误，攻击那些想要禁止举办联谊会派对的人有品格缺陷，而没有提出任何具体的论证。第三个理由似乎受到一厢情愿的思维的影响，第四个理由犯了追求完美解决方案的谬误，第五个理由犯了滑坡谬误，因为完全有可能出台规定禁止举办联谊会派对，而无须将这个规定推广到校园里的其他社交活动上。

2. *Most complaints come from antisocial loners.*
3. *A fraternity member has held several parties with alcohol and nothing went wrong.*
4. *Banning fraternity parties on campus wouldn't solve the problem; there would still be assaults and safety issues at other parties that are not held by fraternities.*
5. *Banning fraternity parties on campus would lead to banning other social events on campus where alcohol is available.*

This essay begins with *ad hominem* and *name-calling* fallacies, attacking the character of those who want to ban fraternity parties rather than addressing any specific arguments. *Wishful thinking* appears to influence the writer's third reason, and the fourth reason commits the fallacy of *search for a perfect solution*. His next reason illustrates the *slippery slope fallacy*, as it is clearly possible to make regulations that ban fraternity parties without necessarily extending such bans to other social events on campus.

第 7 章

证据的效力：个人经历、典型案例、当事人证言和专家陈述

| 学 习 目 标 |

1. 深刻理解事实和见解的区别，以此作为评价证据来源的一种方式。
2. 了解证据的其他来源及其作为结论的证据的价值。

在第 6 章里，通过学习如何找到论证中存在的各种谬误，你已在评价试图说服你的论证方面取得了巨大的进步。本章以及第 8 章会继续关注如何评价论证，主要帮助你学习对交流者用以加强他们的论证的各种证据提一些批判性问题。

CHAPTER 7

The Worth of Personal Experience, Case Examples, Testimonials, and Statements of Authority as Evidence

LEARNING OBJECTIVES

1. Appreciate the fact/opinion distinction as a way of evaluating the sources of evidence.
2. Acquire awareness of alternative sources of evidence and their worth as evidence for conclusions.

In Chapter 6, you made major progress toward evaluating persuasive communications by learning how to detect some fallacies in reasoning. In this chapter and Chapter 8, we continue our focus on evaluation as we learn to ask critical questions about various kinds of evidence communicators use to strengthen their reasoning.

事实还是见解

当证据接近事实而远离纯粹的见解时，它就具有说服力。换句话说，我们希望我们的结论基于关于世界的确凿陈述，而非某种见解，或对于什么真实、什么可靠的异想天开的猜测。证据来自对事实的系统收集和组织。

事实远观似乎无比强大、准确，近观却并非如此。正如玛丽·普维的杰作《现代事实的历史》所阐明的那样，随着人类在新的语境中使用"事实"这个概念，我们所说的"事实"的形式已经发生了多次变化。

想一想事实在不同语境下含义的细微差别吧。例如，在棒球比赛中，裁判是一次投球是坏球还是好球的决定者。但是投球真的是裁判所断言的那样吗？另一个裁判会不会做出不同的断言呢？棒球场上的阴影会不会影响到裁判对投球断言的准确性呢？教练对于之前的一个判罚的抗议是否促使裁判无意或有意地裁定这是一个坏球或一个好球呢？裁判在判断一次投球是好球还是坏球时，其对于击球区的判定标准是否会发生变化呢？

可以肯定的是，像"伊朗拥有大规模杀伤性武器"这样的"事实"至少和棒球投球是好球还是坏球一样值得怀疑。

我们所有人都会经历同样一件事的概率是衡量一个特定事实有多确定和可靠的标准。换句话说，事实的力量有强弱之分，其可能性有大小之分。事实比是非对错更复杂，但这绝不是说它们不重要。强有力的论证是建立在强有力的事实之上的。

见解常常与事实形成对比，本该如此。见解没什么价值，发表见解是不费吹灰之力的。"通心粉比葡萄干麸皮好吃。""美国总统比苏格兰领导人更有常识。"一个表达见解的人表达的是："瞧，不要问我为什么说这样的话。这些只是我的见解而已。"

一个见解可能是一场富有成效的对话的第一步，但在缺乏合乎逻辑、严密周详的理由和证据的情况下，它仅仅只是一个见解。

当你或其他人有值得三思的见解时，我们就可以提出有力的理由和证据，让你的主张更接近事实。

那么，我们该如何定位事实性主张，从而使我们跳出纯粹的见解呢？事实性主张的来源具有不同程度的可靠性。本章和第8章中讨论的每一个来源都会产生一些陈述，这些陈述宣称描述了我们世界的特征，基于提出特定事实性主张的人的看法。作为一个批判性思维者，你需要警惕一些潜在的问题，这些问题可能会削弱来自这些来源的证据。

FACT OR OPINION?

Evidence is stronger when it gets closer to fact and farther away from mere opinion. In other words, we would prefer that our conclusions lean on statements about the world that are not a matter of perspective or whimsical guessing about what is real and dependable. Evidence results from the systematic collection and organization of facts.

But facts are less powerful and accurate than they seem from afar. As Mary Poovey's masterful *A History of the Modern Fact* makes so clear, the idea of what we mean by "a fact" has changed forms multiple times as humans have used the idea in new contexts.

Think about the nuances and contexts that apply to any fact. In baseball, for example, an umpire is the determining voice in terms of whether a pitch is a ball or a strike. But was the pitch what the umpire claimed it was? Would a different umpire have called the pitch the same thing? Would the shadows in the baseball park possibly shape the probability that the pitch was indeed what the umpire claimed it was? Did the recent uproar about a previous call by one of the managers nudge the umpire to unintentionally or intentionally announce that the pitch was a ball or strike? Was the pitch ruled a ball or strike based on changing definitions of the strike zone?

You can be assured that "facts" like "Iran has weapons of mass destruction" are at least as questionable as whether a baseball pitch is a strike or ball.

However, the probability that all of us would experience the same thing is a measure of how certain and dependable a particular fact is. In other words, facts come in various strengths or probabilities. Just because facts are more complicated than being either true or false in no manner reduces their importance. Strong arguments depend on strong facts.

Opinions are often contrasted with facts, and they should be. Opinions are cheap; they require little effort to utter or create. "Macaroni tastes better than Raisin Bran." "American presidents have more common sense than do the leaders of Scotland." A person expressing an opinion is saying, "Look, don't ask me why I say such things. They are simply my opinions."

An opinion might be the initial step in a productive conversation, but in the absence of reasons and evidence for its logic and accuracy, it is simply a *mere* opinion.

And whenever you or anyone else has an opinion that is worth considering, then we can simply produce strong reasons and evidence that moves your claim closer to being factual.

So where do we locate the factual claims that will move us away from mere opinions? Factual claims have sources with varying degrees of reliability. Each of the sources discussed in this chapter and Chapter 8 produces statements allegedly describing the characteristics of our world based on the perceptions of the person making particular factual claims. As a critical thinker you need to be alert to the potential problems that can weaken evidence drawn from these sources.

⚠ **批判性问题**：源自个人经历、典型案例、当事人证言以及专家意见的证据，其效力如何？

我们需要可靠的证据

要评价论证过程，我们就要记住，有些事实断言比别的事实断言更加值得信赖。例如，"大部分美国参议员都是男性"这个断言是真的，你可能对此很有把握，而要说"练习瑜伽降低了罹患癌症的风险"这个断言也是真的，你就有点信心不足了。

因为对大部分断言来说，要证明它们是绝对的真理或绝对的谬误都极其困难，所以与其问它们是不是真的，我们不如问它们是否可靠。其实，我们想问的问题是："我们可以相信这样的信念吗？"支撑一个断言的证据数量越多、质量越高，我们就越可以信赖这个断言，同时也越可以称这样的断言为"事实"。

比如说，有大量证据表明乔治·华盛顿是美利坚合众国的第一任总统，因此，我们可以将这个断言当成事实。相反，"瓶装水比自来水饮用起来更安全"这样的看法，则有许多相互矛盾的证据，因此我们不能将这个信念当成事实。见解和事实之间最大的区别就在于相关证据的情况。支撑一个信念的证据越多，这个信念的"事实程度"也就越高。对于什么是事实，我们也可以换一种说法：它们多多少少有可能是真的。

在判断一则信息的说服力之前，我们要知道哪个事实断言最可靠。怎么确定事实断言的可靠程度呢？我们会问以下这样的问题：

- 你的证据是什么？
- 证据在哪里？
- 你确信它是真的吗？
- 你怎么知道它是真的？
- 你为什么相信它？
- 你能证明吗？

如果你养成经常问这些问题的习惯，那么你离跻身最具判断性思维者之列就不远了。这些问题要求立论者对自己的言论负责，揭示出其论证成立的基础。任何一个想让你认真考虑其观点的立论者，都会毫不犹豫地回答你的这些问题。他们知道自己有实质性的证据来支撑这些断言，因此，他们也想告诉你这些证据，希望你能逐渐认同他们的结论。对展示证据这样简单的要求，如果有人的反应是怒火中烧或退避三舍，那通常是因为他们自己觉得难为情——他们意识到自己缺了证据，他们对自己的信念本该少一分底气的。

我们如果经常提这些问题，就会注意到对于许多信念，总是没有足够的证据来明确支

? *Critical Question:* **How good is the evidence originating from personal experience, case examples, testimonials, and appeals to authority?**

THE NEED FOR DEPENDABLE EVIDENCE

To evaluate reasoning, we need to remember that some factual claims can be counted on more than others. For example, you probably feel quite certain that the claim "most U.S. senators are men" is true, but less certain that the assertion "practicing yoga reduces the risk of cancer" is true.

Because it is extremely difficult, if not impossible, to establish the *absolute* truth or falsity of most claims, rather than asking whether they are *true*, we prefer to ask whether they are *dependable*. In essence, we want to ask, "*Can we count on such beliefs?*" The greater the quality and quantity of evidence supporting a claim, the more we can *depend on it*, and the more we can call the claim a "fact."

For example, abundant evidence exists that George Washington was the first president of the United States of America. Thus, we can treat that claim as a fact. On the other hand, there is much conflicting evidence for the belief "bottled water is safer to drink than tap water." We thus can't treat this belief as a fact. The major difference between claims that are *opinions* and those that are *facts* is the present state of the relevant evidence. The more supporting evidence there is for a belief, the more "factual" the belief becomes. An alternative way to express the nature of facts is that they are more or less probably true.

Before we judge the persuasiveness of a communication, we need to know which factual claims are most dependable. How do we determine dependability? We ask questions like the following:

What is your proof?   How do you know that's true?
Where's the evidence?   Why do you believe that?
Are you sure that's true?   Can you prove it?

You will be well on your way to being among the best critical thinkers when you develop the habit of regularly asking these questions. They require those making arguments to be responsible by revealing the basis for their arguments. Anyone with an argument that you should consider will not hesitate to answer these questions. They know they have substantial support for their claims and, consequently, will want to share their evidence in the hope that you will learn to share their conclusions. When people react to simple requests for evidence with anger or withdrawal, they usually do so because they are embarrassed as they realize that, without evidence, they should have been less assertive about their beliefs.

When we regularly ask these questions, we notice that for many beliefs there is insufficient evidence to clearly support or refute them. For example,

持它们或把它们驳倒。例如，很多证据都能证明"隔一天服一片阿司匹林可以减少患心脏病的风险"这个断言，但也有一些其他的证据可以反驳它。在这种情况下，我们就需要判断哪一方占据可靠证据的数量优势，这样我们才能确定这一事实断言的可靠程度。

证据的来源

什么时候我们才能接受一个事实断言，认为它是可靠的？在以下三种情况下，我们最倾向于认可事实断言：

1. 当这个断言看起来是无可置疑的常识时，比如"举重有助于增肌"这个断言；
2. 当这个断言是从无懈可击的论证中得出的结论时；
3. 当这个断言得到有证据支持的理由充分的支撑时。

本章我们所关心的是第三种情况。要确定证据充分与否，我们需要提出这样的问题："这个证据的效力怎么样？"要回答这个问题，我们首先要问："我们所说的证据到底是什么意思？"

⚠ **注意**：所谓证据，就是立论者告知的明确信息，用来支撑或证明一个事实断言的可靠性。在规定性论证中，需要有证据来证明属于事实断言的那些理由；在描述性论证中，需要有证据来直接证明一个描述性的结论。

展示 7-1　主要的证据类型

√ 个人经历

√ 典型案例

√ 当事人证言

√ 权威或专家意见

√ 个人观察

√ 研究结果

√ 类比

如果运用得当，每种证据都可以成为"有效证据"，有助于支撑写作者的断言。正如淘金者仔细检查自己淘金盘里的石子，以筛选出可能存在的高质量矿石，我们也必须仔细检查证据，判断其质量。我们想知道：作者的证据是否为其断言提供了可靠的支撑？因此，

much evidence supports the assertion that taking an aspirin every other day reduces the risk of heart attack, although some other evidence disputes it. In such cases, we need to make judgments about where the *preponderance of reliable evidence* lies as we decide on the dependability of the factual claim.

SOURCES OF EVIDENCE

When should we accept a factual claim as dependable? There are three instances in which we will be most inclined to agree with a factual claim:

1. when the claim appears to be undisputed common knowledge, such as the claim "weight lifting increases muscular body mass";
2. when the claim is the conclusion from a well-reasoned argument; and
3. when the claim is adequately supported by reasons that are well supported.

Our concern in this chapter is the third instance. Determining the adequacy of evidence requires us to ask, *"How good is the evidence?"* To answer this question, we must first ask, "What do we mean by *evidence?*"

 Attention: *Evidence is explicit information shared by the communicator that is used to back up or to justify the dependability of a factual claim. In prescriptive arguments, evidence will be needed to support reasons that are factual claims; in descriptive arguments, evidence will be needed to directly support a descriptive conclusion.*

EXHIBIT 7-1 Major Kinds of Evidence

- ✓ personal experiences
- ✓ case examples
- ✓ testimonials
- ✓ appeals to authorities or experts
- ✓ personal observations
- ✓ research studies
- ✓ analogies

When used appropriately, each kind of evidence can be "good evidence." It can help support an author's claim. Like a gold prospector closely examining the gravel in her pan for potentially high-quality ore, we must closely examine the evidence to determine its quality. We want to know, "Does an author's evidence provide dependable support for her claim?" Thus,

我们在开始评价证据时就要问："证据的效力怎么样？"我们要一直在脑海深处铭记，没有哪一种证据可以像灌篮那样一锤定音、一劳永逸。你总是在找更好的证据，不可能找到完美的证据。

在本章以及第 8 章里，我们会逐一探讨我们可以针对每种类型的证据（见展示 7-1）提出的各种问题，这些问题能帮助我们确定证据的质量如何。本章将要考察的证据类型包括直觉、个人经历、典型案例、当事人证言和专家意见。

以个人经历作为证据

下面两个论证使用了一种特定类型的证据来支撑一个事实断言。

1. "我的朋友朱迪熬了一个通宵来复习备考，结果考得相当不错，所以在明天考试之前，我觉得我没必要睡觉了。"

2. "吃一大块巧克力蛋糕以后，我常常觉得好过很多，所以我认为，感到郁闷的人只要多吃点巧克力蛋糕就行了。"

这两个论证都把个人经历当成证据。看到"我认识一个人……""以我的经验，我发现……"这样的句子时，你就应该注意这类证据。因为个人经历过的事在我们的记忆里总是十分生动，所以我们总是将它们用作证据来支持某个看法。例如，也许你跟一个汽修工打交道后感觉糟糕透顶，因为他修车的费用实在高得离谱，这让你认为大多数汽修工都喜欢漫天要价。

无论这样概括所有汽修工是否正确，把这样的个人经历作为普遍看法的基础都是个错误！因为单一的个人经历，甚至个人经历的总和，根本不足以构成一个有代表性的经历样本，所以个人经历常会导致我们犯下"以偏概全谬误"。一次突出的经历或者几次同类经历可能说明某个结果有可能出现，例如，你可能遇到过几个人声称他们因为没有系安全带，所以在出车祸时得以保全性命。但这类经历不能说明这样的结果具有代表性或极有可能发生。当你听到自己或者别人说"嗯，以我的经验来看……"的时候，一定要当心。

⚠ **以偏概全谬误**：一个人仅根据群体中极小部分人的经历就得出有关整个群体的结论。

we begin to evaluate evidence by asking, "*How good is the evidence?*" Always keep in the back of your mind that no evidence will be a slam dunk that gets the job done conclusively. You are looking for better evidence; searching for altogether wonderful evidence will be frustrating.

In this chapter and in Chapter 8, we examine the kinds of questions we can ask of each type of evidence to help us decide its quality. Kinds of evidence examined in this chapter are personal experiences, case examples, testimonials, and appeals to authority.

PERSONAL EXPERIENCE AS EVIDENCE

The following arguments use a particular kind of evidence to support a factual claim.

1. "My friend Judy does really well on her tests when she stays up all night to study for them, so I don't see the need for getting sleep before taking tomorrow's test."

2. "I always feel better after having a big slice of chocolate cake, so I think that anyone who is depressed just needs to eat more chocolate cake."

Both arguments appeal to personal experiences as evidence. Phrases like "I know someone who . . ." and "In my experience, I've found . . ." should alert you to such evidence. Because personal experiences are very vivid in our memories, we often rely on them as evidence to support a belief. For example, you might have a really frustrating experience with a car mechanic because he greatly overcharges you for his services, leading you to believe that most car mechanics overcharge.

While the generalization about car mechanics may or may not be true, relying on such experiences as the basis for a general belief is a mistake! Because a single personal experience, or even an accumulation of personal experiences, is not enough to give you a *representative* sample of experiences, personal experiences often lead us to commit the *hasty generalization fallacy*. A single striking experience or several such experiences can demonstrate that certain outcomes are *possible*; for example, you may have met several people who claim their lives were saved because they were not wearing their seat belts when they got into a car accident. Such experiences, however, cannot demonstrate that such outcomes are *typical* or *probable*. Be wary when you hear yourself or others arguing, "Well, in my experience. . . ."

Fallacy: Hasty Generalization: A person draws a conclusion about a large group based on experiences with only a few members of the group.

以典型案例作为证据

一所大学的校长说:"我们的学生当然能找到高收入的工作,也能进入高等学府进一步深造。为什么?因为就在去年我们欢送过一个叫玛丽·尼斯的毕业生前往哈佛大学法学院深造。入学哈佛这一年来,她一直排名班级前5%成绩。因此,我们的学生当然能在一流大学取得令人瞩目的成功。"

有一种类型的论证经常会被用到,即纤毫毕现、栩栩如生地描绘或塑造一个或多个人物(或事件)来证实某个结论。这类描述通常都基于观察或者访谈,其形式也各不相同,从深度探讨到浅层描述都有可能。我们把这样的描述称为"典型案例"。立论者常常在游说报告的开头讲一段活灵活现的故事,生动地描述某个事件,以对听众动之以情。例如,支持禁止在开车时使用手机的一个论辩方法就是讲一些让人肝肠寸断的故事——司机边开车边打电话引发车祸,导致年轻人死于非命。

对我们而言,典型案例常常很有说服力,因为它们如此具体生动而又情节感人,很容易在我们的脑海里产生画面。政治候选人越来越偏爱在演讲中展示典型案例,知道案例中呈现出的丰富细节会激发听众热烈的情感回应。

遇到将引人注目的典型案例用作证据的情况时须审慎!

是否在某些时候,典型案例即使算不上有力的证据,也非常有用呢?当然!如同个人经历一样,它们也展现出各种重要的可能性,在抽象的数据让呈现出真实的人的画面。它们让人们更容易与论题产生联结,因此对它产生更浓厚的兴趣。

以当事人证言作为证据

1. 加油站墙上的一则说明:"我的车老是漏油,送给简修过以后就再也不漏了。因此我强烈建议你将爱车送到简那里维修,不管什么发动机问题她都能修好。"

2. 这本书看起来真不错。在书的封底,有读者的评论说:"这本书我一拿起来就再也放不下。"

商业广告、电影预告、图书封底的各式推荐、超自然现象存在的证明,或对其他有争议或令人意想不到的生活事件的描述,常常都利用了一种特殊类型的诉诸个人经历的方法来说服别人相信它们;它们引述具体当事人的话,尤其是名人的话,用以证明某个想法或某个产品的好坏,或证明那些非比寻常的事情确实发生过,这些都基于他们的亲身经历。这种引用的说法都被称为当事人证言。

CASE EXAMPLES AS EVIDENCE

> President of a large university: "Of course our students can move on to high-paying jobs and further study at large universities. Why, just this past year we sent one of our students, Mary Nice, off to law school at Harvard. In her first year, Mary remained in the top 5 percent of her class. Therefore, our students can certainly achieve remarkable success at elite universities."

A frequently used kind of argument is the use of a detailed catchy description of, or story about, one or several individuals or events to support a conclusion. Such descriptions are usually based on observations or interviews and vary from being in-depth to being superficial. We call such descriptions *case examples*. Communicators often begin persuasive presentations with dramatic descriptions of some event to emotionally involve their audience. For example, one way to argue for the banning of cell phone use in cars is to tell heart-wrenching stories of young people dying in car accidents because the driver was talking on a cell phone.

Case examples are often compelling to us because of their vividness and their interesting details, which make them easy to visualize. Political candidates have increasingly resorted to case examples in their speeches, knowing that the rich details of cases generate an emotional reaction.

> *Be wary of striking case examples as proof!*

Are there times that case examples can be useful, even if they are not good evidence? Certainly! Like personal experiences, they demonstrate important *possibilities* and put a personal face on abstract statistics. They make it easier for people to relate to an issue and thus take more interest in it.

TESTIMONIALS AS EVIDENCE

> 1. Note on service station wall: "Jane did a wonderful job fixing the oil leak my car had. I strongly recommend that you take your car to Jane to fix any engine problem you have."
> 2. This book looks great. On the back cover, comments from readers say, "I could not put this book down."

Commercials, ads for movies, recommendations on the backs of book jackets, and "proofs" of the existence of the paranormal or other controversial or extraordinary life events often try to persuade by using a special kind of appeal to personal experience; they quote particular persons, often a celebrity, as saying that a given idea or product is good or bad, or that extraordinary events have occurred, based upon their personal experiences. Such quoted statements serve as *personal testimonials*.

也许你在选择要报考的大学时也听到过大学生的当事人证言。证言也是一种形式的个人经历，某个人（常常是名人）提供一种说法来证实某个产品、事件或服务的价值，而这种赞同和支持缺乏我们所需以判断应在多大程度上接受其影响的信息。

这类证据到底有多大用处？它的用处通常并不大。多数情况下，我们对这类当事人证言无须过多关注，除非我们能找出它们背后更多相关的专门知识、利益纠葛、价值观和偏见。对下列与当事人证言相关的问题，我们都要特别小心。

- **选择性**。人们的经历各不相同。那些想要说服我们的人，总是小心选择他们要用的证人证言。在书的封底我们最有可能见到的总是最好的溢美之词，而不是最有代表性的读者反馈。我们应该时时刻刻追问这个问题："那些我们没机会听到其反馈的人的经历感受是什么样的？"同样，那些提供证言的人往往对他们关注的信息会有所选择，他们格外关注那些可以证实他们信念的信息，而忽略那些可能证伪他们信念的信息。我们常说眼见为实，相反的情况也常常出现：你相信什么就会看见什么！我们的期盼心理极大影响着我们体验事件的方式。如果我们相信外星人生活在我们当中，或者人类从来就没有真正登上过月球，那么我们就更有可能把模糊不清的影像看成外星人，或者发现美国政府有关登月的阴谋的证据。

- **个人利益**。许多证言，例如图书、电影和电视节目的推荐都来自那些可以从证言中获得一定利益的人。例如，医药公司常常支付给医生一些研究经费，条件是他们给病人开这家公司生产的药。因此，我们需要问一问："作证者是不是与他鼓吹的东西之间有什么利益关系，以致我们可以在他的证言里发现强烈的偏见？"

- **省略信息**。当事人证言很少能提供足够的信息，我们无法将之作为判断的基础。例如，如果朋友劝你去看一部新电影，因为这部电影是"百年一遇的好片子"，那么你应该善意地问一声，是什么让他对这部电影如此印象深刻。我们的评判标准和那些提供证言的人的评判标准可能完全不一样。

- **人的因素**。当事人证言之所以显得如此可信，一个原因就是它们都来自充满激情的人，这些人看起来信得过、心肠好并且很诚实。我们不由自主地想要相信这样的人所说的话。

You may have listened to personal testimonials from college students when you chose your college. Testimonials are thus a form of personal experience in which someone (often a celebrity) provides a statement supporting the value of some product, event, or service and the endorsement lacks any of the information we would need to decide just how much we should let it influence us.

How helpful is such evidence? Usually, it is not very helpful at all. In most cases, we should pay little attention to personal testimonials until we find out much more about the expertise, interests, values, and biases behind them. We should be especially wary of each of the following problems with testimonials:

- **Selectivity.** People's experiences differ greatly. Those trying to persuade us have usually carefully selected the testimony they use. What we are most likely to see on the back of a book jacket is the *best praise*, not the most typical reaction. We should always ask the question, "What was the experience like for those whom we have not heard from?" Also, people who provide the testimonials have often been selective in their attention, paying special attention to information that confirms their beliefs and ignoring disconfirming information. Often, believing is seeing! Our *expectations* greatly influence how we experience events. If we believe that aliens live among us, or that humans never really landed on the moon, then we are more likely to see ambiguous images as aliens or as proof of the government conspiracy regarding the moon landing.

- **Personal interest.** Many testimonials such as those used for books, movies, and television products come from people who have something to gain from their testimony. For example, drug companies often give doctors grants to do research, as long as they prescribe the drug company's brands of medication. Thus, we need to ask, "Does the person providing the testimony have a relationship with what he is advocating such that we can expect a strong bias in his testimony?"

- **Omitted information.** Testimonials rarely provide sufficient information about the basis for the judgment. For example, when a friend of yours encourages you to go see this new movie because it is the "best movie ever," you should ask, with warmth, about what makes the movie so impressive. Our standards for judgment may well differ from the standards of those giving the testimony.

- **The human factor.** One reason that testimonials are so convincing is that they come from very enthusiastic people, who seem trustworthy, well-meaning, and honest. Such people make us *want* to believe them.

以专家意见作为证据

> 根据医生的建议,我应该服用抗抑郁药物,以帮助自己应对最近这段时间的抑郁症状,并且我无须担心药物的副作用。

说这句话的人通过诉诸权威来为他的断言辩护,这些权威向来被认为对某个特定主题比我们大部分常人更为了解,他们就是所谓的专家。当立论者诉诸专家意见时,他们相信专家有渠道接触某些特定事实,并且这些人有特殊的能力,以从这些事实中得出结论。因此,基于专家的背景,这类专家意见比前文中讲的当事人证言能给论证带来更大的吸引力。你每天都会遇到诉诸各种类型的专家意见的论证,你别无选择,只有依赖他们,因为面对纷繁复杂的生活,我们只能在某几个方面有所作为,要深入探究其他方面的信息,我们既没有时间,也没有足够的知识储备。

> 影评人:"这是年度十佳影片之一。"引自《托莱多报》的瓦莱丽影评专栏
> 现场访谈节目嘉宾:"经济正在走向衰退。"
> 某组织:"美国医药协会支持这一立场。"
> 科研人员:"研究表明……"
> 亲戚:"我爷爷说……"
> 宗教:"《古兰经》上说……"
> 杂志:"根据《新闻周刊》的报道……"

我们可以从这些渠道获取专家的建议,了解怎样减肥、怎样获取幸福、如何变成有钱人、怎样降低胆固醇、怎样养育适应能力强的孩子,以及怎样钓一条大鱼,等等。你很容易就能在这个单子上添加更多的条目。

很显然,虽然同样是诉诸专家意见,但有些证据比其他证据更受我们的青睐。为什么呢?因为有些专家在发表观点时比其他专家谨慎得多。

你要记住,专家常常也会犯错误,并且他们内部也常常意见不统一。下面这些取材于《专家如是说》一书的例子就清楚地表明专家观点的不可靠。

> "我想电脑在全球大概只卖得出去 5 台。"
> ——托马斯·沃森,IBM 董事长,1943 年
> "录像机占领任何市场不会超过 6 个月。人们很快就会厌倦每天晚上盯着个夹板箱看。"
> ——达里尔·F. 扎努克,20 世纪福克斯电影公司总裁,1946 年

APPEALS TO AUTHORITY AS EVIDENCE

> According to my doctor, I should be taking antidepressant drugs to help me cope with my recent episodes of depression and I don't need to worry about side effects.

The speaker has defended his claim by appealing to authority—sources that are supposed to know more than most of us about a given topic—so-called experts. When communicators appeal to authorities or experts, they appeal to people who they believe are in a position to have access to certain facts and to have special qualifications for drawing conclusions from the facts. Thus, such appeals potentially provide more oomph to an argument than testimonials, depending on the background of the authority. You encounter appeals to many forms of authority on a daily basis. And you have little choice but to rely on them because you have neither the time nor the knowledge to become adept in more than a few dimensions of our very complicated lives.

> Movie reviewers: "One of the ten best movies of the year." Valerie Viewer, *Toledo Gazette*.
> Talk show pundits: "The economy is heading for a recession."
> Organizations: "The American Medical Association supports this position."
> Researchers: "Studies show . . ."
> Relatives: "My grandfather says . . ."
> Religion: "The Koran says . . ."
> Magazines: "According to *Newsweek* . . ."

We can get expert advice from such sources on how to lose weight, achieve happiness, get rich, lower cholesterol, raise a well-adjusted child, and catch a big fish. You can easily add to our list.

It should be obvious that some appeals to authority should be taken much more seriously as evidence than others. Why? Some authorities are much more careful in giving an opinion than others.

You should remember that *authorities are often wrong*. Also, they often disagree. The following examples, taken from *The Experts Speak*, are clear reminders of the fallibility of expert opinion (Christopher Cerf and Victor Navasky, 1998, Rev. Ed., Villard Books, New York).

> "I think there is a world market for maybe five computers."
> —Thomas Watson, chairman of IBM, 1943.

> "Video won't be able to hold onto any market it captures after the first six months. People will soon get tired of staring at a plywood box every night."
> —Darryl F. Zanuck, Head of Twentieth Century Fox Studios, ca. 1946.

上述引用的话提醒我们，当立论者引述专家意见的时候，我们需要问一些批判性问题。我们要问一问："我们为什么要相信这个专家？"说得更具体一点，我们应该对专家追问下面这些问题。

对于所谈论的主题，该专家拥有的专长、训练或特别知识到底有多少？这个主题是不是他潜心研究多年的主题？或者，这个人有没有与此主题相关的丰富经历？

这个专家所处的地位是否使其有特别好的渠道来获取相关事实？比如说，对于他所断言的事，他是否涉足其中并拥有第一手资料？总体而言，比起通过第二手资料获得信息的专家，你应该对掌握第一手资料的专家（涉足相关的事情并拥有第一手资料）更有信心。例如，《滚石》和《连线》杂志都是第二手的资料来源，而研究型期刊如《美国医学协会杂志》则是第一手资料来源。

有没有较好的理由让人能够相信，相对而言，专家更难遭受各种歪曲的影响？可能影响证据呈现方式的因素非常多，包括个人需求、先前的预期、一般看法、态度、价值观、相关理论和意识形态，等等。例如，如果一所公立大学的校长被问及削减教育经费是否会对大学造成不利影响，他十有八九会回答"有影响"，并给出一大堆过硬的理由。也许他对现状的见解不偏不倚，但由于他所处的职位，我们需要考虑，有没有可能他找出这些理由只是为了证明他自己的偏见。

我们所说的抱有偏见或心存歧视，意思是我们在寻找证据之前，就先对事情的好坏有了强烈的个人感受，以致干扰了我们公正地评价证据的能力。因为几乎在我们所有的判断中都会有很多让我们产生偏差的因素，所以我们很难期望一个专家完全不带任何偏见。不过，我们从某些专家那里可以比从其他专家那里少得到一些偏见，并且尝试通过搜寻关于专家在这个主题上个人利益的信息来确定这类偏见。例如，如果一个专家从他提倡的某些行动中获得较大的经济利益时，我们就要特别当心。

我们不能仅仅因为怀疑专家的个人利益有可能影响到他的公正性，就断然抛弃一个断言。我们可以采取的有效步骤是检查一下，看看那些持不同观点、不同预期、不同价值观和有不同利益的各路专家是否同意他的断言。因此，问一问这个问题会有好处："这个专家是不是因为经常做出可靠的断言而名声在外？"

当你在互联网上遇到事实断言时，你会特别关心专家的可靠程度。在网上，几乎每个人都可以变成潜在的"权威人士"，人们想说什么就说什么，断言满天飞。电脑并没有内置的处理器来评价这些断言。这显然是个"买方注意，上当自负"的情况！

对书籍、家居产品、餐饮、度假胜地、酒店、公司和服务质量的评论日益给人们带来一种关于其可靠性的幻觉，尽管很多评论常常是通过伪造和买卖的手段得来的。研究发现，

These quotes should remind us that we need to ask critical questions when communicators appeal to authority. We need to ask, "*Why should we believe this authority?*" More specifically, we should ask the following questions of authorities.

How much expertise, training, or special knowledge does the authority have about the subject about which he is communicating? Is this a topic the person has studied for a long time? Or, has the person had extensive experience related to the topic?

Was the authority in a position to have especially good access to pertinent facts? For example, was she involved firsthand with the events about which she makes claims? In general, you should be more impressed by an authority who is a *primary source*—someone having firsthand involvement with relevant events—than by *secondary sources*. *Rolling Stone* and *Wired*, for example, are secondary sources, while research journals such as the *Journal of the American Medical Association* are primary sources.

Is there good reason to believe that the authority is relatively free of distorting influences? Among the factors that can influence how evidence is reported are personal needs, prior expectations, general beliefs, attitudes, values, theories, and ideologies. For example, if a public university president is asked whether cuts in funding for education are bad for the university, he will in all probability answer "yes" and give a number of good reasons. He may be giving an unbiased view of the situation. Because of his position, however, we would want to be concerned about the possibility that he has sought out only those reasons that justify his own biases.

By having bias and prejudice, we mean the existence of a strong personal feeling about the goodness or badness of something up front before we look at the evidence, such that it interferes with our ability to evaluate evidence fairly. Because many factors bias us in virtually all our judgments, we cannot expect any authority to be *totally* unbiased. We can, however, expect less bias from some authorities than from others and try to determine such bias by seeking information about the authority's personal interest in the topic. For example, we want to be especially wary when an authority stands to benefit financially from the actions she advocates.

We should not reject a claim simply because we suspect that the authority's personal interests may interfere with her fairness. One helpful step we can take is to check to see whether authorities with diverse attitudes, prior expectations, values, and interests agree. Thus, it is also helpful to ask the question, "*Has the authority developed a reputation for frequently making dependable claims?*"

You will want to be especially concerned about the quality of authorities when you encounter factual claims on the Internet. When we go online, virtually everyone becomes a potential "authority" because people are free to claim whatever they wish, and there is no built-in process to evaluate such claims. It is clearly a "buyers beware" situation!

Reviews of books, home products, restaurants, vacation resorts, hotels, businesses, and services are increasingly offering the illusion of reliability—even

在很多网站，多达 1/3 的评论都是伪造的，换句话说，评论者对自己评论的东西根本就没有直接的知识，或者对它有着强烈的个人偏见。图书的作者可以花钱购买好评。经常出现的情况是，评论的主要目的是推广产品，而不是客观告诉消费者具体的情况。例如，亚马逊网站 60% 的商品评论都是五星级，这显示出一种朝向正面评论的偏差。因此，你需要寻找一些标志，证明那些证言和专家说法是值得相信的，而不是假冒的。

你应该努力了解网站的建设目的、信誉以及相关的发帖者的经历，并了解他们提供的那些用来支持其结论的论证的质量。尤其要注意他们的论证结构。检查一下网站是不是与其他声誉较高的网站有关联或者有链接。

网站可能不可靠的线索还包括帖子没有注明发表日期，网站的外观不专业，模糊不清、一概而论（例如用"总是""从来不"这样的词）情绪化而非小心求证的断言，完全一边倒的观点，缺乏第一手证据，将道听途说的证据拿来就用，以及大量的论证谬误。最后，你应在不同网站搜寻有关同一主题的证据。

使用这个批判性问题

在发现将直觉、个人经历、典型案例、当事人证言和专家意见用作证据所存在的问题后，你就有了正当理由以评判是否接受基于这样的证据得出的结论。知道这些问题让你有了防护，可以抵御一些虚假的论证。但你还是要公平对待人们提出的、希望你考虑的论点。所以，对给你提供一些经不起推敲的证据的人，问问他们能不能多提供一些扎实的证据，这样做很有意义，要给所有论证充分展示的机会。

在本章中，我们主要关注如何评价几类用来支撑事实断言的证据：个人经历、典型案例、当事人证言以及专家意见。依赖这类证据时一定要格外小心。我们已经提出一些问题，你可以依靠这些问题来判断这类证据是否效力充足。在第 8 章中，我们会接着讨论证据的其他来源，我们还要问同样的问题："这个证据的效力怎么样？"凭借有力的证据，我们才更接近事实或接受某人的观点。

思维体操

⚠ **批判性问题：** 源自个人经历、典型案例、当事人证言和专家意见的证据，其效力如何？

though many reviews are often faked, bought, and sold. Studies have found that at many sites as many as one-third of the reviews are faked; in other words, the reviewer has no direct knowledge of what is being reviewed or has a strong personal bias. Book authors can purchase book reviews. Frequently, reviews serve primarily to promote the product than to objectively inform the consumer. For example, 60 percent of Amazon product reviews are five stars, suggesting a bias toward positive reviews. Thus you need to look for signs that testimonials and appeals to authority are trustworthy and not a sham.

You should strive to learn as much as you can about the purpose of Web sites, the credentials and experience of the contributors associated with them, and the nature of the reasoning support provided for their conclusions. Pay very close attention to the reasoning structure. Check to see whether the site is associated with or linked to highly reputable sites.

Further clues that the site may be undependable include a lack of dates associated with postings; an unprofessional look to the site; claims that are vague, sweeping (e.g., "always," "never"), and emotional, rather than carefully qualified; a totally one-sided view; the absence of primary source evidence; the presence of hearsay evidence; and numerous reasoning fallacies. Finally, seek out evidence on the same topic from other sites.

USING THIS CRITICAL QUESTION

When you identify problems with personal experience, case examples, testimonials, and appeals to authority as evidence, you have a proper basis for hesitating to accept the conclusion based on that evidence. Knowing these problems gives you some protection against bogus reasoning. However, you do want to work hard to be fair to the arguments that people present for your consideration. So it makes sense to ask those who provide you with insubstantial evidence whether they can give you some better evidence. Give arguments every chance they deserve.

In this chapter, we have focused on the evaluation of several kinds of evidence used to support factual claims: personal experience and anecdotes, testimonials, and appeals to authorities. Such evidence must be relied on with caution. We have provided you with some questions you should ask to determine whether such evidence is *good evidence*. In Chapter 8, we discuss other sources of evidence, as we continue to ask the question, *"How good is the evidence?"* Depending on the strength of the evidence, we draw closer to fact or just someone's opinion.

PRACTICE EXERCISES

Critical Question: **How good is the evidence: personal experience, case examples, testimonials, and appeals to authority?**

请评价下列三篇文章中的证据。

第一篇

很多青少年和年轻人都开始使用一种名为高伦雅芙的创新产品，来治疗各种粉刺难题。这种治疗产品价格实惠，配方功能强大，可以改善和恢复肌肤的活力，让你的肌肤变得更健康，更年轻。根据生产这款产品的公司的说法，用户"只需在短短三天之内，就可获得更为洁净的肌肤。"很多名人在接受采访时都对这款产品做了评价。

朱莉安·浩夫：现在我的肌肤更洁净，让我觉得新鲜、有活力。我充满了自信。

娜雅·里维拉：你得为肌肤努力做点什么，你得找到有效的产品。幸运的是，我找到了高伦雅芙。

第二篇

对很多烟民来说，电子烟正开始迅速取代传统香烟。然而，一些人声称，使用电子烟会带来一系列健康问题，因为电子烟"香料"中使用的化学物质会产生一种烟雾，容易被人吸入。电子烟真的是传统香烟的安全替代品吗？根据心脏病专家鲍勃·R. 罗伯茨博士的说法，使用电子烟没有任何危险的副作用。"多年来，作为一种帮助戒掉普通香烟的方法，我一直向我的病人推荐电子烟，我从未见过任何一个病人因为使用了电子烟而患病。吞云牌电子烟是我的病人最喜爱的品牌。"吞云是美国最大的电子烟生产商，它最近发布了一份官方声明，声称电子烟从未被证明会对使用者造成不利影响。在最近的一次采访中，吞云总裁斯穆克说："没有人比我们更了解电子烟，我们知道电子烟不仅是普通香烟的安全替代品，用于休闲消遣也是安全的。"

第三篇

吃素是不是真的比吃肉更健康？答案是"确实如此"！已经有好几项研究列举出经常吃素相比于吃肉带来的各种好处。此外，如果你问问素食主义者，他们很快就会向你解释，他们的饮食为何优于非素食主义者的饮食一大截。更重要的是，很多专注于打造素食饮食和菜单项目的饭店和企业纷纷开张。很明显，吃素比吃肉更健康。

Evaluate the evidence in the following three passages.

Passage 1

Many teens and young adults have begun using a new and innovative product called Proactiv® to treat problems of acne. The treatment is an affordable and powerful formula that improves and rejuvenates your skin, and keeps your skin healthier, longer. According to the company that makes the product, users will have "clearer skin in as few as three days." Many celebrities have commented on the product during interviews:

Julianne Hough: "Now that my skin is clear I feel fresh and vibrant. I feel confident."

Naya Rivera: "You've got to find something that works and you've got to fight back. Luckily, I got Proactiv. . . ."

Passage 2

Electronic cigarettes, or e-cigarettes, are quickly starting to replace traditional cigarettes for many smokers. However, some people have claimed that e-cigarette use can come with its own set of health issues due to the chemicals used in the e-cigarette "juice" that creates the inhaled vapor. Are e-cigarettes truly a safe alternative to traditional cigarettes? According to cardiologist Dr. Bob R. Roberts, there are no dangerous side effects associated with e-cigarette use. "I've been recommending e-cigarettes to my patients for years as a way to quit normal cigarettes, and I've never seen a patient become ill from e-cigarette use. In particular, MegaVapor brand e-cigarettes are the ones my patients love the most." MegaVapor, the largest e-cigarette producer in the nation, recently released an official statement claiming that e-cigarettes have never been shown to cause negative effects in users. In a recent interview, MegaVapor president H.V. Smoker said, "No one knows e-cigarettes better than we do, and we know that e-cigarettes are not only a safe alternative to normal cigarettes but are also safe for recreational use."

Passage 3

Is vegetarianism really healthier than eating meat? The answer is yes! There have been several studies outlining the various advantages that being vegetarian has over eating meat regularly. Furthermore, just ask vegetarians and they will quickly explain how their diet is superior to that of non vegetarians. More importantly, many restaurants and businesses are opening that focus on vegetarian options and menu items; so clearly, vegetarianism is healthier than eating meat.

给个提示

第一篇

  结论：使用高伦雅芙可以有效治疗粉刺。

  理由：名人纷纷对它治疗粉刺的效果赞不绝口。

我们不能依赖这些名人证词，把它们当作有效"证明"。这篇文章极好地展现了把当事人证言当成证据的弱点所在，也展现了期望在影响人们的认识方面具有的威力。这些成功的经历到底有多大的代表性？随机选择的用户使用这种产品后会不会对它有这么多的溢美之词？这些被选择的名人是不是耳根比较软？除非我们能收集到更加系统的研究数据，否则我们就不应得出结论，说这种产品可以有效治疗粉刺问题。

第二篇

  结论：电子烟是安全的。

  原因：一位医学专家和一家电子烟主要生产商声称电子烟是安全的。

我们应该在多大程度上相信这些权威的证言？它们的可靠性不高。首先，两个权威都有可能偏颇不公。有可能罗伯茨博士被吞云公司收买，从而为该公司产品提供正面的证言。如果该公司付钱给罗伯茨博士，那么可能即使产品并不安全，他也会声称电子烟是安全的，从而在经济上获利。不管是否从该公司获得经济利益，罗伯茨博士的说法本身也值得怀疑，因为这仅仅依据他个人的经验。他很可能没有去寻找失败的证据。该公司的断言和罗伯茨博士的断言一样值得怀疑，因为该公司生产并销售电子烟，夸大产品的安全性可以让吞云公司从中渔利。如果该公司拿出点像样的系统性研究成果，证明为什么电子烟是安全的，那么也许能使其断言不那么可疑。

Sample Responses

Passage 1

CONCLUSION: *Using Proactiv® will effectively treat acne.*

REASON: *Celebrities rave about the positive impact of the acne treatment.*

We should not rely on these celebrity testimonials as good "proof." This passage illustrates the weaknesses of testimony as evidence, as well as the power of expectations in affecting perceptions. How typical are these success stories? Would randomly selected users of the acne treatment have voiced so much praise? Are these selected celebrities highly suggestible? Until more systematic research data are collected, we should not conclude that the product is effective in treating the problem of acne.

Passage 2

CONCLUSION: *E-cigarettes are safe.*

REASON: *A medical professional and a major e-cigarette producer claim that e-cigarettes are safe.*

How much should we depend on these appeals to authority? Not much. First, both authorities are likely to be very biased. It's possible that Dr. Roberts has been paid by MegaVapor to provide a positive testimonial about MegaVapor's products. If MegaVapor is paying Dr. Roberts, he stands to gain financially by claiming that e-cigarettes are safe even if they are not. Regardless of any potential financial gain from MegaVapor, Dr. Roberts's claim is suspect because it is based on his experience only. He has probably not sought out evidence of failures. The claims of MegaVapor are as questionable as those of Dr. Roberts because MegaVapor produces and sells e-cigarettes. Therefore, MegaVapor has much to gain by promoting their safety. If the company were to provide some sort of systematic research for why e-cigarettes are safe, perhaps its claims would be less suspect.

第 8 章

证据的效力：个人观察和调查研究

| 学 习 目 标 |

1. 了解以个人观察作为证据来源的作用和危险。
2. 理解使用调查研究作为证据的相关问题。

在本章中，我们将继续评价证据的效力。我们集中讨论两种常见的证据：个人观察和调查研究。当我们遇到它们作为证据出现时，需要提出一些问题。

⚠ **批判性问题**：个人观察和研究报告形式的证据效力怎么样？

以个人观察作为证据

> 警察向徒手的人开枪并将其射杀就应该论罪受罚。虽然他宣称自己以为受害者是在伸手拿枪，但目击者报告说受害人根本就没有做出任何有威胁性的举动。

CHAPTER 8

How Good Is the Evidence: Personal Observation and Research Studies?

LEARNING OBJECTIVES

1. Acquire awareness of the role and dangers of personal observation as a source of evidence.
2. Develop understanding of the problems associated with using research studies as evidence.

In this chapter, we continue our evaluation of evidence. We focus on two common kinds of evidence: personal observation and research studies. We need to question each of these when we encounter them as evidence.

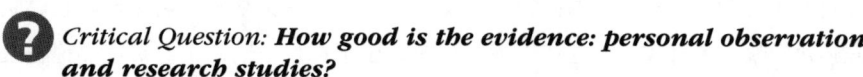

Critical Question: **How good is the evidence: personal observation and research studies?**

PERSONAL OBSERVATION AS EVIDENCE

> The policeman who shot and killed an unarmed man should be charged with a crime. Although he claims he thought the victim was reaching for a gun, onlookers reported that the victim was not making a threatening movement.

我们能在多大程度上依赖目击者的此类观察呢？个人观察是一种有价值的证据，它是很多日常论证和科学研究的基础。例如，我们对亲眼所见的事情会感觉信心十足。因此，我们倾向于依赖目击证人的证词，并将之视作证据。但是，因为很多原因，个人观察常常被证明是不可信赖的证据。

观察者不像镜子，并不能给我们提供"纯粹"的观察。我们所"见"、所说的都是经过一系列价值观、偏见、态度、以往经历、过去接受过的训练以及期待过滤后剩下的东西。我们看到和听到的都是我们愿意看到和听到的东西，对于一次经历，我们挑选和记住的，都是和以往的经历和背景最一致的那些侧面。此外，在很多情况下，会有各种严重障碍阻碍我们看清楚发生的一切，例如注意力不集中、观察的事件快速进行，以及压力重重的环境等。例如，你可以假想一下，如果一个人挥舞手枪指着银行出纳，而你站得离他非常近，你的观察可能会出现偏差。

当报纸、杂志、书本、电视、网络和研究使用观察报告作为证据时，你需要判断有没有过硬的理由来支持我们相信这样的报告。最可信的报告往往根据最近的观察，并且是处在最佳环境里的几个人同时得到的观察结果，他们对观察的事件没有明显而强烈的期待，也不抱任何偏见。

有偏差的调查和问卷

夜幕刚刚降临，你刚吃过晚饭，电话铃声响了。"我们正在做一项民意调查。你能否回答我们几个问题？"如果你回答说"可以"，你将成为成千上万参加年度民意调查的受访者中的一员，调查是我们最常遇到的一种研究方法。想一想你听到下面这个短语的频率有多高："根据最近的民调结果……"

调查和问卷通常被用来测量人们的行为、态度和看法。它们的可靠性如何？得看具体情况！对调查的回答会受到多种因素的影响，所以，我们在解释它们的意义时必须倍加小心。下面，我们就来盘点一下其中的一些影响因素。

首先，要让受访者对调查的回答有意义，这些回答就必须是实话实说。也就是说，口头报告必须反映心中真实的信念和态度。可是，由于种种原因，人们常常要掩盖真相。例如，他们可能会提供自己认为应该提供的答案，而不是反映内心真实想法的答案。他们可能会对问卷调查或其中问题的类型心怀敌意。或者他们可能在做出回答前未经过深思熟虑。如果你曾经接受过此类问卷调查，或许你就能想到很多其他影响因素。

⚠ **注意**：你不能想当然地以为调查得到的回答能准确反映出调查对象的真实态度。

How much can we count on the observation of such onlookers? One valuable kind of evidence is personal observation, the basis for much everyday reasoning as well as scientific research. For example, we feel confident of something we actually see. Thus, we tend to rely on eyewitness testimony as evidence. For many reasons, however, personal observations turn out to be untrustworthy evidence.

Observers, unlike certain mirrors, do not give us "pure" observations. What we "see" and report is filtered through a set of values, biases, attitudes, prior experiences, previous training, and expectations. We tend to see or hear what we wish to see or hear, selecting and remembering those aspects of an experience that are most consistent with our prior experience and background. In addition, many situations present major impediments to seeing accurately, such as poor attention, rapid movement of events observed, and stressful environments. Imagine, for example, possible distortions in your observation if you were standing near a person waving a gun at a bank teller.

When reports of observations in newspapers, magazines, books, television, and the Internet, as well as in research studies, are used as evidence, you need to determine whether there are good reasons to rely on such reports. The most reliable reports will be based on recent observations made by several people observing under optimal conditions who have no apparent, strong expectations or biases related to the event being observed.

BIASED SURVEYS AND QUESTIONNAIRES

It's early evening. You have just finished dinner. The phone rings. "We're conducting a survey of public opinion. Will you answer a few questions?" If you answer "yes," you will be among thousands who annually take part in surveys—one of the research methods you will encounter most frequently. Think how often you hear the phrase "according to recent polls."

Surveys and questionnaires are usually used to measure people's behavior, attitudes, and beliefs. Just how dependable are they? It depends! Survey responses are subject to many influences; so one has to be very cautious in interpreting their meaning. Let's examine some of these influences.

First, for survey responses to be meaningful, they must be answered honestly. That is, verbal reports need to mirror actual beliefs and attitudes. Yet, for many reasons, people frequently shade the truth. For example, they may give answers they think they ought to give, rather than answers that reflect their true beliefs. They may experience hostility toward the questionnaire or toward the kind of question asked. Alternatively, they may give too little thought to the question. If you have ever been a survey participant, you can probably think of other influences.

 Attention: *You cannot assume that survey responses accurately reflect true attitudes.*

其次，很多民意调查中的问题在措辞上显得模棱两可，对这样的问题可以做出多重解读。事实上，不同的人很可能在回答不同的问题！例如，设想一下人们可能对这样的调查问题做出的多种解读："你认为电视上有没有高质量的节目？"调查问题的措辞越含混，调查结果的可信度就越要打折扣。

你应该时刻记住问一下这个问题："这个调查的问题是如何表述的？"通常情况下，一个问题表述得越具体详尽，不同人群就越有可能对其做出相同的解读。

再次，调查本身所包含的很多偏差让它们变得更加可疑。其中两个最重要的偏差是措辞偏差和语境偏差。调查问题中的措辞偏差是个常见的问题，对提出一个问题的方式稍加改变，就会对回答方式产生重大的影响。我们先来看看下面这个基于最近的民意调查得出的结论，然后再回头看看调查中的问题。

> 一位大学教授发现在他任教的大学中有56%的调查对象相信奥巴马的医疗改革方案对美国而言是大错特错的。

现在，我们再来仔细看看民意调查中的问题："你对总统受人误导在全国大张旗鼓地强制推行奥巴马医改式的社会主义有什么看法？"你有没有发现这问题本身存在的偏差？其中引导性的词语是"总统受人误导在全国大张旗鼓"以及"强制推行奥巴马医改式的社会主义"。如果将问题稍做改变，变成"你对总统的医改尝试，即要建立一套覆盖人群更广、成本更低廉而医保覆盖项目大大增加的医疗体系有什么看法"，得到的回答会不会大不相同？因此，这里获得的回答呈现出的人们对新医改方案的态度是失真的。

调查和问卷数据必须不断接受检视，以确保不存在偏差。要小心查看提问问题的遣词用字！

语境对回答所产生的影响也可能非同寻常。对于同样的问题，在不同的民意调查中回答也可能有所不同，主要取决于问卷是如何呈现的，以及问题是如何嵌入调查中的。最近进行的两个调查都包含了下面这个问题："你认为我们是否应该将合法饮酒年龄降低到21岁以下？"在一个调查中，这个问题前面还有另外一个问题："你认为，选举权是否应该像现在这样被赋予18岁的孩子？"而在另一个调查中，这个问题前面没有其他问题。毫不奇怪，两个调查的结果会完全不同。你能明白语境会怎样影响调查对象了吧？

另一个重要的语境因素就是问卷的长度。在比较长的调查当中，人们对后面问题的回答可能与对前面问题的回答截然不同，这只是因为他们做题做烦了。在评价调查结果时一定要警惕语境因素的影响。

Second, many survey questions are ambiguous in their wording; the questions are subject to multiple interpretations. Different individuals may in essence be responding to different questions! For example, imagine the multiple possible interpretations of the following survey question: "Do you think there is quality programming on television?" The more ambiguous the wording of a survey, the less credibility you can place in the results.

You should always ask the question, "How were the survey questions worded?" Usually, the more specifically a question is worded, the more likely that different individuals will interpret it similarly.

Third, surveys contain many *built-in biases* that make them even more suspect. Two of the most important are *biased wording* and *biased context*. Biased wording of a question is a common problem; a small change in how a question is asked can have a major effect on how a question is answered. Let's examine a conclusion based on a recent poll and then look at the survey question.

> A college professor found that 56 percent of respondents attending his university believe that the Obama health care program is a major mistake for the country.

Now look closely at the survey question: "What do you think about the president's misguided efforts to impose Obamacare on the nation?" see the built-in bias? The "leading" words are "the president's misguided efforts" and "impose Obamacare ." Wouldn't the responses have been quite different if the question had read, "What do you think about the president's attempt to provide a health care system that will provide expanded coverage, lower costs, and increased health care coverage to Americans?" Thus, the responses obtained here are a distorted indicator of attitudes concerning the new health care program.

Survey and questionnaire data must always be examined for possible bias. *Look carefully at the wording of the questions!*

The effect of *context* on an answer to a question can also be powerful. Even answers to identical questions can vary from poll to poll, depending on how the questionnaire is presented and how the question is embedded in the survey. The following question was included in two recent surveys: "Do you think we should lower the drinking age from 21?" In one survey, the question was preceded by another question: "Do you think the right to vote should be given to children at the age of 18 as it currently is?" In the other survey, no preceding question occurred. Not surprisingly, the two surveys showed different results. Can you see how the context might have affected respondents?

Another important contextual factor is *length*. In long surveys, people may respond differently to later items than to earlier items simply because they get tired. *Be alert to contextual factors when evaluating survey results.*

Because the way people respond to surveys is affected by many unknown factors, such as the need to please the interviewer or the interpretation of the

人们对于调查的回答方式会受到许多未知因素的影响，例如讨好访谈者的需要，或者对问题的不同解读，那么我们是否应该将调查证据当成有效力的证据？对这个问题，人们有激烈的争论，我们的回答是"应该"，只要我们足够小心，不要在调查证据范围之外去过多概括。有些调查的声誉优于其他的调查。调查的质量越高，你就可以越放心地受它的结论影响。

我们的建议是仔细检查调查的程序，然后再接受调查的结果。一旦你确定程序的质量，就可以选择做出自己的合理概括，这个概括会考虑到你发现的所有偏差。哪怕是有偏差的调查也可能会承载大量的信息，但是你需要知道偏见之所在，防止自己过度被调查结果说服。

以研究报告作为证据

> "研究表明……"
> "研究人员在最近一项调查中发现……"
> "《新英格兰医学期刊》的一份报告显示……"

由训练有素的科研人员系统地收集观察结果所形成的科学研究属于一种权威。它们充分的观察，常常是被高度认可的证据。研究结果的可靠性到底如何？面对研究结果时，应采用与对待更普遍的诉诸权威意见一样的方法，只有等我们问了一些问题之后，才能知道研究结果有多可靠。

我们的社会越来越依赖于科学方法，将其作为重要指导，帮助人们判定事实真相，因为这个世界上各种事件之间的关系错综复杂，人类对于这些事件的观察和理论很容易出错。科学方法力求避免我们在观察世界时以及我们的直觉和常识中带有的很多内在偏见对我们获取真相造成影响。

科学方法有什么特别之处？首先，它追求的信息是以公开证实过的数据的形式出现的——也就是说，它的数据是在一定条件下获取的，其他有资质的人根据同样的条件，可以展开类似的观察并获得同样的结果。因此，如果有研究人员报告说他能在实验室条件下获得冷聚变，只有其他研究人员也能获得同样的结果，这个实验才可信。换句话说，我们可以更加信赖这些科学研究的结果，因为这些结果已经得到复制（即重复）。

其次，科学方法的第二个主要特征就是控制——即使用特别的程序来减少观察和研究成果诠释中出现的错误。例如，如果观察中存在的偏见可能是个主要的难题，那么研究人员就要尝试控制这类错误，如让多个人员一起进行观察，然后看他们的观察结果相互之间

question, should we ever treat survey evidence as good evidence? There are heated debates about this issue, but our answer is "yes," as long as we are careful and do not generalize further than warranted. Some surveys are more reputable than others. The better the quality of the survey, the more you should be influenced by its results.

Our recommendation is to examine survey *procedures* carefully before accepting survey *results*. Once you have ascertained the quality of the procedures, you can choose to generate your own *qualified generalization*—one that takes into account any biases you might find. Even biased surveys can be informative; but you need to know the biases in order to not be unduly persuaded by the findings.

RESEARCH STUDIES AS EVIDENCE

"Studies show . . ."
"Research investigators have found in a recent survey that . . ."
"A report in the *New England Journal of Medicine* indicates . . ."

Research studies that systematically collect observations by people trained to do scientific research are one form of authority. They rely extensively on observation and are often a highly regarded form of evidence. How dependable are research findings? As is true for appeals to authority in general, we cannot know the answers until we ask lots of questions.

Society has turned to the scientific method as an important guide for determining the facts because the relationships among events in our world are very complex, and because humans are fallible in their observations and theories about these events. The scientific method attempts to avoid many of the built-in biases in our observations of the world and in our intuition and common sense.

What is special about the scientific method? Above all, it seeks information in the form of *publicly verifiable data*—that is, data obtained under conditions such that other qualified people can make similar observations and get the same results. Thus, for example, if one researcher reports that she was able to achieve cold fusion in the lab, the experiment would seem more credible if other researchers could obtain the same results. In other words, we can depend more on the results of these scientific studies because they have been replicated (i.e., repeated).

A second major characteristic of the scientific method is *control*—that is, the use of special procedures to reduce errors in observations and in the interpretation of research findings. For example, if bias in observations may be a major problem, researchers might try to control this kind of error by using multiple observers to see how well they agree with one another. Physical scientists frequently maximize control by studying problems in the laboratory so that they can minimize extraneous factors. Unfortunately, control is usually

能在多大程度上取得一致。物理学家经常在实验室里研究问题，以使外部因素的影响达到最小化，使控制达到最大化。可惜的是，在真实社会中实施控制通常比在物理世界中困难得多，因此，要将科学方法成功应用到解决很多复杂的人类行为问题上是很难的。

再次，语言的精确性是科学方法的第三个主要组成部分。许多概念常表意不明、模糊，显得模棱两可。科学方法则力图在语言运用上做到精确和前后一致。

科学还存在很多其他要素，远非我们在这里可以探讨，但我们想让你记住，（做得好的）科学研究是我们获得证据的一个优质的来源，因为科学研究强调可重复性、控制和精确性。

研究发现中存在的一般问题

可惜的是，研究被用于论证一个问题这一事实并不必然意味着研究证据是可靠的证据，或者对于这个证据含义的解释准确无误。如同诉诸任何来源的证据一样，取材于科学研究的证据也需要我们谨慎对待。同时，对于有些问题，尤其是那些关注人类行为的问题，哪怕有最好的证据，我们也只能尝试性地进行回答。因此，对于研究而言，我们还要提出很多重要的问题，这样才能决定它们的结论可靠性到底如何。

当立论者以研究作为证据来源时，你应该记住以下几点：

1. 研究的质量有高有低，差别很大。有的研究细照笃行，有的研究粗制滥造，我们自然更应该相信前者。因为研究过程太过复杂，而且受到太多的外在因素影响，哪怕是训练有素的研究人员，有时所做的研究也难免会存在重大不足，在科学刊物上发表文章并不能确保一项科学研究没有重大缺陷。
2. 研究成果常常会互相矛盾。因此，脱离了调查某一具体问题的科学研究的大环境，单一的研究呈现的常常是误导性的结论。最值得我们注意的研究是那些不止一个人或不止一群研究人员反复做过的研究。有很多断言从来没有被重新验证过，还有很多断言在重新验证后未能重复其原始的结论。比如说，最近发表在一份声望很高的医学杂志上的研究，重新检验了一些宣称产生了成功的医学干预效果的备受重视的研究断言，结果令人信服地表明，原来的断言中有41%都是错误的或被极大地夸大的。（见"Lies, Damned Lies, and Medical Science," November 2010, Atlantic Magazine）。我们需要不断追问："其他的研究人员有没有重复取得这些发现？"
3. 研究发现并不能证明结论，充其量只能支持结论。这些研究发现本身并不足以说明

more difficult in the social world than in the physical world; thus, it is very difficult to successfully apply the scientific method to many questions about complex human behavior.

Precision in language is a third major component of the scientific method. Concepts are often confusing, obscure, and ambiguous. Scientific method tries to be precise and consistent in its use of language.

While there is much more to science than we can discuss here, we want you to keep in mind that scientific research, when conducted well, is an excellent source of evidence because it emphasizes *replication*, *control*, and *precision*.

General Problems with Research Findings

Unfortunately, the fact that research has been applied to a problem does not necessarily mean that the research evidence is dependable evidence or that the interpretations of the meaning of the evidence are accurate. As with appeals to any source, appeals to research evidence must be approached with caution. Also, some questions, particularly those that focus on human behavior, can be answered only tentatively even with the best of evidence. Therefore, we have to ask a number of important questions about research studies before we decide how much to depend on their conclusions.

When communicators appeal to research as a source of evidence, you should remember the following:

1. Research varies greatly in *quality*. There is competent research and there is poorly done research, and we should rely more on the former. Because the research process is so complex and is subject to so many external influences, even those well trained in research practices sometimes conduct research studies that have important deficiencies; publication in a scientific journal does not guarantee that a research study is not flawed in important ways.
2. Research findings often contradict one another. Thus, *single* research studies presented out of the context of the family of research studies that investigate the question often provide misleading conclusions. Research findings that most deserve our attention are those that have been replicated by more than one researcher or group of researchers. Many claims never get retested, and many of those that are retested fail to replicate the original results. For example, a recent study published in a prestigious medical journal found that 41 percent of efforts to retest very highly regarded research claims of successful medical interventions convincingly showed the original claims to be wrong or greatly exaggerated (see "Lies, Damned Lies, and Medical Science," November 2010, *Atlantic Magazine*). We need to always ask the question, "Have other researchers replicated the findings?"
3. Research findings *do not prove* conclusions. At best, they *support* conclusions. Such findings do not speak for themselves! Researchers must

问题。研究人员总是要解释他们的发现的意义，而所有的科学发现都可以找到不止一种解释方法（参见第7章）。因此，研究人员的结论不应该被当成是已证明了的"真理"。当你遇到"研究结果表明……"这样的表述，应该重新将其解读为"研究人员解释，他们的研究发现表明了……"

4. 如同我们大家一样，研究人员也有自己的期待、态度、价值观、训练和需求，这使他们所问的问题、做研究的方法、解释研究发现的方式都烙上偏见的印记。例如，科学家通常都对某个假说投入大量的感情。如果美国食糖研究所为你提供暑期研究经费，那么你就很难发现青少年过量消费食糖的问题。和所有容易犯错的人一样，科学家可能也会发现，客观对待那些与他们所相信的假说相冲突的数据非常困难。科学研究的一个主要优势在于它总是尽力将程序和结果公开化，这样其他人就可以判断研究的优缺点，然后可以尝试重复验证。但是，一份科学报告不论看上去多么客观，仍难免会夹杂严重的主观因素。

5. 发言者和写作者常常歪曲或者简化研究结论。原始研究得出的结论和立论者使用这个证据来支持其信念的方式之间常常有重大差异。例如，研究人员可能在其原始研究报告中仔细限定了他们的结论，只不过当结论到了其他人手里时，这些限定马上就被拿掉了。

6. 研究的"事实"会随着时间的流逝而发生改变，尤其是关于人类行为的断言。比如，下面的研究"事实"都在主流的科学资料来源上报道过，却都被最近的研究证据驳倒。

   • 对大部分抑郁症患者来说，百忧解、左洛复和帕罗西汀的医疗效果要好于安慰剂。
   • 吃鱼肝油、锻炼身体、做智力游戏能有效抵抗老年痴呆症。
   • 麻疹疫苗会引发自闭症。

7. 不同研究的人为程度有很大差异。研究者经常为了达到控制的目的，使得研究失去一部分现实世界的特征。研究的人为因素越多，研究结论就越难推广到外部世界。研究的人为因素问题在研究涉及复杂的社会行为时变得尤其明显。例如，社会科学家会让人们坐在一间有电脑的房间里，玩一些游戏，其中包含测试人们的论证过程。研究人员想要弄明白，为什么人们面对不同的场景时会得出不同的决定。面对这样的研究，我们应当问："坐在电脑前思考假想的情境，这种环境是不是太过人为，根本就不能告诉我们人们在面临真实的两难处境时做出决断的方式？"

always *interpret* the meaning of their findings, and all findings can be interpreted in more than one way (see Chapter 7). Hence, researchers' conclusions should not be treated as demonstrated "truths." When you encounter statements such as "research findings show . . . ," you should retranslate them into "researchers interpret their research findings as showing. . . ."

4. Like all of us, researchers have expectations, attitudes, values, training, and needs that bias the questions they ask, the way they conduct their research, and the way they interpret their research findings. For example, scientists often have an emotional investment in a particular hypothesis. When the American Sugar Institute is paying for your summer research grant, it will be very difficult for you to find that sugar consumption among teenagers is excessive. Like all fallible human beings, scientists may find it difficult to objectively treat data that conflict with their hypothesis. A major strength of scientific research is that it tries to make public its procedures and results so that others can judge the merit of the research and then try to replicate it. However, regardless of how objective a scientific report may seem, important subjective elements are always involved.

5. Speakers and writers often distort or simplify research conclusions. Major discrepancies may occur between the conclusion merited by the original research and the use of the evidence to support a communicator's beliefs. For example, researchers may carefully qualify their own conclusions in their original research report only to have the conclusions used by others without the qualifications.

6. Research "facts" change over time, especially claims about human behavior. For example, the following research "facts" have been reported by major scientific sources, yet have been refuted by recent research evidence:

- Prozac, Zoloft, and Paxil are more effective than a placebo for most cases of depression.
- Taking fish oil, exercising, and doing puzzles helps fend off Alzheimer's disease.
- Measles vaccine causes autism.

7. Research varies in how artificial it is. Often, to achieve the goal of control, research loses some of its real-world quality. The more artificial the research, the more difficult it is to generalize from the research study to the world outside. The problem of research artificiality is especially evident in research studying complex social behavior. For example, social scientists will have people sit in a room with a computer to play games that involve testing people's reasoning processes. The researchers are trying to figure out why people make certain decisions when confronted with different scenarios. However, we should ask, "Is sitting at the computer while thinking through hypothetical situations too artificial to tell

8. 对经济效益、社会地位、人身安全和其他因素的需求可能会影响到研究的结果。研究人员也是人，不是电脑，要他们做到百分百的客观是极其困难的。例如，如果研究人员想通过他们的研究来发现某个特定结果，他们就有可能按照心中所想的结果那样来解释自己的研究发现，以得到他们想要的结果。获得资助、终生教职或其他个人奖励的压力可能会影响研究人员解读数据的方式。例如，一家制药公司出钱赞助一项研究，主要研究使用这家公司的药物进行干预的结果，比起另一项针对同样的药物但是受到与那家制药公司无关的资助（如政府基金的资助）的研究，这项研究更倾向于得出较高比率的阳性结果。

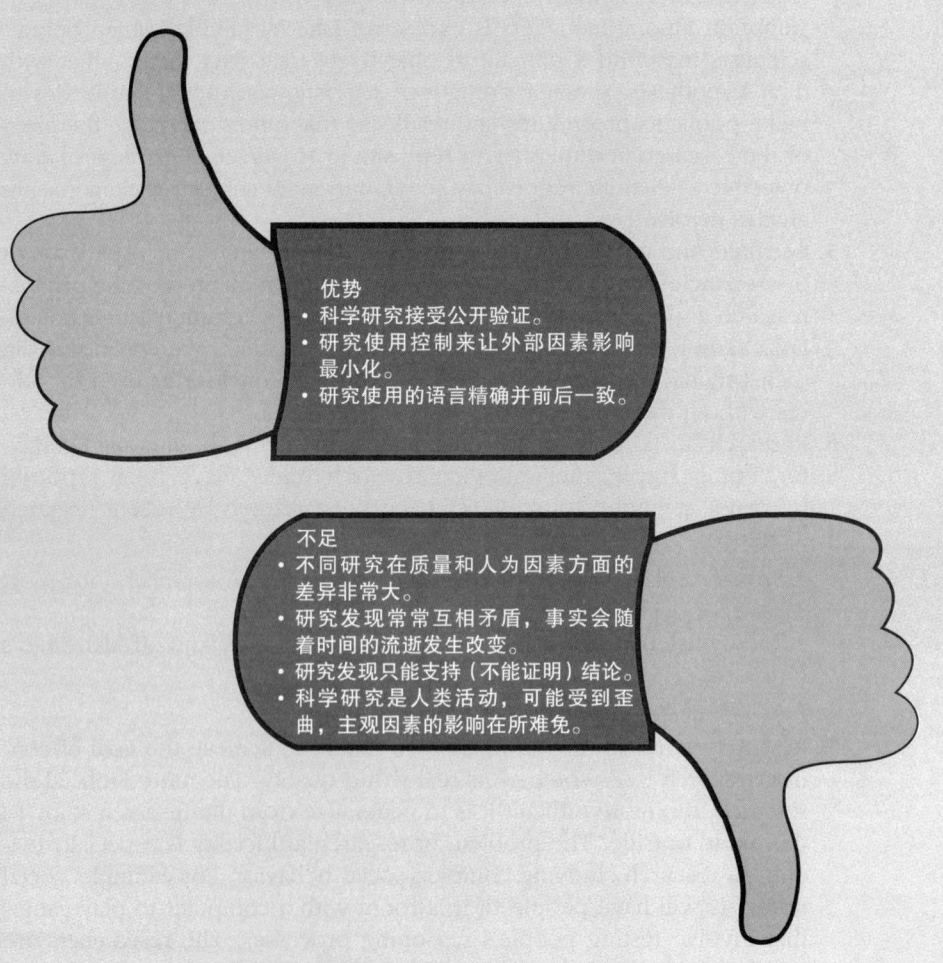

以科学研究作为证据

us much about the way people make decisions when confronted with real dilemmas?"

8. The need for financial gain, status, security, and other factors can affect research outcomes and selection of which studies will be published. Researchers are human beings, not computers. Thus, it is extremely difficult for them to be totally objective. For example, researchers who want to find a certain outcome through their research may interpret their results in such a way to find the desired outcome. Pressures to obtain grants, tenure, or other personal rewards might ultimately affect the way in which researchers interpret their data. For example, research studies funded by a pharmaceutical company tend to have a much higher rate of positive findings for drug interventions using that company's drugs than does research studying the same drugs funded by sponsors not associated with that drug company, such as federal government funding agencies.

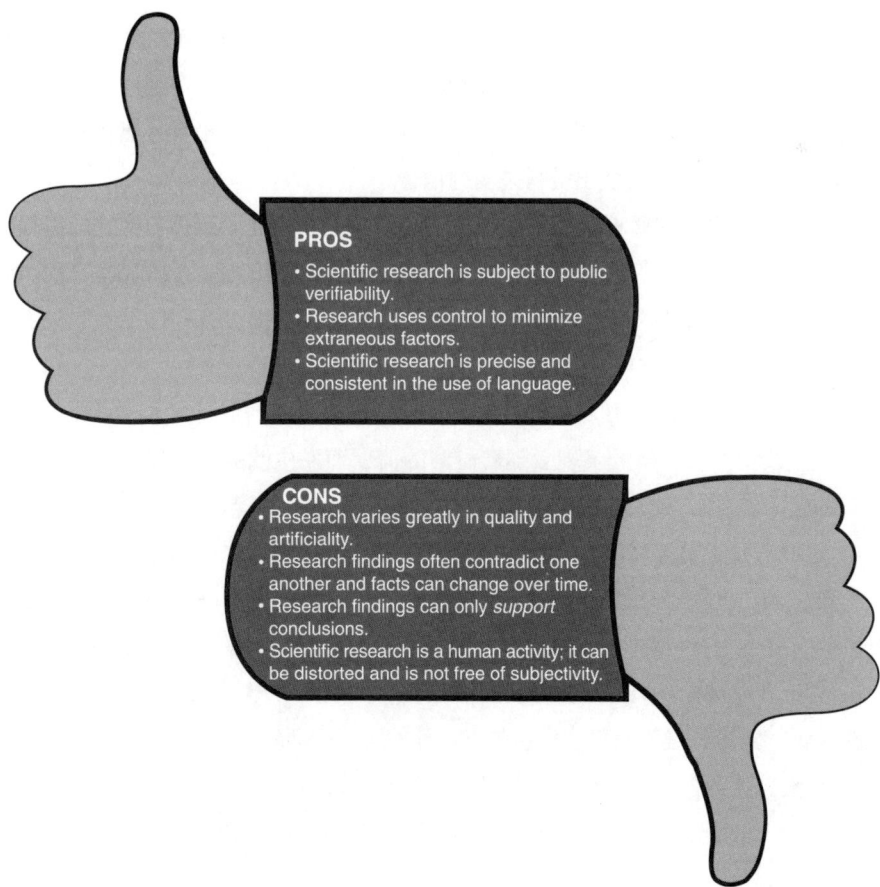

PROS
- Scientific research is subject to public verifiability.
- Research uses control to minimize extraneous factors.
- Scientific research is precise and consistent in the use of language.

CONS
- Research varies greatly in quality and artificiality.
- Research findings often contradict one another and facts can change over time.
- Research findings can only *support* conclusions.
- Scientific research is a human activity; it can be distorted and is not free of subjectivity.

Scientific Research as Evidence

现在你发现，尽管研究证据有很多优点，但我们要避免过早拥抱研究结论。不过，你也不能仅仅因为有一丝疑虑，就武断地抛弃一个建立在科学基础上的结论。"确定性"通常是个不可企及的目标，但并不是所有的结论都同样不确定，我们应该时刻准备去拥抱其中一些结论，而抛弃其他结论。因此，当我们客观地评价研究断言和信念的时候，请小心，不要犯这样的论证错误：在一些结论中强求确定性，其实这些结论中虽然存在一定的不确定性，但并不足以否定这一结论。我们把这一论证错误称为"强求确定性谬误"。

⚠ **强求确定性谬误**：认为一个研究结论如果不是百分百确定，就应该被抛弃。

评价科学研究的一些线索
应用以下问题，以判断研究发现是不是可靠的证据。

1. 报告的资料来源的质量怎么样？通常情况下，最可靠的报告往往出自那些发表在由同行专家评定的期刊上的文章，在这些期刊中，一项研究结果在经过一系列相关专家评价以后才会被接受。通常情况下，资料来源的声誉越好，研究设计也就越好（并不总是如此）。所以，要尽最大努力找出资料来源的信誉。

2. 除了资料来源的质量以外，报告中有没有其他线索显示这项研究完成得很出色？例如，报告有没有详细说明这项研究有什么过人之处？可惜的是，我们在流行杂志、报纸、电视报道和博客里遇到的绝大部分关于研究发现的报告，都没有提供足够的关于研究的细节信息，以确保我们对这项研究质量做出准确评价。

3. 研究实施的时间距离现在有多久，有没有理由让人相信研究发现可能会随着时间的流逝而发生改变？很多研究发现会随着时间的流逝而发生改变。例如，抑郁症、犯罪或者心脏病的起因，在1980年和2014年时可能大不相同。

4. 这项研究的发现有没有被其他研究重复过？如果某种联系总是在精密设计的研究中一致地被发现，比如吸烟和癌症之间的联系，那么我们就有理由相信它，至少在不同意这一结论的人提供较有说服力的证据来证明他们的观点之前，我们都会相信它。

5. 立论者在选择研究的时候是否有选择性？例如，得出相反结论的相关研究有没有被他忽略不计？立论者是不是只选择那些支持他的观点的研究？

6. 有没有什么强势批判性思维的证据？发言者或写作者对于那些支持他的观点的研究有没有表现出一种批判的态度？研究都有局限性，研究得到的大部分结论都需要有所限定。立论者有没有表现出加以限定的意愿？

As you can see, despite the many positive qualities of research evidence, we need to avoid embracing research conclusions prematurely. However, you should not *reject* a scientifically based conclusion just because there is *some* doubt associated with it. Certainty is usually an impossible goal, but all conclusions are not equally uncertain, and we should be willing to embrace some conclusions much more than others. Thus, when critically evaluating research claims and beliefs, be wary of the reasoning error of demanding certainty in some conclusion when some uncertainty is to be expected. We label this reasoning error the *impossible certainty fallacy*.

Fallacy: Impossible Certainty: Assuming that a research conclusion should be rejected if it is not absolutely certain.

Clues for Evaluating Research Studies

Apply the following questions to research findings to determine whether the findings are dependable evidence.

1. *What is the quality of the source of the report?* Usually, the most dependable reports are those published in peer-reviewed journals, those in which a study is not accepted until it has been reviewed by a series of relevant experts. Usually—but not always—the more reputable the source, the better designed the study. So, try to find out all you can about the reputation of the source.
2. Other than the quality of the source, *are there other clues included in the communication suggesting the research was well done?* For example, does the report detail any special strengths of the research? Unfortunately, most reports of research findings encountered in popular magazines, newspapers, television reports, and blogs fail to provide sufficient detail about the research to warrant our judgment of the research quality.
3. *How recently was the research conducted, and are there any reasons to believe that the findings might have changed over time?* Many research conclusions change over time. For example, the causes of depression, crime, or heart disease in 1980 may be quite different from those in 2014.
4. *Have the study's findings been replicated by other studies?* When an association is repeatedly and consistently found in well-designed studies—for example, the link between smoking and cancer—then there is reason to believe it, at least until those who disagree can provide persuasive evidence for their point of view.
5. *How selective has the communicator been in choosing studies?* For example, have relevant studies with contradictory results been omitted? Has the researcher selected only those studies that support his point?
6. *Is there any evidence of strong-sense critical thinking?* Has the speaker or writer showed a critical attitude toward earlier research that was supportive of her point of view? Most conclusions from research need to be qualified because of research limitations. Has the communicator demonstrated a willingness to qualify?

7. 有没有理由让人蓄意歪曲这项研究？我们要当心研究人员亟须找到特定结果的那些情况。

8. 研究的条件是不是人工制造的并因此遭到扭曲？记住，一定要问一声："研究进行的客观条件和研究者总结概括的研究环境到底有多少相似之处？"

9. 根据研究样本，我们概括的范围到底有多大？因为这个问题非常重要，我们将在下一部分深入讨论。

10. 研究人员所使用的调查报告、问卷调查、等级评定或其他测量结果有没有偏见或者歪曲的现象存在？我们应该相信研究人员想要测量的东西，他们一定会测量得准确无误。但是片面的调查报告和问卷调查这一问题在科学研究中简直无孔不入，我们只有在后面部分详细加以讨论。

从研究样本进行概括

写作者和发言者通常用研究报告来支撑他们的普遍概括，也就是对于普遍事件的断言。比如说，"在本研究中，此例药物对 75% 的癌症患者治疗效果明显"并不是一个普遍概括，而"此例药物可治愈胰腺癌"才是普遍概括。我们见到的很多公开发表的普遍概括都需要接受进一步的检查，看看有没有可能出现过度概括的情况。让我们来看看这是为什么。

首先，我们选取样本的方式对判断我们能在多大范围内进行概括至关重要。能否从研究发现中进行概括，主要取决于科研人员所研究的事件或人群的样本数量、覆盖范围和选取的随机性。选取事件或人群进行研究的过程叫作"抽样"。

因为研究人员永远不可能对他们想概括的所有事件或人群都进行研究，所以他们必须选择一些样本来做研究。有些抽样方法比其他抽样方法更加可取。在你评价研究样本的时候，有几个重要的考虑因素必须牢记在心：

1. 样本必须足够大，以支撑进行概括或得出结论。大多数情况下，研究人员观察的事件或人群越多，他们得出的结论就越可靠。如果我们要对大学生在做学期论文时从别人那里获得帮助的频率有多高这一问题概括出一般性的看法，那么研究 1000 名大学生比只研究 100 名大学生自然会更有说服力。

2. 对于研究者将要从中得出结论的所有事件的类型，样本必须覆盖足够的广度，或者说样本应具有足够的多样性。例如，如果研究者想要归纳出大学生普遍的饮酒习惯，那么他们的证据必须建立在对各种不同类型大学的各种不同类型的学生进行抽样的基础之上。

3. 样本的随机性越大越好。研究者随机取样意味着他们尽量保证想要概括的所有事件

> **7.** *Is there any reason for someone to have distorted the research?* We need to be wary of situations in which the researchers *need* to find certain kinds of results.
> **8.** *Are conditions in the research artificial and therefore distorted?* Always ask, "How similar are the conditions under which the research study was conducted to the situation the researcher is generalizing about?"
> **9.** *How far can we generalize, given the research sample?* Because this is such an important issue, we discuss it in depth in our next section.
> **10.** *Are there any biases or distortions in the surveys, questionnaires, ratings, or other measurements that the researcher uses?* We need to have confidence that the researcher has measured accurately what she has wanted to measure. The problem of biased surveys and questionnaires is so pervasive in research that we discuss it in more detail in a later section.

GENERALIZING FROM THE RESEARCH SAMPLE

Speakers and writers usually use research reports to support generalizations, that is, claims about *events in general*. For example, "the medication was effective in treating cancer for 75 percent of the patients in the study" is not a generalization; "the medication cures pancreatic cancer" is. Most publicized generalizations that we encounter need to be closely examined for the possibility of overgeneralizing! Let's see why.

First, how we sample is crucial in determining to what extent we can generalize. The ability to generalize from research findings depends on the *number*, *breadth*, and *randomness* of events or people in the researcher's study. The process of selecting events or persons to study is called *sampling*.

Because researchers can never study all events or people about which they want to generalize, they must choose some way to sample, and some ways are preferable to others. You need to keep several important considerations in mind when evaluating the research sample:

1. The sample must be *large* enough to justify the generalization or conclusion. In most cases, the more events or people researchers observe, the more dependable their conclusion. If we want to form a general belief about how often college students receive help from others on term papers, we are better off studying 1,000 college students than studying 100.
2. The sample must possess as much *breadth*, or diversity, as the types of events about which conclusions are to be drawn. For example, if researchers want to generalize about college students' drinking habits in general, their evidence should be based on the sampling of a variety of different kinds of college students in a variety of different kinds of college settings.
3. The more *random* the sample, the better. When researchers randomly sample, they try to make sure that all events about which they want to generalize have an *equal chance* of getting sampled; they try to avoid a

都有同等的机会得到取样，并尽量避免有偏差的取样。大型的民意调查，如盖洛普民意调查，往往尽量随机地抽取样本。这可以让他们避免特定类型的、特征存在偏差的事件或人群局限了样本的范围。你能不能看出以下每个样本特征有怎样的偏差？

a. 自愿报名接受访谈，讨论他们性行为频率的一群人。
b. 只有有线电话的一群人。
c. 一堂心理学导论课上的学生。
d. 特定电视网的观众，例如福克斯或者微软全国广播公司电视网的观众。

因此，我们要对所有的研究问一问："它们抽取的事件或人群样本有多少，样本的覆盖范围有多广，样本的随机性怎么样？"

没有足够注意到样本的局限会导致对研究发现的过度概括，所表述的概括性结论远远超过此项研究所能保证的范围。在第7章里，我们将这种过度概括称为以偏概全谬误。现在我们来仔细看看一项过度概括的研究。

> 参加在线约会的人极有可能成功找到自己的佳偶。研究人员开展了一项针对229人的在线调查，对象是18~65岁的使用过一个约会网站的人。调查询问他们始于线上约会的关系的情况。调查结果显示：接受调查的人中有94%在第一次见过他们的网络伴侣之后会再次占伴侣约会，这种网络情缘的平均持续时间可达到至少七个月。

抽样的程序不支持这样一个宽泛的概括结论。此项研究报告暗示这一结论可以推广到大多数使用在线约会服务的人，而研究本身却只涵盖一个约会网站和229人这样一个小群体。而且研究并没有交代样本是如何选取的，因此，调查的随机性和覆盖率也无从得知。例如，很有可能那些自愿参加调查的人都是成功找到佳偶的春风得意之人。因此研究报告有瑕疵，它过度概括的程度太大。

⚠ **注意：** 我们只能将研究结果推广到那些和我们在研究中探索过的情况类似的人群或事件。

从测量方法进行概括

所有研究都需要决定怎么测量那些受关注的行为。例如，如果研究者要研究一项教育技术能否促进学生对批判性思维的学习，他就必须决定怎么去测量批判性思维。正如很多

biased sample. Major polls, like the Gallup Poll, for example, always try to sample randomly. This keeps them from getting groups of events or people that have biased characteristics. Do you see how each of the following samples has biased characteristics?

a. People who volunteer to be interviewed about frequency of sexual activity.
b. People who have landline phones only.
c. Students in an introductory psychology class.
d. Viewers of particular television networks, such as Fox or MSNBC.

Thus, we want to ask of all research studies, "How many events or people did they sample, how much breadth did the sample have, and how random was the sample?"

Failure to pay sufficient attention to the limits of sampling leads to *overgeneralizing* research findings, stating a generalization that is much broader than warranted by the research. In Chapter 7, we referred to such overgeneralization as the *hasty generalization fallacy*. Let's take a close look at a research overgeneralization:

> People who join online dating services tend to succeed in finding a good match. Researchers conducted an online survey of 229 people, aged 18 to 65, who had all used an Internet dating site. The survey asked respondents about their primary relationship from the online experience. The research showed that 94 percent of those surveyed saw their "e-partners" again after first meeting them, and the relationships lasted for an average of at least seven months.

Sampling procedures prohibit such a broad generalization. The research report implies that the conclusion can be applied to *most* users of online dating services, when the research studied only one online Web site and only a total of 229 people. The study fails to describe how the sample was selected; hence, the randomness and breadth for this survey are unknown. It is quite possible, for example, that those who volunteered to participate were those who had been most successful in finding a good match. The research report is flawed because it greatly overgeneralizes.

 Attention: We can generalize only to people and events that are like those that we have studied in the research.

GENERALIZING FROM THE RESEARCH MEASURES

All research requires decisions about how to measure the behaviors of interest. For example, if a researcher were studying whether an educational technique promoted the learning of critical thinking, she would have to decide

其他概念，批判性思维也有很多种界定和测量方法。因此，任何关于它的结论，只适用于所使用的具体测量方法。因为概念可以用很多不同的方法来测量，所以研究的结论只适用于所选择的测量方法。人们使用很多不同的手段来测量一些重要的行为，包括问卷调查、检查表、访谈回答以及行为测量。有些测量手段被认为比其他手段更能准确反映一种受关注的行为。例如，某种测量批判性思维技能的方法，比如写批判性的文章，就可能被判断为比选择题测试中的表现更能反映出批判性思维能力的强弱。同样，在对幸福来源的不同研究中，一种测量幸福的方法可能和另一种方法有着截然不同的意义。

因为研究结果只能根据你使用的测量方法来加以概括，所以当我们批判地评价一项研究时，我们需要问："研究者怎样测量所关注的概念，这种测量是否令人满意。"判断性的思考需要我们尽可能地回答这个问题。例如，研究者对儿童延迟满足能力的测量，是通过观察他们愿意选择立刻获得一个棉花糖，还是选择 15 分钟后获得两颗棉花糖。我们需要问："棉花糖的选择是不是测量延迟满足的一种好方法？如果采取不同的测量方法，比如采用父母对孩子延迟满足能力的判断，研究结果会不会有所不同？"

在你评价研究结果的时候，请记住要问这个问题："研究采用了什么样的测量方法，它们是否令人满意？"同时要记住，研究结果只能根据采用的测量方法加以概括。因此，如果一个"结过婚的人比单身的人更快乐"这样的研究结论只建立在完成一份询问"你过得幸福吗"的调查问卷的基础上，那么这个结论最好的表达是：结过婚的人比单身的人更幸福，假设幸福被定义成针对一份自我报告的问卷的一次性回答。如果你不同意这个假设的定义，你就难以接受研究者的结论！在对所有研究的客观评价中，一个重要的组成部分都是判断你认为研究所使用的测量方法在多大程度上抓住了所关注概念的含义。熟悉研究者用来研究某个具体行为的各种测量方法可以帮助你判断测量方法的质量高低。例如，有些研究人员通过让人每天多次给幸福感打分来测量幸福，这种方法就比一次性回答的方法质量更高。

确定一种测量方法是否让人满意的过程可能会包含非常技术性的判断，这超出了本书的讨论范围。但是，只要你识别出研究使用的测量方法，设法想象自己处在被测量者的位置，并记住测量结果只能根据测量方法进行概括，你就能认识到研究结果的局限性。把你自己放在某个参加研究的人的位置，常常能让你获得有益的见解。

how to measure critical thinking. Like many concepts, there are many ways to define and measure critical thinking. Thus, any conclusion about it would only apply to the particular measures used. Because concepts can be measured in many ways, the conclusions of research are only appropriate to the measure of choice. Many different devices have been constructed to measure important behaviors, including questionnaires, checklists, responses to surveys, and behaviors. Some measures can be judged as much more accurate indicators of a behavior of interest than others. For example, one measure of critical-thinking skill, such as a critical essay, might be judged as a much better indicator of that skill than performance on a multiple-choice test. Likewise, one measure of happiness might have a very different meaning than another in studies of the causes of happiness.

Because results of research can only be generalized to the kind of measurements used, when we critically evaluate a research study, we need to inquire, "How did the researchers measure the concepts of interest, and how satisfactory is that measure?" Thinking critically requires us to answer that question as best as we can. For example, researchers have measured children's ability to delay gratification by observing their willingness to choose one marshmallow immediately, versus receive two marshmallows about 15 minutes later. We need to ask, "Is marshmallow choice a good measure of delay of gratification, and might findings be different if a different measure had been used, such as parental judgments of their child's ability to delay gratification?"

When you evaluate research studies, always ask the question, "What were the measures used, and how satisfactory were they?" while remembering that the results of the study can be generalized only to the measures used. Thus, for example, a research conclusion that married people are happier than single people based on completions of a survey that asks, "How happy are you?" is best expressed as married people are happier than single people, *given that happiness is defined as a one-time response on a self-report questionnaire.* If you disagree with that definitional assumption, you will not accept the researcher's conclusion! An important component of critical evaluation of all research is deciding how well you think that the measures in the research captured the meaning of the concepts of interest. Acquainting yourself with various measures that researchers have used to study a particular behavior can help you make judgments about the quality of measures. For example, some researchers have measured happiness by having individuals rate their happiness at multiple times on a daily basis.

The decision about how satisfactory a measure is can include quite technical judgments beyond the scope of this text. However, just identifying the measure used, trying to imagine yourself in the position of those being measured, and remembering that results can be generalized only to those measures will give you important insight into the limitations of the research results. Trying to put yourself into the position of someone participating in the study can often provide you with helpful insight.

请经常对遇到的科学研究提出这个问题：测量方法是否让人满意？同时还要问"有没有其他方法可以用来测量这个概念，如果有，它们是否可能提供不同的结果？"

专家观点什么时候最可信

我们已经讨论了在接受专家的结论之前应该向他们提出的问题，从中可以清楚地看到，专家经常会犯错，或者会误导我们。（可以参看《大错特错：为什么专家老是靠不住》）。

在专家的专业性相当可疑的这种大环境下，专家的观点什么时候最可信呢？我们给出以下建议。

- 你已避免用"系统1思维"来判断专家的意见。
- 你已让思维领先，情感随后。你之所以相信一个观点是因为它已经被理由和证据所证明，而不是因为你感觉它是对的，因为它挺有意思，因为它是新颖的，因为它确认或推翻了别的信念，或因为它帮你赢了一场辩论，让你的观点占了上风。如果你感觉自己立刻就想把这个观点发布到社交媒体上，那么这种感觉就是个警告，你需要等一等，先转变到"系统2思维"的模式。
- 它通过了你学会的这一系列批判性问题的盘问。例如，这个意见得到你断定为精心设计的研究的支持，并且不会受到利益冲突的不当影响。
- 它包含了有限定性的表述。
- 它并非适用于一切情况，认识到它在适用范围方面存在的局限，例如这个结论适用的人有什么特点。它不会过度概括。
- 它是在研究的大背景里提出来的，证据绝非凭空而来，并且它的发现并不适用于每个人。它在已经开展的很多其他研究的大背景下出现，这些研究有些支持它，有些与它的结论或结论的某几个方面相悖。最值得信赖的结论并非从某一个吸引人眼球的研究中浮现出来的，而是从很多专家在很长时间里所做的研究中得出来的。仅仅基于单一的研究得出的结论非常值得怀疑，并且常被证明是错误的。
- 它经受住了其他对相关领域颇有研究的专家的批判性审查。
- 你已找到了对研究断言进行深入分析的讨论资源，例如杂志、书籍、期刊、网站和博客等，它们都鼓励充分的批判性讨论。

Always ask of a research study, "How satisfactory were the measures?" as well as "Are there other ways to measure the concept, and if so, might they provide different results?"

WHEN YOU CAN MOST TRUST EXPERT OPINION

It should be clear from our discussion of the questions that need to be addressed before accepting experts' conclusions that experts are often wrong or misleading. (See, for example, *Wrong: Why Experts Keep Failing Us and How to Know When Not to Trust Them*, David H. Freedman, 2010, Little Brown & Company, New York.)

In the context of much questionable expertise, when can you most trust expert opinion? We suggest the following tips.

- You have avoided System 1 thinking in judging the advice.

- You have let thinking lead and emotion follow. You trust the opinion because it has been justified by reason and evidence, not because it feels right, or it's interesting, or it's novel, or it confirms or disconfirms other beliefs, or it helps you win an argument for "your side." A feeling that you immediately want to share the opinion on Facebook, or Twitter, should be a warning to you to WAIT and to shift into System 2 mode.

- It passes the scrutiny of the critical questions that you have learned to ask. For example, the advice is supported by studies that you have judged to be well designed and not unduly influenced by conflicts of interest.

- It includes qualifying statements.

- It is not universal and recognizes limitations to the breadth of its application; such as, What is the nature of the individuals to which the conclusion applies? It does not overgeneralize.

- It is presented within a broad research context. Evidence does not come out of the blue, and its findings don't apply to everyone. It appears in the context of many other studies that have been conducted, some of which support and some which refute its conclusions or aspects of them. The most reliable conclusions do not emerge from single, attention-grabbing studies but from the studies of many experts over a period of time. Conclusions based only on single studies are highly suspect and often wrong.

- It has held up to the critical scrutiny of other well-informed experts.

- You have sought out sources that discuss in-depth analysis of research claims, such as magazines, books, journals, and Web sites and blogs that encourage lengthy critical discussion.

显然，想要深入评价所有专家的观点是不可能的。一个专家观点与你的生活越相关，你就越要对可获得的证据做深入的研究。

科学研究和互联网

现在是 21 世纪。我们怀疑你和《辛普森一家》里的爸爸霍默·辛普森那样的技术愚人之间的差距要用光年来衡量才行，他甚至会大吃一惊地问："现在他们的电脑上都有互联网了？"如果你在准备写作的时候不利用互联网，那么我们也会像霍默一样大吃一惊。互联网研究从根本上改变了我们大部分人收集证据的方式，让我们获得的信息呈现几何级数的增长。这种信息的唾手可得需要我们付出什么代价？我们不得不仔细考虑收集到的证据，面对这座证据的宝库时保持高度的怀疑精神。记住这些建议，它们可以帮助你解决这些随互联网研究而来的特殊难题。

在本章的前面部分，我们讨论过调查作者背景的重要性。我们督促你判断一下各种潜在的相关偏见或利益冲突。为了权衡一个专家观点的分量，我们首先得知道那个人的身份和可能存在的偏见。"洋葱网"这个流行的讽刺新闻网站展示了互联网怎样让这个任务变得异常困难。在 2008 年一篇题为《愚人网上发评论》的调侃文章里，作者引用了"愚人"对自己的计划的描述："今天晚上，我打算看看大家热议的那部视频，然后点击视频上方专为用户评论预留的那个'回复'链接，草拟一个回复。我要尽量小心，保证一点脑筋都不动，同时确保所有字都用大写字母，标点乱用一气，……虽然现在还不清楚我的评论具体包括哪些内容，但我可以百分百肯定地说，它一定蠢到家了。"要是所有网络上的意见提供者都这样直言不讳就好了！

当论证里包含互联网资源的时候，调查资料来源的可信度就显得尤为重要。互联网常常被人比作大开发前的美国西部荒原。小镇上根本就没有治安官来确保只有真实而公正的声明会由负责任的人发布出来。现在，互联网变得相对不受管束。随便什么人都能建一个网站或者博客。有的网站貌似值得信任，实际上发布信息的人暗地里有所企图。

哪怕你判断出一个网上的写作者是可信的，你也应该多提一些问题。因为网上并没有一个治安官，可疑或虚假的证据很容易得到发布。美国喜剧频道的讽刺专家斯蒂芬·科尔伯特说明了错误信息多么容易就能在互联网上发布。在他的一期脱口秀节目《科尔伯特有话说》里，他编辑了互联网公共百科全书维基百科。有五个小时，维基百科的词条里显示说乔治·华盛顿根本没有蓄养过奴隶，非洲大象的数量在过去的六个月里增长了三倍。（对

Obviously, it is impossible to evaluate all expert opinions in great depth. The more relevance that expert opinion may have in your life, the more in depth you will want to study available evidence.

Research and the Internet

It's the 21st century. We suspect that you are light-years ahead of technological half-wits like Homer Simpson, who marveled, "They have the Internet on computers, now?" We'd be surprised if you were not taking advantage of the Internet when you prepare to write. Internet research has fundamentally changed evidence gathering for most of us, making information exponentially more accessible. What's the trade-off for this unprecedented level of availability? We have to consider the evidence we gather, this cornucopia of evidence, with heightened levels of skepticism. Keep these tips in mind to help you address the particular difficulties that arise with Internet research.

Earlier in this chapter, we discussed the importance of investigating an author's background. We urged you to determine potential biases or conflicts of interests. To weigh the opinion of an authority, we need to know that person's credentials and potential biases. The *Onion*, the popular satirical news site, illustrates how the Internet makes this task particularly difficult. In its 2008 mock article "Local Idiot to Post Comment on Internet," it quotes the "local idiot" as he divulges his plans: "Later this evening, I intend to watch the video in question, click the 'reply' link above the box reserved for user comments, and draft a response, being careful to put as little thought into it as possible, while making sure to use all capital letters and incorrect punctuation. [. . .] Although I do not yet know exactly what my comment will entail, I can say with a great degree of certainty that it will be incredibly stupid." If only all contributors to the Internet were so honest with us!

The importance of investigating a source's credibility is even greater when we add Internet sources to the equation. The Internet often draws comparisons to the Wild West. There is no sheriff in town making sure that only true and fair statements are published by responsible folk. In its current form, it is relatively unrestricted. Anyone can create a Web page or a blog. Web pages can appear to look trustworthy when they are actually published by someone with a hidden agenda. Take a look at https://teachbytes.com/2012/11/01/test-website-evaluation-with-10-hilarious-hoax-sites/ for illustrations of hoax sites and suggestions about how to avoid being fooled by them.

Even after you decide that a Web-based author is reliable, you should ask more questions. Because the Web does not have a sheriff, evidence that is questionable or untrue can easily be posted. Comedy Central's satirical pundit Stephen Colbert wanted to demonstrate how easily false information can be posted on the Internet. In one episode of his *Colbert Report*, he edited the public Internet encyclopedia Wikipedia. For five hours, Wikipedia entries stated that George Washington did *not* own slaves and the population of African elephants tripled in the previous six months. (For another satire of

这个真实问题的另一个讽刺，请查看洋葱网 2002 年的文章《互联网上发现的事实性错误》，文章开篇就说"信息时代在本周一被狠狠击了一记闷棍，有桩事实错误在互联网上被人发现。"）为了和这样的问题做斗争，我们就要避免写出没有注明具体来源的证据。花点时间去查看一下资料的原始来源。当某一篇文章的片段被贴出来或被人引用，贴出文章片段的作者也许自己就误解了原作者的意思，或者将此消息剥离了原来的语境加以使用。

像批判性思维者那样写作和发言

写文章或准备发言最令人兴奋的部分之一就是收集和组织证据来证明我们的观点。证据越充分，主张就越有说服力。在第 7 章和本章中，我们看了如何评估他人论证中的证据，以帮助我们判断哪些论证是合理的。让我们来复习一些我们所学到的关键观点，它们在我们自己的写作和发言中也是很重要的。

1. 证据是立论者为支持一项主张而提供的明确信息。
2. 所有说服力强的论证都需要高质量的证据。高质量的证据具有相关性、充分性和代表性。
3. 假设也需要证据支持。
4. 几乎没有证据是完美的，我们希望找到有更好证据的论证。
5. 大多数个人证据，无论是证词、案例研究还是直觉，都需要额外的支持才能令人信服。
6. 科学证据最有可能是较好的证据。

但是优秀的写作者不只会使用质量最好的证据。在我们的写作和发言中，我们想在列表中添加另一个观点：当与其他证据结合起来时，诉诸情感能够恰到好处地发挥作用。因此，有时候我们使用的证据的质量可能并非最高，但它具有强大的修辞效果。感性证词、个人故事、隐喻、叙述和引文可以起到吸引读者的作用，避免论证成为枯燥乏味的辩论。所以，我们不应放弃使用这样的证据，我们只是想把它和更高质量的证据有效地结合起来。

作为写作者和发言者，我们要在上面的条目下加上这一点：

7. 小心使用旨在引起读者注意的证据。

在第 7 章和本章中，我们评估了许多不同类型的证据，所指出的注意事项与我们的写作和发言也有关。不过，在写作和发言的时候，还有一些额外的问题我们必须牢记心头。

this very real concern, check out the *Onion*'s 2002 article "Factual Error Found on Internet," which begins "The Information Age was dealt a stunning blow Monday, when a factual error was discovered on the Internet.") To combat this problem, avoid writing about evidence that has not been credited to a specific source. Take the time to look up the original source. When a snippet of another article is posted or cited, the author who posted the snippet may have misunderstood or taken the information out of context.

WRITING AND SPEAKING AS CRITICAL THINKERS

One of the most exciting parts of writing an essay or preparing a speech is gathering and organizing the evidence to prove our point. The better the evidence, the more convincing the claim. In Chapter 7 and this chapter, we looked at how to evaluate evidence in other people's arguments to help us determine just which arguments are well reasoned. Let's go over some of the key ideas we learned that will also be essential in our own writing and speaking.

1. Evidence is explicit information provided by the communicator to back up a claim.
2. All good arguments need high-quality evidence. High-quality evidence is relevant, sufficient, and representative.
3. Assumptions also require evidence.
4. Almost no evidence is perfect; we want to find arguments with better evidence.
5. Most individual evidence, whether a testimonial, a case study, or intuition, requires additional support to be convincing.
6. Scientific evidence offers the best possibility of being better evidence.

But good writers don't just use the best-quality evidence. So, in our writing and speaking, we want to add another idea to our list: Emotional appeals can be appropriate when joined with other evidence. Thus, sometimes we might use evidence that is not the highest quality but has great rhetorical impact. Emotional testimonials, personal stories, metaphors, narratives, and quotations can serve the purpose of engaging the reader in what otherwise might be a dry argument. So, we don't want to discard the use of such evidence; we just want to combine it effectively with higher-quality evidence.

So, to the above list, we will add, as writers and speakers,

7. Carefully use evidence intended to engage the reader.

In Chapter 7 and this chapter, we evaluated many of the different types of evidence. And the cautions indicated are also relevant for our writing and speaking. However, there are a few additional issues we must keep in mind when we are writing and speaking.

首先，很多人发现，当我们评估其他人的证据质量时，我们可以很好地将他们的个人经历或他人的案例历史排除出证据之列。然而，当面对我们自己或与我们亲近的人的经历时，我们更倾向于相信这些个人的证据是高质量的，即使我们已经拒绝接受来自他人的同类证据的价值。我们必须对这种倾向保持警觉。如果我们的经验和直觉相比于其他人更加成熟、敏锐，那可就太棒了，但遗憾的是，没有证据证明这是真的。我们必须找到有说服力的证据来支持我们的主张，这些证据绝不能仅限于我们的个人经验、我们朋友的经验或我们的直觉。

其次，我们大多数人都不是我们要在大学里写作和发言的学科的专家。所以，我们倾向于严重依赖那些我们认为是专家的人。这种倾向导致我们严重依赖研究和权威。虽然我们有时可以从研究和权威人士那里找到高质量的证据，但情况并不总是如此，甚至通常都不是如此。

我们先来看看用科学研究作为我们的论据的情况。正如我们前面提到的，如果能找到科学研究，那么做得好的科学研究是最可靠的。但"做得好"这一限定值得反复推敲。科学研究中也有欺骗性的、结构紊乱的研究，浮夸的主张，更不用说高度可疑的资金来源了。而且，正如我们所指出的，找到与其他发现相矛盾的研究结果是相当容易的。不过，作为写作者和发言者，我们可以使用一些工具来克服科学研究的部分局限性。

此外，虽然我们不能确保研究的质量是最高的，但大多数科学出版来源试图为读者过滤掉这些缺陷，并且，使用受尊重的科学出版来源的研究作为证据通常是安全的。当你为自己的论点充实证据的时候，为了增加你选择好的研究证据的可能性，随身携带一份第212页的表格"评估科学研究的一些线索"。而且，请记住，并非符合本章所列标准的研究都可选择，选择的研究还必须与我们的论点相关联。

除了研究，我们很可能会求助于那些我们认为是某个领域的专家权威的人来支持我们的主张。想要不被貌似某个领域的专家权威的人劝服是很困难的。但是，在互联网时代，分辨清楚谁是专家权威变得越来越难了。任何人都可以声称自己是权威。

更糟糕的是，研究表明，如果被贴上专家的标签，我们实际上在自己的观点上就会变得更加固执和偏狭。如果这个研究是真实的（在我们决定它是否是真实的之前，我们需要看到证据！），这就意味着专家是不可靠的，因为他们可能不太愿意考虑不同的观点。可悲的是，我们的文化对那些改变观点的人没有广博的宽容心，我们总是称他们为"墙头草"或"伪君子"，而没有认识到，当新的证据出现时，改变我们的观点不仅是自然而然的事情，而且是负责任、有担当的表现。

First, many of us find that when we are evaluating the quality of other people's evidence, we have no problem dismissing their personal experience or someone else's case history as evidence. However, when it comes to our own experiences or those of people who are close to us, we are much more likely to believe that this personal evidence *is* of high quality, even though we've rejected the value of the same type of evidence coming from others. We have to be very watchful of this tendency. It would be wonderful if our experiences and intuitions were more likely to be high quality than other people's, but, sadly, there is no evidence that this is so. Instead, we too have to find good, supporting evidence for our claims that goes well beyond our own personal experiences, our friend's experiences, or our intuition.

Second, most of us are not experts on the subjects that we will be writing and speaking about in college. So, our tendency is to rely heavily on those we think are experts. This tendency leads us to lean heavily on research studies and authorities. And, while we can find good-quality evidence in studies and from authorities, this is not always or even usually the case.

Let's take a look first at using scientific research as our evidence for our arguments. As we mentioned earlier, where possible, scientific research, when done well, is the most reliable. But that caveat, *when done well*, is worth repeating. Scientific research has its share of fraud, poorly structured studies, and grandiose claims, not to mention highly suspect funding sources. And, as we noted, it is fairly easy to find research findings that contradict other findings. However, we have some tools as writers and speakers that can overcome some of the limitations of scientific research.

Also, even when we can't ensure that research is of the highest quality, most scientific publishing sources attempt to filter out these limitations for readers, and, in general, using research as evidence that comes from respected science sources is usually a safe bet. To increase the likelihood that you have selected good research evidence, keep a copy of the chart on page 213, "Clues for Evaluating Research Studies," handy while you are developing the evidence for your argument. And, keep in mind that even research that meets the criteria outlined in this chapter must also be relevant to our argument.

In addition to research studies, we are likely to turn to those we think are the expert authorities on a topic to help support our claims. It's very hard not to be convinced by someone who seems to be an authority on the topic, but, particularly in the era of the Internet, it is becoming harder and harder to tell just who is and who is not an expert or an authority. Anyone can refer to himself as an authority.

And to make matters even more difficult, research suggests that if we are labeled an expert we actually become more rigid and intolerant in our beliefs. If this research is true (and before we decide if it is, we'd need to see the evidence!), it means that experts are suspect because of their likely unwillingness to consider alternative viewpoints. Sadly, as a culture we don't have a high tolerance for people who change their minds, calling them "flip-floppers" or "hypocrites" instead of recognizing that as new evidence emerges it is not just

求助于专家的另一个局限是，我们倾向于相信那些和我们同声相应、意气相倾的专家。作为写作者和发言者，我们必须特别警惕这种倾向。为了使自己免受偏见的影响，我们必须寻找与自己观点相悖的证据，就像我们热切寻找支持我们观点的证据一样。接下来，我们要用同样的标准来评估证据的质量。与我们意见相左的证据实际上可能比与我们意见一致的证据更有说服力，我们往往很难接受这一点。

一旦我们确定了我们的证据来自某个领域有思想深度的专家或权威，而且他的结论没有重大弱点，我们就可以将此作为证据。显然，我们应该对证据的接受建立高标准。

找到高质量的证据，自然皆大欢喜，但现实情况是，很多时候很难找到真正的研究报告，也很难确定写报告的人是不是真的专家。我们可能不得不退而求其次。在很多写作或发言的情境中，我们可能无法确保找到了最好或最可靠的证据来源。因此，重要的是找到使用起来感觉舒服、放心的门户网站，从它们那里获得的资料源可能比其他的更可靠。例如，如果你可以访问大学图书馆的在线目录，那么使用图书馆而不是通过谷歌或其他搜索引擎来进行检索。大学图书馆已经为我们完成了一些筛选过滤工作。

在收集研究资料时，我们可以使用的其他过滤方法包括避开个人网页或博客，除非我们熟知并尊重作者的背景。此外，在网上查找作者的出版物可以让你了解他对这个专题的了解程度。如果作者写作的话题是处方药、运动、文学和技术，他的作品可能生动有趣，会广泛传播，但他很可能不是这些领域的专家。另一方面，如果作者就一个专题发表了六七项成果，并且其中一些可以通过图书馆的门户检索到，那么我们很有可能找到了一个可靠的人。

显而易见，对于写作者和发言者来说，证据可能是一个困难的领域。我们怎么知道我们的证据是否足够好呢？我们如何处理论证中的模棱两可、含糊不清呢？正如我们之前提到的，几乎没有什么事情是我们可以绝对确定的，所以在我们尽可能地将批判性思维技巧运用到我们的写作和发言之后，我们就很幸运地拥有了另一种对我们有所帮助的策略。对反驳论证进行研究能够让我们避免很多愚蠢的争论。

在我们的写作和发言中，我们有机会思考一些相互矛盾的研究或不同的发现，并有说服力地论证为什么我们最终放弃了其他观点，选择了某一个结论和理由。我们选择的观点可能是我们论证中最有说服力的部分之一，因为我们预见到读者的批评，有效地向他们展示为什么用来批评我们论证的观点不够有力。如果做得足够好，文章或发言的这一部分会变得铿锵有力。

natural but indeed responsible to change our opinions when the evidence demands it.

Another limitation of going to the experts is that we tend to believe the experts who believe what we believe. As writers and speakers, we have to be particularly on the alert for this tendency. To protect ourselves from our own bias, we must seek out evidence that contradicts our opinions just as eagerly as evidence that supports it. And then, we have to evaluate the quality of that evidence with the same criteria. It is often difficult to accept that evidence we do not agree with might actually be better than the evidence with which we do agree.

Once we've determined that our evidence has come from someone who is a thoughtful expert or an authority on a topic *and* that her conclusions are free of major weaknesses, we can use that source as evidence. Clearly, the bar for what evidence we should be using is high.

Finding high-quality evidence is the ideal scenario. But, the reality is that many times it is difficult to find the actual research study or to determine if the person writing it is an actual expert. We may have to settle for less. In many writing or speaking situations, we might not be able to ensure we've found the best or most reliable source. So it is important to get comfortable with portals for source material that tend to be more reliable than others. For example, if you have access to your university library's online catalogue, start there rather than with Google or another search engine. The library has already done some of the filtering for us.

Other filters we can apply when collecting research include bypassing personal Web pages or blogs unless we know and respect the background of the author. Also, looking up an author's publications online can give you some idea of how well he knows the topic. If the author writes about prescription drugs, sports, literature, and technology, he might be widely read and very interesting, but chances are pretty high that he is not an expert in any of these areas. On the other hand, if the author has six or seven publications on one topic and some of these are accessible through your library portal, there's a higher likelihood that we've found someone reliable.

Clearly, evidence can be a difficult area for writers and speakers. How do we ever know that we have better evidence? And how do we deal with that ambiguity in our arguments? As we mentioned earlier, there is very little we can be absolutely certain about, so once we've applied our critical-thinking skills to our writing and speaking as best as we can, we are lucky to have another strategy that can also help us. Studying counterarguments can save us from a lot of silly arguments.

In our writing and speaking, we have the opportunity to address some of the contradictory research or alternative findings and to argue, convincingly, why we ultimately discarded these alternative views for the conclusion and reasons we selected. This idea is potentially one of the most convincing sections of our argument because we get to anticipate our readers' criticisms, effectively showing them why what could have been used to critique our

最后，随着写作的进展，你可能已经注意到了一些东西。你可能可以提出一个主张、定义模糊的术语、提供理由、检验假设、进入证据部分，却无法找到可靠的证据来支持你的主张。对于写作者来说，这种可能出现的情况会让人突然感觉十分沮丧：我们相信并准备为之辩护的东西，竟没有足够好的证据支持。我们就是绞尽脑汁也无济于事！对此，我们会做什么呢？尽管没有足够的证据来支持论证，但很多学生想继续试图用贫瘠脆弱的证据来证明他们的论点，通常他们会用很多个人的事例和证言来作为证据，因为他们不想放弃他们到目前为止所完成的所有工作，也不想放弃他们想在结论中体现的信念。

事实上，研究表明，面对大量背道而驰的证据，我们中的许多人都想要坚持自己的观点。所以，如果这是你认为你会做的，那么你并非唯一会这么做的人。但是，你可以想象，这样的反应不是一个批判性思维者应做出的行为。在提高写作技能的同时，我们也在检验自己的观点和信念。一些观点和信念经不起真火的淬炼，这不足为奇。没有关系，观点和信念形成过程的一部分就是抛弃那些无效的东西。大多数人希望自己持有经得起检验和推敲的信念。所以，不管你对一个结论的感受有多强烈，或者你多么喜欢你的理由，如果没有足够好的证据来支持你的观点，最好的选择就是把它当成一次学习经历，然后重整行装再出发。

思维体操

⚠ **批判性问题**：证据的效力怎么样？

评价下列的练习文章，检查作者所提供的证据的质量。

第一篇

酒鬼生的孩子是不是也更容易变成酒鬼？为了回答这个问题，研究人员在匿名戒酒会里抽取了 451 个人，看看其中多少人会说他们父母中有一个或者双双都是酒鬼。这项研究中涉及的戒酒会成员现在都在参加俄亥俄州、密歇根州或者印第安纳州某处的戒酒会，负责当地戒酒会项目的人要求他们自愿填写这样的调查表。

调查发现 77% 的调查对象把他们父母中的至少一方归入酒鬼之列。这项研究还从上述那些州里随机调查了 451 名宣称不酗酒的人。在这些酒瘾不大的人当中，23% 的人将他们父母中的至少一方归为酒鬼。

argument is not strong enough to do so. Done well, this section of an essay or speech can be very powerful.

Finally, you may have noticed something as our writing sections have progressed. It's possible to propose a claim, define the ambiguous terms, provide reasons, examine the assumptions, and get to this evidence section and be unable to find the evidence to reliably support your claim. This possibility can be a very frustrating moment for writers: Something we believe and were ready to argue for does not have enough good evidence to stand up to scrutiny. We might want to pull our hair out! What do we do at this point? Many students are tempted to continue on valiantly and attempt to prove their argument anyway, using poor evidence, usually lots and lots of personal examples, and testimonials, because they don't want to discard all the work they have done up to this point and/or they don't want to discard their belief in their conclusion, even though there is not sufficient evidence to support it.

In fact, research suggests that many of us desire to hold on to an opinion in the face of all the evidence against it. So if that's what you are thinking you would do, you are not alone. But, as you can imagine, such a response is not the behavior of a critical thinker. As we develop our writing skills, we are also testing our own ideas and beliefs. Not surprisingly, some of them just might not hold up under fire. That's okay. Part of the process of developing opinions and beliefs is discarding those that are not valid. Most people would prefer to hold beliefs that hold up under scrutiny. So, no matter how strongly you felt about that conclusion or how much you liked your reasons, if there simply is not enough good evidence to support your point, your best option is to chalk it up to a learning experience and start over.

PRACTICE EXERCISES

 Critical Question: **How good is the evidence?**

Evaluate each of these practice passages by examining the quality of the evidence provided.

Passage 1

Are children of alcoholics more likely to be alcoholics themselves? In answering the question, researchers sampled 451 people in Alcoholics Anonymous (AA) to see how many would say that one, or both, of their parents were alcoholics. People in AA used in the study currently attend AA somewhere in Ohio, Michigan, or Indiana and were asked by people in charge of the local AA programs to volunteer to fill out a survey.

The research found that 77 percent of the respondents had at least one parent they classified as an alcoholic. The study also surveyed 451 people randomly from the same states who claim not to be heavy

第二篇

为什么不允许18岁的美国人喝酒？18岁的人被允许做21岁的人能做的所有其他事情：投票、当兵打仗、驾车以及独自生活。

第三篇

医用大麻在有效治疗癌症患者方面可能带来巨大的飞跃。虽然有人认为医用大麻合法化将会增加对毒品的娱乐性消费，但是我要说："这样的主张有什么根据呢？"上个月，在密歇根州底特律市，有75人接受问卷调查，被问到他们是否认为让医用大麻合法化会增加本州毒品的娱乐性消费。有93%的人回答说他们不相信医用大麻合法化会增加毒品的娱乐性消费。因此，美国的立法者应该通过一项法律，让医用大麻合法化。

给个提示

第一篇

结论：酒鬼的子女比常人的子女更容易变成酒鬼。

理由：比起常人，更多的酒鬼报告说他们父母当中至少有一方酗酒。

注意这里所呈现的结果来自一份并没有说明其结果有多大代表性的研究。我们不知道这则信息到底发表在什么地方，所以也无法评估这份研究在发表之前是经过了严格的评定。尽管如此，对这份研究我们可以问几个有用的问题。样本的数量很大，可是它的覆盖面有问题。尽管有好几个州被抽样，这些州的戒酒会项目中的人又能在多大程度上代表全国的酒鬼呢？同样，戒酒会中的酒鬼和那些没有寻求帮助的酒鬼相比又怎么样呢？也许样本存在的最大问题在于缺乏随机抽样。虽然那些自诩不酗酒的人是在三个州里随机选取的，但戒酒会中的调查对象却是自愿参与的。那些自告奋勇地谈论自己父母的人，是不是有别于那些不愿公开信息的人？如果自愿的人和不自愿的人之间有区别，那么这个样本的选取就是有偏差的。

drinkers. Of the nonheavy drinkers, 23 percent would label at least one of their parents as alcoholic.

Passage 2

Why shouldn't 18-year-olds be permitted to consume alcohol? They are permitted to do all other things that 21-year-olds do: vote, enroll in the war, drive a vehicle, and live on their own.

Passage 3

Medical marijuana could be a huge step toward more effectively treating cancer patients. While some people argue that legalizing medical marijuana will increase recreational use of the drug, I argue, "Where is the proof for such an assertion?" Last month, 75 people were surveyed in Detroit, Michigan, and were asked if they believe that legalizing medical marijuana will increase recreational use of the drug in their state. Ninety-three percent responded that they did not believe legalizing medical marijuana would increase recreational use of the drug. Therefore, our national lawmakers should pass a law legalizing medical marijuana.

Sample Responses

Passage 1

CONCLUSION: *Children of alcoholics are more likely to become alcoholics than are children of nonalcoholics.*

REASON: *More alcoholics than nonalcoholics reported having at least one alcoholic parent.*

Note that the results presented are from one study without reference to how typical these results are. We also do not know where this information was published, so we can make no assessments regarding how rigorously the study was reviewed before publication. However, we can ask some useful questions about the study. The sample size is quite large, but its breadth is questionable. Although multiple states were sampled, to what extent are the people in the AA programs in these states typical of alcoholics across the nation? Also, how do alcoholics in AA compare to alcoholics who have not sought help? Perhaps the most important sampling problem was the lack of a random sample. While the self-reported nonalcoholics were randomly selected in the three states, the respondents in AA were selected on a voluntary basis. Do those who volunteered to talk about their parents differ greatly from those

调查中的测量有多准确？首先，除了说明调查对象来自戒酒会，文章并没有提供酒鬼的明确定义。此外，我们也不知道调查对象将他们的父母界定为酒鬼的标准是什么。因此，我们无从确定调查对象判断某人是不是酒鬼的准确程度如何。同样存在问题的还有以下事实：调查者对所谓的对照组中不酗酒的人的选取全是基于他们的自我评定。我们知道大家普遍会更赞许不酗酒的人，人们在知道一个答案会受到普遍赞许时都会倾向于给出这个答案。这种倾向也会使所谓的对照组里的取样过程产生偏差。关于这些评定标准的准确度如何，我们希望知道更详细的信息，然后才能对这一结论更有信心。

第二篇

结论：应该允许18岁的人喝酒。

理由：21岁的人可以喝酒，而18岁的人和他们并没有什么不同。

首先我们要注意这个论证建立在比较的基础上。我们对21岁的人所享有的权利很熟悉，它被用来帮助我们更好地理解一件在某种程度上和它有些类似的事情：18岁的人和21岁的人有很多相同的权利。但是，两者之间也有个极大的区别，即很多18岁的人不像21岁的人那样心理得到了更全面的发展，或者能够承担更多社会责任。如果我们假设这个差异会影响到18岁的人为喝酒负责的能力，那么这个差异足以让我们舍弃这个用来证明结论的类比。

who did not volunteer? If there is a difference between the volunteers and nonvolunteers, then the sample is biased.

How accurate are the rating measurements? First, no definition for *alcoholic* is given beyond those answering the survey currently being in AA. In addition, we are not told of any criteria given to the research participants for rating parents as alcoholic. Thus, we are uncertain of the accuracy of the judgments about whether someone was an alcoholic. Also, problematic is the fact that the selection of the supposed control group of nonalcoholics is based on self-assessment. We know that there is a socially acceptable answer of not being an alcoholic, and people tend to give socially acceptable answers when they know them. This response tendency could also bias the sampling in the supposed control group. We would want to know more about the accuracy of these ratings before we could have much confidence in the conclusion.

Passage 2

CONCLUSION: *Eighteen-year-olds should be permitted to consume alcohol.*

REASON: *Eighteen-year-olds are no different from 21-year-olds who are permitted to consume alcohol.*

First, we note that the reasoning is based on a comparison. Something we are familiar with, the privileges of 21-year-olds, is used to help better understand an event that is similar in some ways: Both 18-year-olds and 21-year-olds have many of the same privileges. A significant difference, however, is that many 18-year-olds are not as psychologically developed, or socially responsible, as 21-year-olds. If we assume that this difference could affect an 18-year-old's ability to consume alcohol responsibly, then this difference is sufficient for us to reject the analogy as proof for the conclusion.

第 9 章

有没有替代原因

| 学 习 目 标 |

1. 提高对替代原因普遍性的认识。
2. 强化对如何发现替代原因的理解。
3. 培养对混淆相关性和因果关系这一现象的频发和危险的敏感意识。
4. 了解在讨论替代原因时，如何与他人有效互动。

我们以一个故事开始本章。

一个爱问为什么的孩子注意到太阳早上出现在天空，到夜里就不见了。孩子想知道太阳到底去哪儿了，于是想方设法地要凑近看看日落。可是，不管怎么努力，他还是不明白太阳到底去哪儿了。后来，孩子又注意到他的保姆也是早上出现在家里，到夜里就不见了。有一天，他好奇地问保姆，她晚上去哪儿了。保姆回答说："我回家去了。"孩子将保姆的来去和日夜的循环联系在一起，得出结论：保姆的离去导致太阳也一起回家了。

这个故事清楚地表明了使用证据时一个常见的难题：试图弄明白一件事情的起因。我们只有理解引发某个现象的因果模式，才能选择一个明智的办法来避免问题的发生，或者促进某个特别正面的结果的出现。比如说，我们想知道是什么引起了2008年的金融危机。

CHAPTER 9

Are There Rival Causes?

LEARNING OBJECTIVES

1. Develop awareness of the pervasive nature of rival causes.
2. Strengthen the understanding of how to uncover rival causes.
3. Develop sensitivity to the frequency and dangers of confusing correlation and causation.
4. Become sensitive to how to interact effectively with others when discussing rival causes.

e begin this chapter with a story.

> An inquisitive little boy noticed that the sun would show up in the sky in the morning and disappear at night. Puzzled by where the sun went, the boy tried to watch the sunset really closely. However, he still could not figure out where the sun was going. Then, the boy also noticed that his babysitter showed up in the mornings and left at night. One day, he asked his babysitter where she went at night. The babysitter responded, "I go home." Linking his babysitter's arrival and departure with the coming of day and night, he concluded that his babysitter's leaving caused the sun to also go home.

This story clearly illustrates a common difficulty in the use of evidence: trying to figure out what caused something to happen. We cannot determine an intelligent approach to avoiding a problem or encouraging a particular positive

或者，为什么过去的十年里肥胖的比例会有如此急剧的上升。

这个故事同时还显示出在使用证据证明一件事引发另一件事时常见的一个难题，即替代原因的难题。那个虚构的小男孩为他观察到的问题提供了一个解释："太阳晚上下山是因为我的保姆回家了。"他的"原因"好像很合理，并且能讲得通，但我们希望你能明白，还有一个对太阳为什么会落山更可靠的解释。

尽管替代原因很少会像在我们这个故事里表现得那样明显，但你经常会遇到专家提出一个假说来解释某件事或某个研究发现，而同时还有其他言之成理的假说同样能解释它们。通常情况下，专家不会把这些替代原因透露给你，因为他们不想让你分心，从而显得他们那些断言不那么确凿无疑，所以你得自己去发掘出这些替代原因。这样做特别有助于你判断"证据的效力怎么样"。有些事件存在着各种各样言之成理的替代原因，这就大大削弱了我们对写作者提供的原因的信心。

⚠ **批判性问题**：有没有替代原因？

⚠ **注意**：所谓替代原因，即言之成理的替代解释，它能够说明为什么特定的结果会发生。

何时该寻找替代原因

当你有足够的理由相信写作者或发言者在使用证据支持他对某件事的起因的一个断言时，你就需要寻找一些替代原因。"原因"这个词的意思是"引起，让某件事发生，或影响"。立论者可以用很多种不同的方式指出因果思维。下面我们仅列出一些供你参考。

导致……
影响了……
与……有关
阻止了……
增加了……的可能性
决定了……
与……有联系
有……的效果

outcome until we understand the causal pattern that gave rise to the phenomenon in the first place. For example, we want to know what caused the financial crisis of 2008. Or, *why* the rate of obesity has been increasing so sharply over the last 10 years.

The story also shows a very common difficulty in using evidence to prove that something caused something else—the problem of *rival causes*. The fictional little boy offered one interpretation of his observations: "The sun sets at night because my babysitter goes home." His "cause" seems reasonable; it makes sense. However, we expect that you can see another very plausible explanation for why the sun sets.

Although rival causes will rarely be as obvious as they are in our story, you will frequently encounter experts presenting one hypothesis to explain events or research findings when other plausible hypotheses could also explain them. Usually, these experts will not reveal rival causes to you because they do not want to distract you from the sound of certainty associated with their claims; you will have to produce the rival causes yourself. Doing so can be especially helpful as you decide, "How good is the evidence?" The existence of multiple, plausible rival causes for events reduces our confidence in the cause originally offered by the author.

 *Critical Question: **Are there rival causes?***

 Attention: *A rival cause is a plausible alternative explanation that can explain why a certain outcome occurred.*

WHEN TO LOOK FOR RIVAL CAUSES

You need to look for rival causes when you have good reason to believe that the writer or speaker is using evidence to support a claim about the cause of something. The word *cause* means "to bring about, make happen, or affect." Communicators can indicate causal thinking to you in a number of ways. We have listed a few.

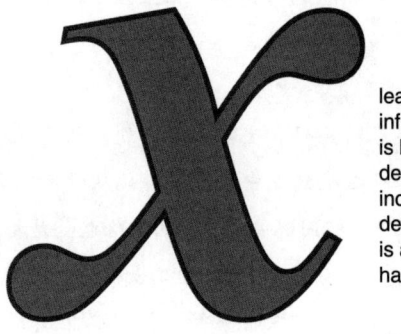

leads to...
influences...
is linked to...
deters...
increases the likelihood...
determines...
is associated with...
has the effect of...

这些因果思维的线索应该能帮助你在立论者做一个因果断言时辨认出来。一旦注意到这样的断言，你就一定要警惕存在替代原因的可能性。

替代原因的普遍性

找出替代原因有助于我们更好地对自己遇到的关于因果关系的结论进行回应，遇到这类结论的情况主要有三种：1. 我们日常的人际交往；2. 以往或现在的世界大事；3. 科学研究的结果。

下面我们来看几个例子。

> 例一 日常人际交往中的论证
>
> 大学生和朋友间的对话：都已经过了24小时了，我男朋友还没有回我的短信。他肯定被我气疯了。
>
> 替代原因：也许他忙着学习以准备考试，或者他落下了手机。
>
> 例二 世界大事
>
> 根据新闻报道，2012年12月14日，20岁的亚当·兰扎在位于美国康涅狄格州纽顿的桑迪胡克小学枪杀了20名儿童和6名成年教职员工，制造了大规模的校园血案。在开车去学校之前，兰扎已经在纽顿的家中将母亲杀害。正如特大新闻事件中一贯的情形，每个人都有自己偏好的因果解释。在血案发生后不久，新闻和电视访谈节目就给出兰扎杀人的好几个可能的动机。
>
> 1. 杀手的行为很可能是由于他的母亲坚持要把他送到精神病院，这引发了他的怒火并酿成血案。
> 2. 玩暴力游戏引发了这场血腥残杀。据报道，警察在枪杀案后搜查兰扎的家时，发现了价值好几千美元的暴力游戏设备。
> 3. 兰扎一直在服用一种治疗精神病的药物，记录显示，这种药物会使一些患者出现冲动－控制障碍和重度抑郁症。
>
> 例三 科学研究
>
> 最近一项研究显示母乳喂养对妈妈和宝宝都有好处。研究发现一生中哺乳期超过一年的女性，在绝经以后患心脏病或中风的风险比从没有哺乳过的女性低大约10%。她们患糖尿病、高血压、胆固醇偏高等病的风险也会大大降低。这项研究发表在五月号的《妇产科》杂志上，分析了139 681名女性的数据，这些女性都登记参加了妇女健康启动项目——一项长期跟踪调查绝经后女性的国家级研究项目。

These clues to causal thinking should help you recognize when a communicator is making a causal claim. Once you note such a claim, be alert to the possibility of rival causes.

THE PERVASIVENESS OF RIVAL CAUSES

Detecting rival causes can help us better react to causal conclusions encountered in (a) our everyday personal relationships, (b) past or ongoing world events, and (c) results of research studies.

Following are several examples.

Example 1. Reasoning in interpersonal relationships.

College student talking to a friend: It's been over 24 hours and my boyfriend hasn't returned my text message. He must be mad at me.

Rival causes: Maybe he's busy studying for a test, or perhaps he has misplaced his cell phone.

Example 2. A major world event.

According to news reports, on December 14, 2012, Adam Lanza, 20, fatally shot 20 children and 6 adult staff members in a mass murder at Sandy Hook Elementary School in Newtown, Connecticut. Before driving to the school, Lanza had killed his mother at their Newtown home. As is very typical in major news events, everyone had his favorite causal explanation. Shortly after the killings, several possible motives for the killings had been suggested on the news and in talk shows.

1. The shooter's actions may have been triggered by his anger at his mother's insistence on having him committed to a psychiatric facility.
2. Playing violent video games triggered the killing rampage. It was reported that when police searched Lanza's home after the shooting, they found thousands of dollars' worth of violent video games.
3. Lanza had been prescribed an antipsychotic drug, with documented links to causing impulse-control disorder and major depression in some users.

Example 3. A research study.

A recent study suggests that breast-feeding benefits mothers as well as babies. The study found that women who had breast-fed for more than a year in their entire lifetimes were almost 10 percent less likely to have had a heart attack or a stroke in their postmenopausal years than those who had never breast-fed. They were also less likely to have diabetes, hypertension, and high cholesterol. The research, published in the May issue of the journal *Obstetrics and Gynecology*, analyzed data on some 139,681 women who had enrolled in the Women's Health Initiative, a long-term national study of postmenopausal women.

在这项研究中，研究人员很可能带着一个假设——母乳喂养给母亲带来健康上的好处开始研究，结果发现了与假设相一致的证据。但是让我们来看看其他不同的或者说替代性的原因，它们同样可以支持这个研究发现。

1. 哺乳的女性可能只是平均来说比没有哺乳过的女性过着更健康的生活。比如说，她们可能比没有哺乳的女性锻炼得更多，或者饮食更为多样化。
2. 选择不用母乳喂养孩子的女性可能在外工作更长的时间，这可能会带来更大的生活压力，因此也会带来更多的健康问题。
3. 一些女性之所以拒绝母乳喂养，可能是因为她们本身比那些选择母乳喂养的女性有更多的健康问题。例如，那些服药的母亲或抽烟的母亲可能会担心母乳喂养的安全性。

在接下来的部分，我们就会为批判性思维者探讨这些教训的启示。

寻找替代原因

寻找替代原因就像做一名出色的侦探。当你发现可能存在替代原因的种种情况，你就要问一下自己这些问题：

- 我能不能想到这个证据的其他解读方式？
- 还有什么别的因素可能会引发这个行动或者导致这些发现？
- 如果我从另一个角度来看这个事件，那么我可能会把哪些因素当成重要的原因？
- 如果这个解读是不正确的，还有什么别的解读可以说得通？

唯一的原因还是原因之一

小学学龄儿童中抑郁症的发病率有了惊人的升高。现场访谈节目的主持人开始采访各路专家，问他们引发这一现象的（唯一）原因。他们可能说是遗传造成的，可能说是同龄人中流行的取笑戏弄造成的，可能说是父母的疏忽大意造成的，可能说是电视新闻里太多的恐怖主义和战争信息造成的，可能说是缺乏宗教信仰造成的，可能说是压力过大造成的。专家们可能宣称他们有答案，但他们不太可能真的知道。这是因为人们常犯的一个错误就是想找出一件事的简单、唯一的原因，而事实上这个"原因"是许多共同起作用的原因结合在一起的产物，即这些原因共同起作用，创造了事件发生所需要的整体环境。比如说，很可能是许多原因以独特方式结合在一起的作用促成了前文例子中提到的校园血案的发生。

In this study, the researcher probably began with the hypothesis that breast-feeding *causes* health benefits for mothers, and she found evidence consistent with this hypothesis. But let us offer different, or rival, causes for the same findings.

1. Women who breast-feed may simply on average lead more healthful lives than those who do not. For example, they may exercise more, or eat differently, than women who do not breast-feed.
2. Women who choose not to breast-feed may work outside the home more hours, possibly causing more life stress and thus incurring more health problems.
3. Reasons that women reject breast-feeding may be related to having more health problems than those who choose to breast-feed. For example, mothers who are taking medications or are smokers may be concerned about the safety of breast-feeding.

In the following sections, we explore the implications of these lessons for the critical thinker.

DETECTING RIVAL CAUSES

Locating rival causes is much like being a good detective. When you recognize situations in which rival causes are possible, you want to ask yourself questions like the following:

- Can I think of any other way to interpret the evidence?
- What else might have caused this act or these findings?
- If I looked at this event from another point of view, what might I see as important causes?
- If this interpretation is incorrect, what other interpretation might make sense?

THE CAUSE OR *A* CAUSE

There is an alarming increase in the rate of depression among elementary-aged children in one particular local school. Talk show hosts begin to interview the experts about *the* cause. It is genetic. It is the prevalence of teasing among peer groups. It is parental neglect. It is too much TV news coverage of terrorism and wars. It is lack of religion. It is stress. The experts may *claim* to have the answer, but they are not likely to *know* it. That is because a frequently made error is to look for a simple, single cause of an event when "the" cause is really the result of a combination of many *contributory* causes—causes that help to create a total set of conditions necessary for the event to occur. For example, the impetus to commit mass murder likely results from unique combinations of many contributory causes.

在涉及人类特征或活动的情况下，多个原因共同起作用比单一原因起作用的情况出现得更加频繁。最好的因果解释常常结合了多种原因，只有这些原因共同作用才能导致某件事的发生。所以，针对现场访谈节目主持人的问题，专家给出的最佳答案是："我们不知道这类事件发生的确切原因，但是我们可以推测出促成这类事件发生的多种可能原因。"我们在寻找替代原因时，必须要记住：我们找出的任何一个单独的原因，都极有可能是导致事件发生的原因之一，而不是其唯一的原因。

对很多事件来说，除了多种原因共同起作用的可能性，我们还需要认识到，不同的人做出同样一个行为的原因可能完全不同。一个人的抑郁症可能主要由某种生理功能障碍引发，而另一个人的抑郁症则可能主要由一件压力很大的生活事件（例如痛失爱子）引发。我们必须要警惕过度概括。常常最合适的问题并不是什么原因引发了抑郁症，而是什么原因引发了这个人的抑郁症。不同的抑郁症病例，不同的大屠杀事件，常常有不同的诱因。

如果人们不能考虑到原因的复杂性，他们就犯了"过度简化因果关系谬误"。

⚠ **过度简化因果关系谬误：** 依赖并不足以解释整个事件的因果因素来解释一个事件，或者过分强调这些因素中的一个或多个因素的作用。

在某种意义上，几乎所有的因果解释都是过度简化，因此，面对他人提供的并不包含某件事的每一种可能原因的解释，你也得公平对待。不过，因果性的结论应该包括足够多的因果因素，以说服你相信它们并不是过度简化，或者写作者应该向你说明，他在结论中强调的因果因素仅仅是很多可能的促成原因之一，即只是一个原因，而不是唯一的原因。

将多种视角作为寻找替代原因的指导

当我们试图理解为什么人们会以某种的方式行事时，不同的视角或观点往往会影响我们选择何种原因加以考虑。我们选择去寻找的事物会影响我们能够看到的事物。我们熟悉的视角越多，在找到可能的替代原因方面就会越有创造力。例如，社会学家、社会工作者、心理学家、心理医生、生物学家、神经科学家、营养学家、环境主义者、警察、共和党人和民主党人，以及商人，他们在思考各种可能的原因时，可能会青睐不同类型的原因。你越能尽快熟悉多种不同的视角，就越能找到事件可能的替代原因。当你在作业中遇到各种

Multiple contributory causes occur more often than do single causes in situations involving the characteristics or activities of humans. The best causal explanation is often one that combines many causes that *only together* are sufficient to bring about the event. So, the best answer experts can give to the talk show hosts' question is, "We don't know *the* cause for such events, but we can speculate about possible causes that might have contributed to the event." Thus, when we are searching for rival causes, we need to remember that any single cause that we identify is much more likely to be a contributory cause than *the* cause.

In addition to the likelihood of multiple contributory causes for many events, we need to recognize the possibility of different people having very different causes for the same behavior. Thus, one person's depression may be primarily caused by some biological dysfunction while another person's may be primarily caused by a very stressful life event, such as the death of a child. We need to be wary of overgeneralizing. The appropriate question often is not what causes depression; it is what are the causes of depression for this particular individual? Different cases of depression and different cases of mass murder often have different contributing causes.

When people fail to consider the complexity of causes, they commit the *causal oversimplification fallacy*.

Fallacy: Causal Oversimplification: Explaining an event by relying on causal factors that are insufficient to account for the event or by overemphasizing the role of one or more of these factors.

In some sense, almost all causal explanations are oversimplifications; therefore, you have to be fair to those who offer explanations that do not include *every* possible cause of an event. Causal conclusions, however, should include sufficient causal factors to convince you that they are not too greatly oversimplified, or the author should make clear to you that the causal factor she emphasizes in her conclusion is only one of a number of possible contributing causes—*a* cause, not ***the*** cause.

MULTIPLE PERSPECTIVES AS A GUIDE TO RIVAL CAUSES

Different perspectives or points of view influence our choices of causes to consider when we are trying to understand why people behave the way they do. What we choose to look for influences what we are able to see. The more perspectives with which we are familiar, the more creative we can be in generating possible rival causes. For example, sociologists and social workers, psychologists and psychiatrists, biologists, neurologists, nutritionists, environmentalists, police officers, Republicans and Democrats, and business people each are likely to favor different kinds of causes when they contemplate possible causes. The more familiar you can become with multiple perspectives,

不同的视角时,一定要努力让自己尽快熟悉各种可能的原因。同时,在你努力找到各种原因时,要警惕专家和你都具有的一个倾向——沉浸在"确认偏误"当中,只努力寻找并依赖和我们的信念相一致的证据。

混淆相关性和因果关系

我们生来就有一种倾向:把相互之间有关系的事件,或者一起发生的事件,"看"成彼此之间有因果关系的事件。也就是说,我们会得出如下结论:因为特征X(如吃的能量棒的数量)与特征Y(如在一场体育比赛中的表现)之间有关系,我们就说X导致了Y。

如果我们这样思考问题,那么我们常常会大错特错。为什么?通常情况下有多种假设可以解释为什么X和Y两者同时发生。实际上,至少有四种不同的假设可以解释任何一种这样的关系。知道这些假设是什么可以帮助我们找到替代原因。我们举一个研究的例子来看看这四种假设分别是什么。

> 最近的一项研究指出"吸烟可抵抗流感"。研究人员分析了525个烟民,结果发现67%的烟民在过去三年里从没有得过一次流感,他们推测说香烟燃烧时产生的尼古丁杀死了流感病毒,让它无法传播并引发疾病。

在身体微恙的人纷纷开始吸烟来预防流感侵袭之前,他们应该仔细考虑下面四个对于这项研究发现的可能解释。

- 解释一:X是Y的一个原因。(吸烟确确实实杀死了流感病毒。)
- 解释二:Y是X的一个原因。(从来不受流感病毒侵扰的人更有可能继续吸烟。)
- 解释三:X和Y有关系是因为第三种因素——Z。(吸烟和不患流感都是由相关因素引起的,例如吸烟以后经常洗手,这种行为进而阻碍了流感病毒的传播。)
- 解释四:X和Y相互影响。(不常患感冒的人有吸烟的倾向,而吸烟有可能影响到一些潜在的疾病。)

记住:相关性或关联性并不能证明因果关系!

大多用来证明因果关系的证据都仅仅建立在相关性或关联性的基础之上。当你发现一个写作者指出某两者之间有联系,以此来支持他的一种假设,一定记住要问一问:"有没有可以解释这种联系的其他原因?"你可以用下面这个研究来考一考自己。

the more you will be able to generate possible rival causes for events. As you encounter varied perspectives in your course work, strive to expand your familiarity with possible causes. Also, when striving to identify causes, be wary of the tendency of experts and yourself to engage in the confirmation bias tendency to seek and rely on only that evidence that is consistent with what we already believe.

CONFUSING CAUSATION WITH ASSOCIATION

We have an inherent tendency to "see" events that are associated, or that "go together," as events that cause one another. That is, we conclude that because characteristic X (e.g., amount of energy bars consumed) is associated with characteristic Y (e.g., performance in an athletic event), X causes Y.

When we think this way, we are, however, often very wrong. Why? Usually multiple hypotheses can explain why X and Y "go together." In fact, there are at least four different kinds of hypotheses to account for any such relationship. Knowing what these are will help you discover rival causes. Let's illustrate each of the four with a research example.

> A recent study reported that "smoking combats flu." The researchers studied 525 smokers and found that 67 percent of the smokers did not have the flu once over the last three years and hypothesized that the nicotine in the smoke from cigarettes destroys the flu virus before it can spread and cause sickness.

Before people who are feeling sick start smoking to prevent the onset of the flu, they should consider each of the following four potential explanations for the research findings.

> Explanation 1: *X is a cause of Y.* (Smoking does indeed kill the flu virus.)
>
> Explanation 2: *Y is a cause of X.* (Being free from viruses makes it more likely that people will keep smoking.)
>
> Explanation 3: *X and Y are associated because of some third factor, Z.* (Smoking and being without the flu are both caused by related factors, such as frequent washing of the hands after smoking, which in turn prevents the spread of the flu virus.)
>
> Explanation 4: *X and Y influence each other.* (People who do not usually catch the flu have a tendency to smoke, and the smoke may affect some potential illnesses.)

> *Remember:* Association or correlation does not prove causation!

Yet most evidence used to prove causation is based only on association or correlation. When an author supports a hypothesis by pointing to an association between characteristics, always ask, "Are there other causes that explain the association?" Test yourself on the following study.

最近一项研究指出"冰淇淋会导致犯罪"。研究人员研究了美国十大城市过去五年的冰淇淋销量和犯罪率，结果发现随着冰淇淋销量的增长，犯罪率也呈现出上升的趋势。他们由此推测吃冰淇淋诱发人脑中的一种化学反应，增加了人们的犯罪倾向。

我们希望你现在能明白，吃冰淇淋的人根本无须担心他们马上就要以身试法变成罪犯。你能想到哪些替代原因？夏天逐年升高的气温是否能解释冰淇淋销量（X）和犯罪率（Y）二者间的联系？

混淆因果关系和相关性这一现象可以理解，但非常危险。虽然原因确实先于结果出现，但先于结果出现的还有很多种其他因素，其中很多都不是引发结果的原因。

仔细分析上面提到的对事件之间如何产生联系的四种可能的解释，你现在应该能够辨认出两种常见的因果论证谬误："因果混淆谬误"和"忽略共同原因谬误"。

⚠ **因果混淆谬误**：将事件的起因和结果相混淆，或未能认识到两件事之间可能是相互影响的关系。

⚠ **忽略共同原因谬误**：未能认识到两件事之间之所以有联系，是因为第三种因素在起作用。

混淆"在此之后"与"因此"

我们常常按照这样的模式解释事件：因为 B 事件发生在 A 事件之后，所以 A 事件引发了 B 事件。这样的论证之所以会发生，是因为人类都有一种强烈的倾向，愿意相信如果两件事前后紧随着发生，那么第一件事肯定导致了第二件事。

但是很多事件紧随在其他事件之后发生，却并不是由前面的事件所引发。如果我们错误地得出结论说第一件事引起第二件事，因为它发生在前，我们就犯了"事后归因谬误"（拉丁文的意思是"在此之后，所以是因此"）。例如，你可能在写出一篇极出色的论文时戴了某一顶帽子，所以现在你每次写论文都坚持非要戴同一顶帽子不可。

⚠ **事后归因谬误**：假设 B 事件是由 A 事件所引发的，仅仅因为 B 在时间上紧随 A 之后发生。

> A recent study reported that "ice cream causes crime." Researchers studied ice cream sales and crime rates over the last five years in the ten largest U.S. cities and found that as ice cream sales increase, so does the crime rate. They hypothesized that eating ice cream triggers a chemical reaction in one's brain that results in an inclination toward crime.

We hope you can now see that ice cream eaters need not be concerned that they are about to commit a crime. What rival causes did you think of? Couldn't the increased summer heat account for the association between ice cream sales (X) and crime (Y)?

This confusion between correlation and causation is as understandable as it is dangerous. A cause will indeed precede its effect. But many things preceded that effect. Most of them were not causal.

You should now be able to identify two common causal reasoning fallacies, *confusion of cause and effect fallacy* and *neglect of a common cause fallacy*, by attending to the four possible explanations of why events might be associated that we mentioned earlier.

Fallacy: Confusion of Cause and Effect: Confusing the cause with the effect of an event or failing to recognize that the two events may be influencing each other.

Fallacy: Neglect of a Common Cause: Failure to recognize that two events may be related because of the effects of a common third factor.

CONFUSING "AFTER THIS" WITH "BECAUSE OF THIS"

Often, we try to explain a particular event as follows: Because event B *followed* event A, event A *caused* event B. Such reasoning occurs because human beings have a strong tendency to believe that if two events occur close together in time, the first one must have caused the second one.

Many events that occur after other events in time are not caused by the preceding events. When we wrongly conclude that the first event caused the second because it preceded it, we commit the *post hoc, ergo propter hoc* (meaning "after this, therefore because of this") *fallacy*, or, for short, the *post hoc fallacy*. For example, you may have written an excellent paper while wearing a particular hat, so now you always insist on wearing the same hat when you write papers.

Fallacy: Post Hoc: Assuming that a particular event, B, is caused by another event, A, simply because B follows A in time.

下面这个例子进一步阐释了这种论证方式带来的问题。

> 我昨天找到的那个25美分硬币肯定是我的幸运币。找到它以后，我在一门特别难的课程考试考了个A，我最不喜欢的一门课停了一次，而我最喜欢看的电影昨天晚上在电视上播出了。（不必在意我为了准备考试埋头苦读，教授六岁的孩子最近得了流感，电视节目早在我找到硬币之前就已经定好了。）

你也许能够猜到，政治领导人和商界领袖都喜欢使用事后归因这种论证，特别是当这对他们有利的时候。比如说，他们喜欢把自己走上领导岗位之后发生的一切好事都揽到自己头上，而把发生的一切坏事都推给别人。

记住：一件事紧接在另一件事之后发生这一发现本身并不能证明两者之间有因果关系，这可能只是一个巧合。当你看到这种论证方式的时候，你要问问自己"有没有替代原因能解释这个事件"以及"除了一件事紧随在另一件事之后发生这个事实以外，还有没有什么其他过硬的证据"。

解释单个事件或行为

是什么原因引起2010年冰岛火山爆发？为什么社交媒体这样流行？

和我们对桑迪胡克小学校园血案提出的问题一样，这些问题都在寻求对单个历史事件的解释。首先，就像我们在桑迪胡克小学校园血案中见到的那样，对于同一事件，有多种不同版本的故事都能说得通。其次，我们解释事件的方式深受各种社会力量和政治力量的影响，同时还受到和信念有关的个人视角的影响。我们还受到一种常见的偏见——基本归因错误的影响，这种错误指我们在解释他人的行为时普遍高估了个人倾向的重要性而低估了环境因素的作用。也就是说，我们总喜欢认为别人的行为来自其内部因素的作用（他们个人的性格特点），而不是来自外部因素的作用（环境的力量）。因此，当发现有人偷窃时，我们很可能将偷窃行为一下子就看成小偷骨子里没廉耻或没良心的结果，或认为他们做了糟糕的选择。然而，我们还应该考虑一下外部环境的作用，考虑是否贫困或者来自同龄人的压力等因素也发挥了作用。

构建过去事件的各种原因还存在一个重大的困难，即很多证据都依赖于人们的记忆，而大量研究显示记忆常常会出现极大的扭曲。

我们怎么知道我们是不是有了对某件事或某些事的合理解释呢？我们永远也不可能有百分百的把握。但是通过问一些批判性的问题，我们可以取得一些进步。

The following example illustrates the problem with this kind of reasoning.

> The quarter I found yesterday must be lucky. Since I have found it, I got an A on a really hard test, my least favorite class was canceled, and my favorite movie was on TV last night. (Never mind the fact that I studied really hard for my test, my professor has a 6-year-old who recently had the flu, and the TV schedule is made far in advance of my finding a quarter.)

As you might guess, political and business leaders are fond of using the post hoc argument, especially when it works in their favor. For example, they tend to take credit for anything good that takes place after they assumed their leadership role and to place blame elsewhere for anything bad that happens.

Remember: The finding that one event follows another in time does not by itself prove causation; it may be only a coincidence. When you see such reasoning, always ask yourself, "Are there rival causes that could account for the event?" and, "Is there any good evidence other than the fact that one event followed the other event in time?"

EXPLAINING INDIVIDUAL EVENTS OR ACTS

What caused the 2010 volcanic eruptions in Iceland? Why is Facebook so popular?

Like our question about the Sandy Hook massacre, these questions seek explanations of individual historical events. First, as we saw in the Sandy Hook case, so many different stories for the same event can make sense. Second, the way we explain events is greatly influenced by social and political forces, as well as by individual perspectives regarding beliefs. Also, a common bias is the *fundamental attribution error*, in which we typically overestimate the importance of personal tendencies relative to situational factors in interpreting the behavior of others. That is, we tend to see the cause of others' behavior as coming from within (their personal characteristics) rather than from without (situational forces). So, for example, when people steal, we are likely to view the stealing initially as a result of their tendency to be immoral, to their lacking a conscience, or to the bad choices they made. However, we should also consider the role of outside circumstances such as poverty or peer pressure.

A further major problem in constructing the causes of past events is that much evidence relies on the memories of people, and abundant research suggests that memories are often greatly distorted.

How can we know whether we have a "good" explanation for a particular event or set of events? We can never know for sure. But we can make some progress by asking critical questions.

Be wary of accepting the first interpretation of an event you encounter. Search for rival causes and try to compare their credibility. Consider other perspectives from which the event of interest might be viewed. Read multiple

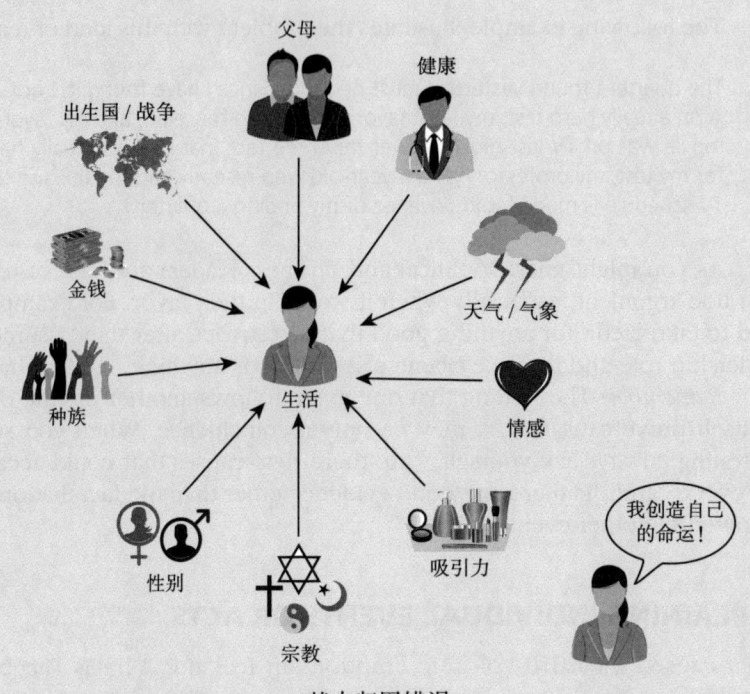

基本归因错误

一定要当心，千万不要贸然接受你所遇到的事件的第一个解释。要寻找替代原因，并努力去比较不同原因的可信度。要考虑采取看待同一事件的不同视角。阅读事件的多种不同叙述版本，以帮助你扩大见解的范围。我们必须接受这一事实：很多事件并非只有一种简单的解释。

评价替代原因

你想出来的替代原因越显得言之有理，你对刚开始见到的那个解释的信心就越会打折扣，至少在获得进一步的证据以供认真考量之前是这样的。作为一个批判性思维者，你必须竭尽所能地评估各种不同解释，看看每一种解释与当前的证据的契合度，尽量对你的个人偏见保持敏感和警惕。

替代原因与你的表达和交流

因果论证对于写作者而言是最难写的论证之一，你必须要筛查大把可能存在的原因，其中一些货真价实，而另一些则可能以假乱真。然后你必须要展示一种实实在在的因果关

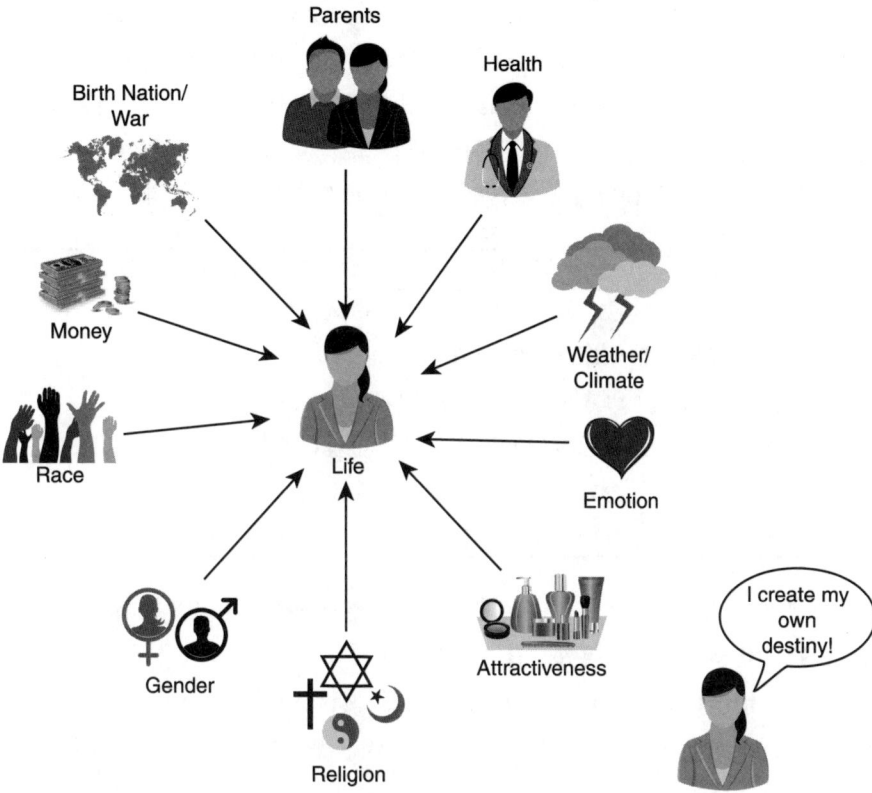

Fundamental Attribution Error

versions of events to help expand your viewpoints. We must accept the fact that *many* events do not have a simple explanation.

EVALUATING RIVAL CAUSES

The more plausible the rival causes that you come up with, the less faith you can have in the initial explanation offered, at least until further evidence has been considered. As a critical thinker, you have to assess as best you can how each of the alternative explanations fits the available evidence, trying to be sensitive to your personal biases.

RIVAL CAUSES AND YOUR OWN COMMUNICATION

Causal arguments are among the most difficult for writers to construct. You have to sift through a bunch of possible causes, some that are legitimate and others that are falsely attractive. Then you must show that an actual causal relationship exists. This problem is illustrated in a classic clip of PBS's *Sesame*

系的存在。这个问题在美国公共广播公司的《芝麻街》的一个经典桥段中得到了展现，在这个桥段里木偶伯特发现阿尼把一根香蕉举到耳朵边。伯特问他为什么有这样奇怪的举动，阿尼回答说："听着，伯特，我是用这根香蕉来驱赶短吻鳄的。"有点生气的伯特指出芝麻街上压根就没有短吻鳄，阿尼骄傲地回答说："对啊，这根香蕉无意间起了大作用，是不是，伯特？"阿尼错误地推断两个同时发生的事件相互之间有关系。

展示 9-1　比较原因时，使用下列标准

✓ 它们在逻辑上的合理性。哪些原因对你而言最说得通
✓ 它们和你所学的其他知识之间的一致性
✓ 它们以前在解释或预测类似事件上的成功率
✓ 和其他解释相比，大量已接受的事实支持一个解释的程度
✓ 和其他解释相比，是否只有较少已接受的信念不支持一个原因
✓ 和其他解释相比，它解释大量以及各种类型的事实的效力

在你证明了某种关系确实存在之后，接下来你必须要说明这种关系会朝着你提出的那个方向发展。也就是说，A 导致了 B，而不是 B 导致了 A，或者 C 导致了 A 和 B。也不能是另外的什么情况，例如在 J. K. 罗琳的《哈利·波特》系列著作中，作者再现了那个有关因果先后方向的经典的鸡和蛋之谜："凤凰和火焰，哪个先有？"卢娜·洛夫古德——小说主角们的一个古怪的朋友，正确地回答道："一个圆圈根本就没有起点。"

最后，你也许想说明你关注的因果关系在解释这个现象时比其他的替代解释效力更强。这整个过程可能很难一下子实现，我们建议你将其拆成一个个小步骤。第一步就需要用到一些创造性思维。

发掘潜在的原因

开始关于因果关系的写作时，你要做的和写其他的论证时一样。先选定一个自己感兴趣的论题。在这种情况下，你寻找的论题主要探索的是因果关系。这样的论题可能明确提到"原因"这样的词语，例如"美国 AMC 有线电视台播放的《行尸走肉》的观众人数打破了有线电视台的收看记录，这是什么原因造成的"或者"是什么原因造成疾病对治疗有了一定的抵抗力"。同样，这个论题也可以明确使用"后果"这个词语："勒布朗·詹姆斯决定离开这座城市前往迈阿密热火队效力，这可能为克利夫兰的经济带来怎样的后果？"

在选定了一个论题后，下一步就是竭力思考这个问题可能的答案。这个过程可以充满创造力。处理这个任务的一种最佳方法就是像一个五岁的淘气小孩一样一直问个不停，

Street, during which the Muppet Bert found Ernie holding a banana to his ear. Bert asked him about his peculiar behavior and Ernie responded, "Listen, Bert, I use this banana to keep the alligators away." An irritated Bert pointed out that there are no alligators on Sesame Street, to which Ernie proudly replied, "Right. It's doing a good job causally, isn't it, Bert?" Ernie mistakenly reasoned that two simultaneous events were related.

EXHIBIT 9-1   In Comparing Causes, Apply the Following Criteria

✓ Their logical soundness. Which ones make the most sense to you.
✓ Their consistency with other knowledge that you have.
✓ Their previous success in explaining or predicting similar events.
✓ The extent to which the explanation is implied by a greater variety of accepted truths than other explanations.
✓ The extent to which it has been disconfirmed by fewer accepted beliefs.
✓ The extent to which it explains a larger number and variety of facts than competing explanations.

After you prove that a relationship exists, you must then demonstrate that the relationship moves in the direction you suggest. That is, A caused B, not B caused A, or C caused both A and B. Or something else entirely—in J.K. Rowling's *Harry Potter* series, the author recreates the classic chicken and the egg riddle about causal direction as "Which came first, the phoenix or the flame?" Luna Lovegood, a whimsical friend of the main characters, answered correctly: "A circle has no beginning."

Lastly, you may want to demonstrate that the causal relationships you focused on explain the phenomenon better than the alternatives. This entire process can be overwhelming. We suggest you break it down into steps. The first of these steps involves some creative thinking.

Exploring Potential Causes

You start the writing process like any other argument. You decide on a particular issue that interests you. In this instance, you are looking for an issue that explores causality. Such an issue may mention the term *cause* explicitly, such as "What were the causes of AMC's *The Walking Dead*'s record-breaking viewership on cable television?" or "What causes diseases to become resistant to treatment?"

Once you decide on an issue, your next step is to brainstorm potential answers to the question. The process can be a creative one. One excellent way to approach the task is to adopt the questioning attitude of an annoying 5-year-old. Keep asking *why*. Let's return to our example of AMC's *The*

不断追问为什么。我们回到《行尸走肉》这部电视剧的例子来做一下说明。"为什么《行尸走肉》能打破有线电视台的收视纪录?""也许是因为18～49岁的人喜欢看僵尸题材。"让我们采取五岁儿童的态度一路追问:"他们为什么喜欢僵尸题材?"你会怎么回答这个问题?

> **学会的教训**
>
> 1. 很多类型的事件都可以由各种替代原因来解释。
> 2. 专家可能检查同一个证据并发现不同的原因来对它加以解释。
> 3. 大部分立论者都只提供那些他们偏好的原因,具备批判性思维的读者或听众必须自己找出替代原因。
> 4. 想出替代原因是个创造性的过程,通常情况下,这类原因不会一目了然。
> 5. 最后,一个因果断言的确定性和言之成理的替代原因的数量成反比。因此,找到多个替代原因,可以让批判性思维者获得适当的理智上的谦逊。

我们心中的"小孩"接下来可能还会问什么问题?"还有什么原因呢?"因为《行尸走肉》填补了任何一家其他电视台都不曾填补过的空白。"还有什么原因呢?"因为表演、剧本和导演都完美无缺。"还有什么原因呢?"

你现在该明白了。朋友、同学以及生活中的其他人在你开动脑筋的时候都可以帮助你。他们可能想出你从来就没有想到过的原因。

思维体操

下面每个例子都提供了支持一个因果断言的论证。请尽量为这些断言想出替代性的原因。然后看一看通过了解这些替代原因,你在多大程度上削弱了作者原来的断言的分量。

第一篇

父母没有大学学历的孩子更容易贫穷吗?为了找到答案,研究人员最近对552名接受政府援助的人进行了抽样调查,看看有多少人会说他们的父母中一方或双双没有大学学历。取样是在自愿原则下进行的,样本范围包括俄亥俄州、肯塔基州和西弗吉尼亚州接受政府援助的人员。调查显示,85%的受访者的父母中至少有一方没有大学学历。研究人员还随机调查了这三个州中552个没有接受政府援助的人。在这一样本中,只有40%的人声称他们父母中至少有一方没有大学学历。

Walking Dead to demonstrate. Why did *The Walking Dead* break cable records? Well, maybe because 18- to 49-year-olds like zombies. Enter the 5-year-old's attitude: "Why do they like zombies?" How would you answer that question?

> **Lessons Learned**
>
> 1. Many kinds of events are open to explanation by rival causes.
> 2. Experts can examine the same evidence and discover different causes to explain it.
> 3. Most communicators will provide you with only their favored causes; the critical reader or listener must generate rival causes.
> 4. Generating rival causes is a creative process; usually, such causes will not be obvious.
> 5. Finally, the certainty of a particular causal claim is inversely related to the number of plausible rival causes. Hence, identifying the multiple rival causes gives the critical thinker the proper sense of intellectual humility.

What might our inner youngster ask next? "Why else?" *The Walking Dead* filled a niche that no other network filled. "Why else?" The acting, writing, and directing were well executed. "Why else?"

You get the picture. Your friends, classmates, and other people in your life can help you during the brainstorming stage. They might think of a cause that had not crossed your mind.

PRACTICE EXERCISES

Each of the following examples provides an argument to support a causal claim. Try to generate rival causes for such claims. Then try to determine how much you have weakened the author's claim by knowledge of rival causes.

Passage 1

Are the children of parents without college degrees more likely to be poor? To find out, researchers recently sampled 552 people who were on government assistance to see how many would say that one, or both, of their parents lacked college degrees. The sample and results were obtained by sending out a voluntary survey to people on government assistance in the states of Ohio, Kentucky, and West Virginia. The survey revealed that 85 percent of the respondents had at least one parent who lacked a college degree. The researchers also randomly surveyed 552 people who did not receive government assistance in the same three states. Of this sample, only 40 percent claimed that at least one parent lacked a college degree.

第二篇

为什么这位公司高管要从自己的企业里偷钱呢？细细查看一下他的生活就可以找到清晰而令人信服的答案。这个高管来自一个非常成功的家庭，父母都是医生，兄弟姐妹都是律师。作为公司高管，他挣的钱不如家人多。同时，他坚信美国梦和以下思想：一个人只要努力工作，最终总会成功。但是，尽管工作十分卖力，但他最近还是经历了许多生意上的挫败，包括在股市里赔掉相当一大笔钱。更糟糕的是，他的孩子需要做手术。为了不让家人失望，做个成功人士，为家庭提供稳定的收入，这位高管不得不从自己的公司里偷钱。

第三篇

大学校园里日益增加的细菌数量导致大学生的发病率不断攀升。大学生不太可能在校园的生活区和公共区域里进行消毒，这导致大量的细菌依附在物体表面，飘散到空中，导致更多的学生生病。

给个提示

第一篇

结论：父母没有大学学历的孩子比父母有大学学历的孩子更容易贫穷。

理由：接受政府援助的人比没有接受政府援助的人更多报告说，他们父母中至少有一方没有大学学历。

请注意，这里呈现的结果只来自一项研究，并未提及这些结果具有多大的典型意义。我们也不知道这些信息是在哪里发表的，因此我们无法评估该研究在正式发表之前是否经过了严格的审稿程序。

不过，我们可以问一些关于此项研究的问题，这些问题对我们很有帮助。样本量相当大，但其范围广度令人质疑。虽然对多个州进行了抽样调查，但这些州接受政府援助的人能在多大程度上代表全美国贫困人口？例如，不同的州对一个人在寻求政府援助之前必须满足的条件有不同要求。此外，依靠政府援助的贫困人口与没有寻求援助的贫困人口相比较，情况如何？

Passage 2

Why did the corporate executive steal funds from his business? A close look at his life can provide a clear and convincing answer. The executive comes from a very successful family where his parents are doctors and his siblings are lawyers. As a corporate executive, he was not making as much money as his family members. Also, the executive believes heavily in the American dream and the idea that if one works hard enough that person will succeed. However, despite his hard work, the executive has had a number of recent business failures, including losing a substantial sum of money in the stock market. To make matters worse, his children need braces. To live up to expectations, become a success, and provide for his family, the executive had to steal the money from his business.

Passage 3

Increased amounts of germs and bacteria on college campuses cause higher rates of illness in college students. College students are less likely to sanitize living areas and common areas on campus, which in turn creates excessive germs on surfaces and in the air, leading to more sickness in students.

Sample Responses

Passage 1

CONCLUSION: *The children of parents without college degrees are more likely to be poor than the children of parents with college degrees.*

REASON: *More people on government assistance reported having at least one parent without a degree than those not on government assistance.*

Note that the results presented are from one study without reference to how typical these results are. We also do not know where this information was published, so we can make no assessments regarding how rigorously the study was reviewed before publication.

However, we can ask some useful questions about the study. The sample size was quite large, but its breadth is questionable. Although multiple states were sampled, to what extent are the people using government assistance programs in these states typical of poor people across the nation? For example, different states have different requirements a person must meet before a person can seek government assistance. Also, how do poor people on government assistance compare to poor people who have not sought help?

也许抽样中最重要的问题在于缺乏随机样本。虽然没有享受政府援助的人是在三个州中随机选择的，但接受政府援助的人是在自愿的基础上选择的。自愿参加调查的人和没有自愿参加的人是否有很大区别？例如，有可能男性参加调查的可能性比女性高80%。这就使得样本将不成比例地偏向男性，使样本无法准确地代表广大贫困人口。研究人员必须向我们提供更多关于样本的信息，以让我们确信调查取样是没有偏差的。

测量评定准确性如何？首先，除了接受政府援助这一信息，调查没有给出穷人的定义。但是人们接受政府援助的原因有很多。例如，医疗保险可以被视为政府援助，在美国，只要年满65岁并有资格享受社会保障，就有资格享受医疗保险。因此，我们不仅不确定抽样调查的人群是否接受了公平取样，也不确定参与者是否真的贫困。

同样值得质疑的是，所谓的不贫困人群对照组的选取依据是自我评估。我们知道不贫困是受到社会赞许的，人们知道后往往会给出受社会赞许的答案。此外，如前所述，有人可能很穷，但他们不寻求帮助。如果一些对照组参与者其实很贫穷，但他们没有报告自己的贫困状况，那么这种情况可能会使对照组进一步失之偏颇。在我们对结论抱有信心之前，我们要对这些评定的准确性知道更多。

第二篇

结论：高管从自己的公司偷钱是为了和自己家人竞争，显示他不是失败者，同时也为了养家糊口。

理由：高管很可能关心上面提到的所有因素。

有可能上面所有的因素都是这位高管从自己公司偷钱的重要原因。但是社会上还有很多其他人士肩上也背负同样的压力，他们却没有诉诸非法手段来获得钱财。有没有其他可能的原因导致这位高管的这种行为？就像我们在恐怖行为的案例中看到的一样，可能存在多种言之成理的解释。例如，我们想要多了解他的童年，了解他生活中最近发生的事。

Perhaps the most important sampling problem was the lack of a random sample. While the people who did not use government assistance were randomly selected in the three states, the respondents who were on government assistance were selected on a voluntary basis. Do those who volunteered to take the survey differ greatly from those who did not volunteer? For example, it's possible that males were 80 percent more likely to take the survey than females were. This would make the sample disproportionately male and prevent the sample from accurately representing the poor population at large. The researchers would have to provide us with more information about the sample to assure us that it was not biased.

How accurate are the rating measurements? First, no definition for *poor* is given beyond those answering the survey who were on government assistance at the time. But people use government assistance for a variety of reasons. For example, Medicare could be considered government assistance, and one can become eligible for Medicare simply by being over 65 and being eligible for Social Security. Thus, not only are we uncertain of whether the population was fairly sampled, but we are also uncertain if a participant was actually poor.

Also questionable is the fact that the selection of the supposed control group of people who were not poor is based on self-assessment. We know that there is a socially acceptable answer of not being poor, and people tend to give socially acceptable answers when they know them. Also, as previously mentioned, it's possible for someone to be poor and not seek help. This occurrence could further slant the control group if any of the participants were indeed poor but did not report themselves as poor. We would want to know more about the accuracy of these ratings before we could have much confidence in the conclusion.

Passage 2

CONCLUSION: *The executive stole money from his company to compete with his family members, to show that he is not a failure, and to provide for his family.*

REASON: *The executive was probably concerned with all of the above elements.*

It is possible that all the above factors were important in causing the corporate executive to steal from his company. But many other people in society have the same pressures put upon them, and they do not resort to illegal means to obtain money. Are there other possible causes for such behavior? As in the case of any act of terrorism, there may be many alternative plausible explanations. For example, we would want

- 这位高管最近有没有和老板吵架？
- 他有没有服药？
- 他最近有没有承受高强度压力的经历？
- 他有没有偷窃的历史？

通过事后观察，我们常常可以发现儿童时期的经历能部分解释成人的行为。但在我们做出因果结论之前，必须要寻找更多证据来证明是一系列事件引起了另一系列事的发生，仅仅一系列事件先于另一系列事件发生这样一个简单的事实是不够的。我们还必须要小心不被基本归因错误误导，确保自己在考虑内部因素的同时，也考虑外部因素。

to know more about his childhood and more about recent events in his life.

- Has the corporate executive had any recent disagreements with his boss?
- Has he been using drugs?
- Has he had any recent highly stressful experiences?
- Does he have a history of stealing?

After the fact, we can always find childhood experiences that make sense as causes of adult behavior. Before we draw causal conclusions, however, we must seek more evidence to prove that one set of events caused the other than the mere fact that one set of events preceded the other set. We must also be wary not to fall victim to the fundamental attribution error and be certain to consider external causal factors, as well as internal ones.

第 10 章

数据有没有欺骗性

| 学 习 目 标 |

1. 认识到有偏差的和不知来历的统计数据的危险。
2. 加强对各种形式的平均值的重要性的理解。
3. 了解测量误差的危险。
4. 认识到一个使用统计数据的人得出的结论可能会与统计数据本身显示的情况大相径庭。

下面这篇简报能在多大程度上说服你?

新闻简报:经济获得了长足发展。仅上个月我们的失业率就下降了一个百分点。

你不该被上面的论证打动。这个论证很可能用数据欺骗了我们!

写作者给出的证据当中最为常见的一种就是"统计数据"。你可能经常听到人们使用下面这句话来支撑他们的论证:"我有统计数据来证明。"我们使用统计数据(通常以不合适的方式)来揭示战争伤亡人数的增加或减少,提醒公众注意发病率的变化,测量一种新

CHAPTER 10

Are Any Statistics Deceptive?

LEARNING OBJECTIVES

1. Recognize the danger of biased and unknowable statistics.
2. Increase understanding of the importance of alternative forms of averages.
3. Become aware of the dangers of measurement errors.
4. Recognize that a person using a statistic may conclude something quite different from what the statistic itself suggests.

How much should you be persuaded by the following passage?

> News bulletin: The economy is greatly improving. Last month alone our unemployment rate decreased by 1 percent.

You should not be very impressed by the above reasoning. The argument might *deceive us with statistics!*

One of the most frequent kinds of evidence that authors present is "statistics." You have probably often heard people use the following phrase to help support their argument: "I have statistics to prove it." We use statistics (often inappropriately) to reveal increases or decreases in war casualties, to alert the public to changing disease rates, to measure the sales of a new product, to

产品的销量，判断某只股票的前景，确定下一张牌是 A 的概率，衡量不同大学的毕业率，记录不同年龄段的人性生活的频率，并为很多其他问题提供信息。

统计数据就是用数字表达出来的证据。这样的证据可能看起来非常有说服力，因为数字让证据显得非常有科学性，非常精确，似乎它就代表了"事实"。但是，统计数据可能（而且经常会）撒谎！它们并不一定能证明它们想要证明的观点。

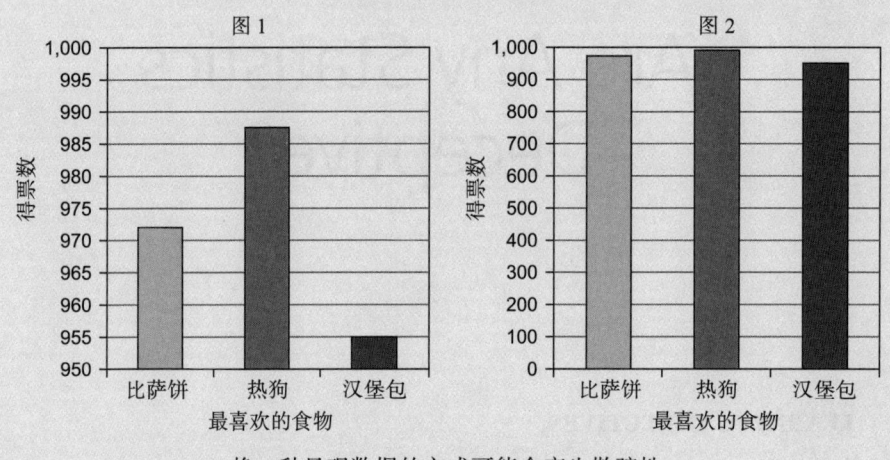

换一种呈现数据的方式可能会产生欺骗性

⚠ **注意**：统计数据可能（而且经常会）撒谎。它们并不一定能证明表面上证明的观点。

作为一个批判性思维者，你应该努力辨别错误的统计数据式的论证。我们无法在短短几页中向你全面展示人们"用统计数据撒谎"的所有不同方法。不过，本章将为你提供一些基本策略，你可以用这些策略来发现其中骗人的小伎俩。同时，本章还通过展示写作者错误使用统计数字来当证据的最常见的方法，以提醒你注意数据论证中存在的缺陷。

⚠ **批判性问题**：数据有没有欺骗性？

不知来历和有偏差的数据

> 最近的一个新闻标题：40% 的大学生饱受注意力缺陷障碍的折磨！

你发现自己有时候注意力不集中，你该不该担忧呢？你怎么知道自己可以相信眼前的统计数字？

judge the moneymaking capabilities of certain stocks, to determine the likelihood of the next card being the ace, to measure graduation rates for different colleges, to record the frequency of different groups having sex, and to provide input for many other issues.

Statistics are evidence expressed as numbers. Such evidence can seem quite impressive because numbers make evidence appear to be very scientific and precise, as though it represents "the facts." *Statistics, however, can, and often do, lie!* They do not necessarily prove what they appear to prove.

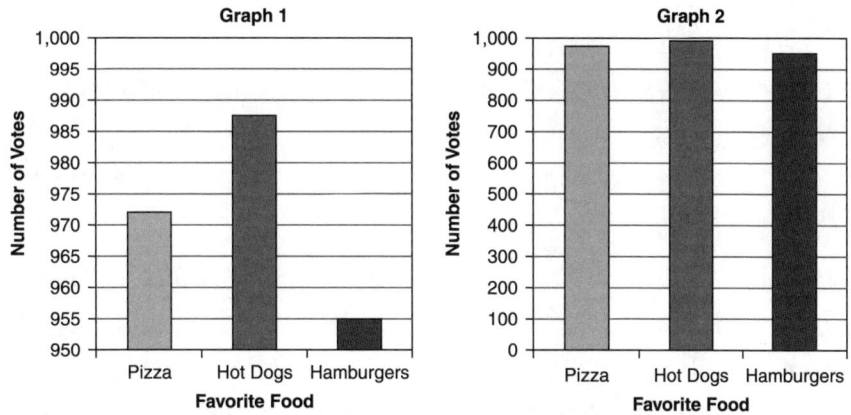

Alternative Ways to Provide Statistics That Potentially Deceive

 Attention: *Statistics can and often lie. They do not necessarily prove what they appear to prove.*

As a critical thinker, you should strive to detect erroneous statistical reasoning. In a few pages, we cannot show you all the different ways that people can "lie with statistics." However, this chapter will provide some general strategies that you can use to detect such deception. Also, it will alert you to flaws in statistical reasoning by illustrating a number of the most common ways that authors misuse statistical evidence.

 Critical Question: ***Are the statistics deceptive?***

UNKNOWABLE AND BIASED STATISTICS

Recent headline: 40 PERCENT OF COLLEGE STUDENTS SUFFER FROM ATTENTION DEFICIT DISORDER!

Should you be alarmed that you sometimes find your attention wandering? How do you know that you can trust that statistic?

任何统计都要求发生在某地的某些事件能被界定并准确识别出来，但是这样的要求常常得不到满足。因此，要找出欺骗性的数据，第一个策略就是尽可能多地找到关于这些数据如何采集的信息。我们能不能准确地知道美国到底有多少人在报税单上作弊，多少人每晚阅读他们喜爱的经文，多少人开车时打电话，或多少人使用违禁药品？

想象一下做这些统计的细节，你的答案肯定是"不太可能"。为什么？因为要为特定的目的得到精确的数据，你常常会遇到各种各样的拦路虎和绊脚石，其中包括关键词语的模棱两可，识别相关人员或事件的种种困难，人们不愿意提供真实的信息，不能准确报告各种事件，以及观察事件时种种身体上的障碍等。因此，统计数据往往是以有根据的推测的形式呈现的。这些估计有时候很有用，但它们也可能有欺骗性。记住总要问一问："作者是怎样得出这个估计的？"得到的细节越多越好。

不知来历的数据最常见的用处就是借助很大的数字让别人加深印象或警示别人，这些呈现出来的数字的精确性常常是可疑的。例如，一些报道可能会用很大的数字来提醒公众注意日见增长的生理或心理障碍的发生率，例如癌症、饮食障碍或幼儿多尿症等。如果我们能知道这些数据确定的过程有多谨慎，我们更能相信这些数据。例如，你的朋友报告说，他们对 90% 的邻居的访谈结果显示大多数人对住在他们的社区感到开心，此时，你必须非常小心，不要轻易相信这些结果反映了人们对其社区的真实感受。记住：在对这样的数据做出反应之前，我们先要问一问它们是怎么得来的。

令人困惑的平均值

请检查下面的陈述：

1. 一个快速致富的方法就是做一名职业橄榄球员，2015 年美国国家橄榄球联盟球星的平均收入是 220 万美元。

2. 为在大学里取得好成绩，学生需要付出的努力越来越少了。根据最近一项调查，大学生每周平均花在学习上的时间是 12.8 小时，和 20 年前大学生的学习时长相比，前者大概只有后者的一半。

两个例子当中都使用了"平均"这个词，但是实际上有三种不同的方法来测定平均值，而且在大多数情况下，每种方法都会给出不同的数值。

第一种方法是把所有数值相加，然后用总数除以相加的数值的数目。这种方法所得的结果就是平均数。第二种方法是将所有数值从高到低排列，然后找到位于最中间的数值，

Any statistic requires that some events somewhere have been *defined* and *accurately identified*, but these conditions are often not met. Thus, the first strategy for locating deceptive statistics is to try to find out as much as you can about how the statistics were obtained. Can we know precisely the number of people in the United States who cheat on their taxes, read their favorite sacred text each night, talk on their cell phones while driving, or use illegal drugs?

If you imagined the details of doing such counts, we suspect your answer is "Not likely." Why? Because there are a variety of obstacles to getting accurate statistics for certain purposes, including ambiguity of key terms, difficulties in identifying relevant persons or events, people's unwillingness to provide truthful information, people's failure to report events, and physical barriers to observing events. Consequently, statistics are often in the form of educated guesses. Such estimates can be useful; they can also be deceiving. Always ask, "How did the author arrive at the estimate?" The more detail you get, the better.

One common use of unknowable statistics is to impress or alarm others with large numbers, often presenting them with suspicious precision. For example, large numbers may be used to try to alert the public to the increasing incidence of physical or mental disorders, such as cancer, eating disorders, or childhood diabetes. We want to be most impressed by these numbers if we know how carefully they were determined. For example, when your friends report that their interviews with 90 percent of their neighbors determined that most people enjoy living in their neighborhoods, you must be very careful about believing that such results capture how people actually feel about their neighborhoods. *Remember:* Before reacting to such statistics, we need to ask how they were determined.

CONFUSING AVERAGES

Examine the following statements:

1. One way to make money fast is to become a professional football player. An **average** NFL football player made $2.2 million in 2015.
2. Making the grade in college classes is requiring less work for students; according to a recent survey, college students are studying an **average** of 12.8 hours per week, about half of the hours studied 20 years ago.

Both examples use the word "average." But there are three different ways to determine an average, and in most cases, each will give you a different value.

One way is to add all the values and divide this total by the number of values used. The result is the *mean*. A second way is to list all the values from highest to lowest, then find the one in the middle. This middle value is the *median*. Half of the values will be above the *median*; half will be below it. A third way is to list all the values and then count each different value or each range of values. The value that appears most frequently is called the *mode*, the third kind of average.

这个中间数值就是中位数。一半的数值在中位数之上，另一半在中位数之下。第三种方法是将所有数值排列好，计算每个不同数值出现的次数或每个不同数值范围出现的次数，出现频率最高的数值就叫作众数，这是第三种平均值。

展示 10-1　平均值的种类
√ 平均数　通过把所有数值相加然后用总数除以相加的数值的数目来测定
√ 中位数　通过将所有数值从高到低排列，然后找到位于最中间的数值来测定
√ 众数　通过计算不同数值出现的次数，然后找出出现频率最高的数值来测定

一个写作者所用的术语"平均值"谈论的是平均数、中位数还是众数？这会产生很大的区别。

在第一个例子中，哪一种平均值最能说明问题？请考虑一下职业化运动当中大牌球星的收入与一般球员收入的对比。最大牌的球星，比如说橄榄球明星四分卫，收入比球队里大部分其他球员要高出很多。事实上，2015年薪酬最高的橄榄球运动员年收入超过3500万美元——远远高于平均值。这样高的收入将会急剧拉高平均数，但是对于中位数或众数而言影响不大。举例来说，美国国家橄榄球联盟的球员2015年工资平均数是220万美元，但是其工资中位数却只有83万美元。因此，对于大部分职业运动，运动员工资平均数要比中位数或者众数高出很多。所以，如果有人想让工资水平显得非常非常高，他就会选择平均数作为平均值。

现在让我们来仔细看看第二个例子。如果这里列举的平均值是中位数或众数，我们就有可能低估了平均学习时间。有些学生很可能花了极多的时间学习，比如一周30或40个小时，这会提高平均数的数值，但是不影响中位数或者众数的数值。学习时间的众数数值可能远低于或远高于中位数，主要取决于多长的学习时间对学生而言最为常见。

当你见到平均值的时候，一定要记得问一下："这是平均数、中位数还是众数？平均值的含义不同会不会产生什么影响？"在回答这些问题时，请想一想平均值的不同含义会给信息的意义带来怎样的变化。

不仅判断一个平均值是平均数、中位数还是众数非常重要，判定最小数值和最大数值之间的差距（即全距）以及每个数值出现的频率（即数值分布），常常也很重要。

下面我们来看一个例子，在这个例子里知道数值的全距和数值分布就非常重要。

> 医生对20岁的病人说：你所患癌症的预后不容乐观。患同样癌症的病人存活时间的中位数是10个月。所以剩下来的这几个月你想做什么就做点什么吧，不必有什么顾虑了。

EXHIBIT 10-1 Types of Averages	
✓ Mean	Determined by adding all the values and dividing by the total number of values
✓ Median	Determined by listing all the numbers from highest to lowest and finding the one in the middle
✓ Mode	Determined by counting the frequency of different values and then finding the value that appears most frequently

It makes a big difference whether a writer who uses the term "average" is talking about the mean, median, or mode.

What average makes the most sense in the first example? Consider the salaries of the stars versus those of the average players in professional sports. The biggest stars, such as the star quarterback, will make much higher salaries than most other players on the team. In fact, the highest-paid football player for the year 2015 made more than $35 million—well above the average. High salaries will increase the mean dramatically, but will have no major effect on the median or mode. For example, the mean salary for NFL players in 2015 was $2.2 million, but the median salary was "only" $830,000. Thus, in most professional sports, the mean salaries will be much higher than the median or modal salaries. Consequently, if one wished to make the salaries seem extremely high, one would choose the mean as the indicator of the average.

Now, let's look carefully at the second example. If the average presented is either the mode or the median, we may be underestimate the average amount of study time. Some students likely put in a very high number of study hours, such as 30 or 40 hours per week, thus raising the value of the mean but not affecting the value of the mode or median. The modal study hour value could be significantly lower or higher than the median, depending on what number of study hours is most frequent for students.

When you see "average" values, always ask, "Does it matter whether it is the mean, the median, or the mode?" To answer this question, consider how using the various meanings of average might change the significance of the information.

Not only is it important to determine whether an average is a mean, median, or mode, but it is also often important to determine the gap between the smallest and largest values—the *range*—and how frequently each of the values occurs—the *distribution*.

Let's consider an example in which knowing the range and distribution would be important.

> Doctor speaking to 20-year-old patient: The prognosis for your cancer is very poor. The median length of survival is 10 months. You should spend the next few months of your life doing those things that you have always wanted to do.

病人听到医生给出这样的诊断结果，对自己的未来该做出怎样可怕的预期呢？首先，我们确定知道的是获得这种诊断的病人有一半不到10个月就去世了，还有一半人存活时间超过了10个月。但是我们并不知道活下来的那部分人的存活时间的全距和数值分布。也许这些信息会显示，有些人甚至很多人存活的时间远远超过了10个月。其中有些人甚至很多人可能活到了80岁以上呢！知道病人存活情况的完整分布可能会改变这个癌症患者对未来的看法。

一般来说，病人应该考虑不同的医院对于他的疾病的存活率记录是不是有不同的全距和数值分布。这样，他应该考虑选择在有最乐观的数值分布情况的医院就诊。

当你遇到平均值的时候记住全距和数值分布的一个总体好处，就是提醒你大多数人或事并不符合确切的平均值，与平均值差异极大的结果也在预料之中。

测量误差

统计结果来自测量，测量难免会有误差。新英格兰爱国者队于2015年被指控在对阵印第安纳波利斯小马队的一场重要比赛中给球泄了气，以更易于抓住传球。他们真的这么做了吗？

裁判在中场休息时对橄榄球进行了测量，爱国者队的球测量压力要比小马队的球测量压力小。但是等等，球不是在同一时间测量的：小马队的球是在爱国者队的球之后测量的。那又怎样呢？小马队的球放在一间温暖的房间里一些时间之后，才接受测量，球有更多时间对温度做出反应，以致球的压力增大了。因此，小马队的球被测量并被发现有更大的压力，其解释可能是"测量误差"，法院也同意爱国者队提出的这一论点。

另一个测量误差的例子是饮酒量的测量。辩护律师在这种情况下使用的一个成功策略是要求起诉方提供证据，证明用于判断血液酒精含量的设备最后一次校准的时间，然后要求提供校准设备本身的可靠性数据。换句话说，有两种装置被用作醉酒驾驶的依据，但只有在测量仪器保持准确和一致的情况下，该指控才有效。当它们不可靠时，血液酒精统计数据就不能为检察官的指控提供充分的依据。

错把一件事的结论用来证明另一件事

有些数据确实能证明一件事，而立论者往往欺骗我们，宣称这些数据证明了另一件性质完全不同的事。这些数据压根就证明不了它们被用来证明的事！有两种策略可以帮你找

How dire should the patient view his future after receiving such a diagnosis? First, all we know for sure is that half of the people with this diagnosis die within 10 months, and half live longer than 10 months. But we don't know the range and distribution of how much longer the surviving half lives! The range and distribution of people who live more than 10 months could reveal that some or many people live well beyond 10 more months. Some, or even many, may live past 80! Knowing the complete survival distribution could change how this cancer victim views his future.

In general, a patient should consider whether different hospitals in the country have different ranges and distributions of survival for his disorder. If so, he should consider choosing treatment at the hospital with the most favorable distribution.

A general benefit of keeping the range and distribution in mind when encountering averages is that doing so should remind you that most people or events will not match the exact average value and that outcomes quite distant from the average are to be expected.

MEASUREMENT ERRORS

Statistics result from measurement. But measurements are always subject to error. The New England Patriots were charged in 2015 with deflating the footballs in an important game against the Indianapolis Colts so that passes could be caught more easily. But did they?

When the footballs were measured at halftime, there was less measured pressure in the Patriots' balls than in those in the possession of the Colts. But wait. The balls were not measured at the same time; the Colts' balls were measured *after* those of the Patriots. So what? The Colts' balls were in a warm room before they were measured. The balls had more time to react to the warmth and increase in pressure. So when the Colts' balls were measured and found to have more pressure in them, the explanation could be "measurement error." A court agreed with this argument made by the Patriots.

Another illustration of measurement error is in the measurement of levels of alcohol consumption. One successful strategy used by defense lawyers in such cases is to require the prosecution to provide evidence of when the device used to make judgments about blood alcohol levels was last calibrated and then to require reliability data for the calibration device itself. In other words, here you have two devices being used as the basis for the claim of driving while intoxicated, but that claim has validity only to the extent that the instruments doing the measuring are accurate and consistent. When they are undependable, the blood alcohol statistic does not provide a proper basis for supporting the prosecutor's charge.

CONCLUDING ONE THING, PROVING ANOTHER

Communicators often deceive us when they use statistics that prove one thing but then claim to have proved something quite different. The statistics

出这类欺骗。

一种策略是对立论者提供的数据视而不见，然后问自己："以什么样的统计数据作证据，在证明他的结论时会有帮助？"然后，将"所需"的数据和给出的数据进行比较。如果两者之间难以吻合，你可能就发现了一个数据上的欺骗。下面的例子为你提供一个应用这种策略的机会。

> 如果你乘坐我们这座城市的地铁，十有八九你的手机会被人偷走。我刚读到一份统计数据，说地铁盗窃中的失窃物70%都是小型电子产品。

需要做什么样的研究才能比较准确地获知自己乘地铁时电子产品被人偷走的可能性到底有多大？你自然想知道乘地铁被盗窃的概率，而不是失窃物是电子产品的概率。这个数据证明了一件事：地铁系统的大部分偷窃行为都是奔着小型电子产品去的。但它并没有证明这类偷窃行为发生的概率有多大。要回答这个问题，你需要这样问：在坐地铁的时候被偷的概率到底是多少？有可能总体的偷窃行为非常少，但是其中大部分都牵涉小型电子产品。这个例子给我们的最重要的教训就是：我们一定要加倍注意统计数据和结论的措辞，看看二者是不是指的同一件事情。如果不是，写作者或发言者就很可能是在用数据说谎。

知道什么样的数据证据应被用来支持一个结论是很难的。因此，另一种策略就是不急于看作者的结论，而先仔细检查作者的数据，然后问自己："从这些数据中我们可以得出什么合适的结论？"然后拿你的结论和作者的结论相比较。请用这种策略来检验下面这个例子。

> 大约半数美国人对另一半不忠。研究人员最近在一家购物中心采访了很多人。在接受采访的75人当中，有36人直白地说他们有朋友曾承认对自己的交往对象不忠。

你有没有想到这个例子一开始得出的结论？大约一半人在某个特定地点承认有朋友告诉过自己，他们在和他人交往的过程中至少有过一次不忠行为。你有没有看出数据所证明的东西和写作者的结论之间存在着巨大的差异？如果你看出来了，那你就发现了这位写作者是如何利用数据来进行欺骗的。

通过省略信息欺骗

统计数据经常因为不完整而欺骗我们。因此，还有一种在数据论证中找到缺陷的非常

don't prove what they seem to! We suggest two strategies for locating such deception.

One strategy is to *blind yourself to the communicator's statistics* and ask yourself, "What statistical evidence would be helpful in proving her conclusion?" Then, compare the "needed" statistics to the statistics given. If the two do not match, you may have located a statistical deception. The following example provides you with an opportunity to apply that strategy.

> You are highly likely to have your smart phone stolen if you ride on our city's subway. I just read a statistic that said that electronic gadgets account for 70 percent of the thefts on the subway system.

What study needed to be done to acquire a good idea of the likelihood of being robbed of a gadget while on the subway? You would want to know the likelihood of being robbed, not the likelihood of being robbed of an electronic gadget. The data have proven one thing—most of the thefts in the metro system are of electronic gadgets. It has not proven how likely such thefts are. To answer that question you need to ask the question, "What are the odds of being robbed at all while riding the subway?" It is possible that there are very few total thefts, but most of them involve electronic gadgets. The important lesson to learn from this example is to *pay very close attention to both the wording of the statistic and the wording of the conclusion* to see whether both are referring to the same thing.

Knowing just what statistical evidence should be provided to support a conclusion is difficult. Thus, another strategy is to examine the author's statistics *very closely* while *blinding yourself to the conclusion*; then ask yourself, "What is the appropriate conclusion to be drawn from those statistics?" Then, compare your conclusion with the author's. Try that strategy with the following example.

> Almost half of all Americans cheat on their significant others. A researcher recently interviewed people at a shopping mall. Of the 75 people responding to the survey, 36 reported having friends who had admitted to cheating on someone they were seeing.

Did you come up with the following conclusion? Almost half of the people *in one given location* admit to *having friends* who report having cheated, *at least once*, on someone whom they were dating or were otherwise involved with. Do you see the difference between what the statistics proved and what the author concluded? If so, you have discovered how this author has lied with statistics.

DECEIVING BY OMITTING INFORMATION

Statistics often deceive us because they are incomplete. Thus, a further helpful strategy for locating flaws in statistical reasoning is to ask, "What further information do you need before you can judge the impact of the

有用的策略就是问一问:"在判断数据的影响力之前,还需要什么进一步的信息?"让我们先看看下面的例子,这些例子展示了这个问题所起的作用。

> 1. 大公司正在将市中心地带的小镇气息破坏殆尽。就在去年,城里的大公司数目增长了 75%。
> 2. 艾滋病预防项目需要较大的资金增幅。2009 年,有 54 000 人饱受艾滋病的折磨。

在第一个例子中,75% 这个数字很吸引眼球,但是缺少了一些信息:这个百分比所依据的绝对数值。假如我们知道这种增长是从 4 家增长到 7 家,而不是从 12 家增长到 21 家,我们还会如此惊讶吗?在第二个例子里,我们倒是有数字了,但是不知道比率。难道我们不需要知道这一数字对感染艾滋病的人所占人口百分比来说意味着什么吗?

第二个例子展示了我们社会中一种常见的情况:试图通过聚焦全国范围内受折磨的总人数,以引起公众对某个社会问题的关注。尽管有很多人受艾滋病折磨显然是个亟待解决的问题,但当我们将 54 000 除以美国的总人口约 3 亿人,我们只得到一个大概 0.02% 的数值。

当你遇到听起来让人格外留意的数字或者百分比时,一定要当心!你可能需要其他信息来判定这些数字到底多值得人格外留意!当只有绝对数值摆在眼前的时候,问一问知道百分比是否可能帮你做出更好的判断;当只有百分比出现在眼前的时候,问一问绝对数值是否会丰富它们的含义。

当你遇到统计数字的时候,一定要记得问一问:"缺少了什么相关信息?"

评估数据的一些线索

1. 尽量找出关于数据如何获得的信息,越多越好。问一下:"这位写作者或发言者是怎么知道的?"在立论者想要用很大的数字来让你动心或者让你惊心的时候,你尤其要警惕。
2. 要对论证中描述的平均值类型感到好奇,分析一下知道事件的数值全距和数值分布是否能为你提供一个有用的视角。
3. 如果数据使用者拿指向一件事的数据来证明另一件事,你就要特别当心。
4. 先不看写作者或发言者使用的数据,把所需的统计证据和实际提供的数据做比较。
5. 从数据中得出你自己的结论。如果得出的结论和写作者或发言者的结论不一致,那么很可能论证中有什么地方出错了。
6. 判断缺少了什么信息。对于误导性的数字和百分比以及缺少的比较,你要特别当心。

statistics?" Let's look at the following examples to illustrate the usefulness of this question.

1. Large businesses are destroying the small town feel of our downtown area. Just last year, the number of large businesses in the city increased by 75 percent.
2. AIDS prevention programs need major funding increases. In 2009, 54,000 people were afflicted with AIDS.

In the first example, 75 percent seems quite impressive. But something is missing: the *absolute numbers* on which this percentage is based. Wouldn't we be less alarmed if we knew that this increase was from 4 businesses to 7, rather than from 12 to 21? In our second example, we have the number, but we don't have the *percentage*.ABouldn't we need to know what this number means in terms of the percentage of people having AIDS?

The second example illustrates a common occurrence in our society, an attempt to arouse the public's attention to a societal problem by spotlighting the number of people afflicted nationally. Although clearly an urgent problem, when we divide 54,000 by the approximate population of the United States, 300 million, we get a value of about 0.02 percent.

When you encounter impressive-sounding numbers or percentages, beware! You may need to get other information to decide just how impressive the numbers are. When only *absolute numbers* are presented, ask whether *percentages* might help you make a better judgment; when only *percentages* are presented, ask whether *absolute numbers* would enrich their meaning.

When you encounter statistics, be sure to ask, "What relevant information is missing?"

Clues for Assessing Statistics

1. Try to find out as much as you can about how the statistics were obtained. Ask, "How does the author or speaker know?" Be especially vigilant when the communicator tries to impress or alarm you with large numbers.
2. Be curious about the type of average being described; analyze whether knowing the range and distribution of events would add a helpful perspective to the statistic.
3. Be alert to users of statistics concluding one thing, but proving another.
4. Blind yourself to the writer's or speaker's statistics and compare the needed statistical evidence with the statistics actually provided.
5. Form your own conclusion from the statistics. If it doesn't match the author's or speaker's conclusion, then something is probably wrong.
6. Determine what information is missing. Be especially alert for misleading numbers and percentages and for missing comparisons.

在自己的写作中使用统计数字

我们希望你将统计数字融入自己的写作。如果运用得当，这会是极有价值的工具，可以帮助我们描述和理解趋势和模式，可以帮助我们进行预测，还可以强化我们的论证。不过，本章也说明了在论证中引用数据可能存在的一些非常严重的风险。对于没有受过训练的读者而言，统计数据看起来像权威事实，但是你知道事实多么容易被人操纵。作为一个心怀批判性思维的写作者，你需要实现一个重要的平衡。你必须尽量避免欺骗手法，同时以一种明白易懂的方式呈现出那些常常复杂万端的数据。

为了做出妥当的、带有数据的论证，你可能要从论证中专门拿出点时间来解释这些数据是怎么得出的，有什么含义，以及存在的局限。这样做会增加你在读者心目中的可信度，向他们表明你不是在偷偷摸摸地给他们灌输"私货"。你同时也在鼓励他们成为强势批判性思维者，对数据的质量得出自己的结论。你可以将这些解释呈现在论证的文本当中，也可以将它们放到脚注、尾注或者附录里，这很可能取决于你所在领域的通行做法和你写作的正式程度。

思维体操

⚠ **批判性问题：数据有没有欺骗性？**

在下列每篇练习文章里，找出证据的不当之处。

第一篇

美国政府部门职位的竞选和角逐变得越来越失控。金钱开始在越来越多的选举中起到关键作用。现在要赢得一个参议院席位，平均花在竞选活动中的支出要超过 800 万美元，而总统候选人平均花费超过 3 亿美元。是时候做出一些实质性的改变了，我们不能允许政客通过广告宣传的大笔花费来购买他们的职位。

第二篇

家正变成越来越危险的栖身之所。和家相关的伤害案件的数量不断增加。2000 年，大约有 2300 名 14 岁及以下儿童死于家中发生的事故。此外，每年有 470 万人被狗咬伤。更糟糕的是，电视机这个相对安全的家用电器也开始变成危险源。事实上，每年有 42 000 人

USING STATISTICS IN YOUR WRITING

We hope that you incorporate statistics into your writing. When used appropriately, they are a valuable tool. They help us describe and understand trends and patterns. They help us to predict. Statistics can strengthen our arguments. Even so, this chapter has illuminated some of the very serious risks of incorporating statistics into an argument. To the untrained reader, statistics look like authoritative facts, but you know how easily the facts can be manipulated. As a writer concerned with critical thinking, you are faced with an important balancing act. You must attempt to avoid deceptive techniques, but also present often-complicated statistics in a clear and understandable way.

To make careful arguments with statistics, you may have to take some time away from your argument to explain how the statistics were produced, the implications of the statistics, and the limitations of them. Doing so will improve your credibility with your readers. You are showing them that you are not trying to sneak something by them. You are also encouraging them to be strong-sensed critical thinkers and draw their own conclusions about the quality of the statistics. You may decide to include these explanations in the text of your argument, or you could choose to include them in a footnote or endnote or in an appendix. This decision will likely be based on the habit of your field and formality of the writing.

PRACTICE EXERCISES

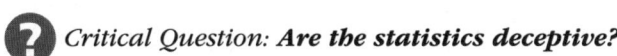

*Critical Question: **Are the statistics deceptive?***

For each of the practice passages, identify inadequacies in the evidence.

Passage 1

Campaigns for national office are getting out of hand. Money is playing a central role in more and more elections. The average winner in a senate race now spends over $8 million on his campaign, while typical presidential candidates spend more than $300 million. It is time for some serious changes, because we cannot simply allow politicians to buy their seats through large expenditures on advertisements.

Passage 2

The home is becoming a more dangerous place to spend time. The number of home-related injuries is on the rise. In 2000, approximately 2300 children aged 14 and under died from accidents in the home. Also, 4.7 million people are bitten by dogs each year. To make matters worse, even television, a relatively safe household appliance, is becoming dangerous. In fact, 42,000 people are injured by televisions and television

因误撞电视机或者电视机架而受伤。既然家里会发生这么多的事故，也许人们需要花更多的时间待在户外。

第三篇

　　一项新的研究表明，发送过多短信的青少年有较高做出不健康行为的风险。这项研究聚焦于密歇根州一个城镇的学生样本，其中大多数学生来自社会经济地位较低的家庭，是少数族裔群体的成员，家里没有父亲。研究人员发现，相比于其他青少年，过度发短信的青少年尝试吸烟的可能性高 50%，与他人发生肢体冲突的可能性高 45%，酗酒的可能性高 35%。这项研究给父母敲响了警钟，提醒他们阻止孩子发过多的短信。

给个提示

第一篇

　　结论：有必要改变政府部门职位的竞选方式。
　　理由：政客们花在竞选活动上的钱太多了。平均每位参议员在竞选活动中花费超过 800 万美元。总统候选人花费超过 3 亿美元来进行竞选活动。

　　竞选活动是不是太费钱了？"平均"这样的词提醒我们应该注意其潜在的欺骗性。我们需要了解这些数据中的平均值的种类。这些数据是平均数、中位数还是众数？例如，在参议院席位竞选数据中使用平均数可能会得出一个歪曲的数字，因为某些参议员，尤其是和竞选对手势均力敌的参议员在竞选中花费的钱确实数额巨大。但是，很多参议员都注定要重新当选，这些竞选活动很可能花费并不多。我们知道只有极小一部分参议员在竞选活动中与对手实力极其接近。因此，如果平均数被用来代表平均值，那么多数参议员的花费可能不及报道中提到的那么多。换句话说，如果用中位数或者众数作为平均值，就可能得到低得多的数值。同样，知道数值的全距和分布也会让你更加了解所关心的竞选活动的花费到底有多少。

　　此外，重要的参照数据也缺失了。当前竞选活动的花费和过去类似活动的花费比起来怎么样？其他部门的竞选花费情况又如何？很可能竞选活动的花费近几年实际上一直在走下坡路。

stands each year. With so many accidents in the home, perhaps people need to start spending more time outdoors.

Passage 3

A new study shows that teens who send excessive numbers of text messages may be at risk for unhealthy behaviors. The study focused upon a sample of students in an urban Michigan county, most of whom were from a lower socioeconomic status, were a member of a minority group, and had no father in the home. Researchers found that teens who excessively text are 50 percent more likely to have tried cigarettes, 45 percent more likely to be in a physical fight, and 35 percent more likely to be binge drinkers. This study should serve as a wake-up call to parents and prompt them to discourage their children from texting excessively.

Sample Responses

Passage 1

CONCLUSION: *A change in campaigning for national office is necessary.*

REASON: *Politicians are spending too much on campaigns. The average senator spends more than $8 million on his campaign. Presidential candidates spend more than $300 million on their campaigns.*

Are campaigns costing too much money? The words *average* and *typical* should alert us to a potential deception. We need to know the kind of average used for these statistics. Was it the mean, the median, or the mode? For example, using the mean in the senate race data could potentially lead to a figure that is skewed because of certain particularly close, senate races where candidates spent large sums of money. However, because many senators are basically guaranteed reelection, these races probably involve less spending. We know that only a few senate race elections are unusually close. Therefore, most probably do not spend as much as was reported, if the mean was used to present the average. In other words, the median or the mode would probably show a lower value. Also, knowing the distribution and range would give you a better idea of how much you would want to be concerned about campaign spending practices.

Additionally, important comparison figures are missing. How does campaign spending compare to similar spending in the past? What about for other offices? It is possible that campaign spending has actually gone down in recent years.

第二篇

　　结论：花时间待在家里正变得越来越危险。
　　理由：和家相关的伤害事件数量正在日渐增加。
　　支撑：一年里，2300名儿童死于家中发生的事故。
　　　　　每年有470万人被狗咬伤。
　　　　　每年有42 000人被电视弄伤。

　　要评价这个论证，我们首先需要确定用什么样的证据来回答下面的问题最合适："现在家里比起从前是不是更不安全了？"在我们看来，用来回答这个问题的最佳数据，是现在每年家中重大事故的发生率和过去几年家中的事故发生率的比较。同样有关的是每小时在家里发生的伤害事件数量相比于过去几年的情况。可能家中发生更多伤害是因为人们比过去花更多时间待在家里。如果他们待在家里的时间更多，在家中发生的伤害的数量相应增加，这就显得符合逻辑了。

　　这个论证中提供的证据还有一些可疑之处。第一，关于发生在家中的伤害的总数，作者根本没有提供任何数字。我们知道作者说这一数量不断上升，但是根本没有证据证明这样的上升。第二，作者没有披露儿童在家庭事故当中丧生的具体细节。这个数据和过去儿童在家中身亡的数据相比怎么样？是什么类型的事故导致这些儿童死亡？第三，狗咬人的数字具有欺骗性。我们根本不知道这些狗咬人事件是不是发生在家中。更重要的是，狗咬人的数字似乎并不能引导我们得出结论说待在家里是不安全的。第四，有关电视机的数据也很可疑。作者是从哪里得到这一令人印象深刻的数据？同样，这些被电视所伤的人中，大部分到底伤得有多严重？

Passage 2

CONCLUSION: *It is becoming increasingly dangerous to spend time in one's home.*

REASON: *Household-related injuries are on the rise.*

SUPPORT: *In one year, 2300 children died in household accidents.*
4.7 million people are bitten by dogs every year.
42,000 people are injured by televisions each year.

To evaluate the argument, we need to first determine what the most appropriate evidence is to answer the question, "Are households more unsafe than they used to be?" In our opinion, the best statistic to use to answer this question is a comparison of the current rate of serious household accidents per year with the rate in past years. Also relevant is the number of injuries per hour spent in the house versus the same statistic for past years. It is possible that more household injuries occur because people are spending more time in their houses than they used to spend. If they are inside the house more, it is only logical that the number of injuries occurring in the house would also rise.

The evidence presented in the argument is questionable for a number of other reasons. First, no number is given at all regarding the total number of household injuries. We know the author says they are on the rise, but no evidence is provided demonstrating a rise. Second, no details are given regarding the deaths of children in household accidents. How does this statistic compare to children's deaths in the home in the past? What types of accidents are causing these children's deaths? Third, the number of dog bites is deceptive. We do not know whether these dog bites occur in the home. More importantly, the number of dog bites does not seem to move us toward the conclusion that being at home is unsafe. Fourth, the statistic regarding televisions is questionable. Where does the author get the impressive-sounding statistic? Also, how serious are most of these injuries?

第 11 章

有什么重要信息被省略了

| 学 习 目 标 |

1. 认识到论证中信息缺失的必然性。
2. 养成提问以澄清缺失信息的习惯。

下面这个广告是不是让人难以抗拒？

试试"欢乐时光"，它是医嘱治疗背痛的头号特效药。

这则广告的目的当然是劝说你购买更多广告中的产品。在你的批判性思维能力发展到现在的水平之前，你一定就知道这类广告所说的话不能全部当真。例如，如果这家"欢乐时光"制药公司比其他制药公司给了医生更低的折扣，给医生提供更多的免费试用药，或者给那些使用他们药品的医生提供豪华游轮旅游的机会，你是不太可能在广告中看到此类信息的。虽然你看不到这些信息，但是它们对于你决定吃什么药来治疗背痛很重要。

批判性思维者追求自主思考，但如果他们做决定的基础是极为有限的一点信息，那么他们很难真正思考。几乎所有的结论或者产品都有一些积极的特点。那些只想告诉我们他

CHAPTER 11

What Significant Information Is Omitted?

LEARNING OBJECTIVES

1. Recognize the inevitability of missing information in an argument.
2. Develop the habit of asking questions to illuminate missing information.

How compelling is the following advertisement?

> Try HappyTyme, the number one doctor-prescribed treatment for backache.

The purpose of the advertisement is, of course, to persuade you to buy more of the designated product. Even before your critical-thinking skills developed to their current level, you knew that such advertisements tell less than the whole truth. For example, if the HappyTyme Company gives a bigger discount to doctors than do other pharmaceutical companies, provides doctors with greater numbers of free samples, or provides cruises for doctors who use their product, you are unlikely to see this information included in the advertisement. You will not see that information, but it is quite relevant to your decision about what to take for your sore back.

While critical thinkers are seeking autonomy, they are in a weak position to do so if they are making decisions on the basis of highly limited information. Almost any conclusion or product has some positive characteristics.

们想让我们知道的信息的人，会告诉我们所有这些积极的特点，而且不厌其详，活灵活现。但是他们会隐藏那些结论的消极方面。制造商可能会告诉我们他们的薯片味道鲜美、酥脆，但他们不太可能透露这些薯片含有高热量和防腐剂。因此，有用的自主思考需要我们坚持不懈地寻找写作者到底隐瞒了什么信息，不论写作者是无心省略还是有意隐瞒。

通过追问前面几章里我们学习的那些问题，诸如涉及歧义、假设和证据的问题，你能找出大量非常重要的省略信息。本章旨在帮助你对那些没有明说出来的信息的重要性变得更加敏感，以此作为重要提示来提醒自己，如果仅仅评价那些摆在面上的信息，我们就不能窥见论证的全貌。因此本章专门论述一个极为重要的问题，你必须要问这个问题才能判断出论证的质量：有什么重要信息被省略了？

⚠ **批判性问题**：有什么重要信息被省略了？

找到省略信息的益处

你应该记住，几乎每个你遇到的信息都有一个目的。换句话说，这个信息的组织结构是由别人精心挑选和呈现的，目的就是让它能影响你的思维方式。例如，每一个统计数据都经过精挑细选，精心组织，被用来达到一定的目的。达到谁的目的？谁给你分享这个数据，就是要达到谁的目的。因此你的任务是判断自己是否想成为某一目的的傀儡。这个目的常常就是说服你。

广告商、教师、政客、作家、演说家、研究员、博主和父母都精心组织信息来影响你的判断和决定。那些尽力要说服你的人，几乎总是将他们的立场置于最强的光线之下。所以，当你发现你相信较具说服力的理由——那些你正在努力勘探的金块时，较为明智的做法是犹豫片刻，想想写作者可能没有告诉你的信息，那些你的批判性问题尚未揭示出来的信息。

这里所说的重要的省略信息，是指那些将会决定你该不该被写作者或者发言者的论证所影响，即影响你的论证过程的信息。本章中给出了很多说服力不够强的论证的例子，它们之所以说服力不够，并不是因为说出来的信息不顶用，而是因为省略掉的信息太关键。仔细研读这些例子，注意在每个例子中找不到省略掉的信息会怎样导致你做出仓促的、有可能错误百出的判断。

⚠ **注意**：重要的省略信息就是那些影响论证过程的遗漏信息。

Those who have an interest in telling us only the information they want us to know will tell us all of these positive characteristics in lengthy and vivid detail. But they will hide the negative aspects of their conclusions. We may be told by manufacturers that their potato chips are tasty and crispy, but they are not likely to reveal that those same chips are high in calories and loaded with preservatives. Thus, useful autonomy requires our persistent search for what is being hidden, either accidentally or on purpose.

By asking questions learned in previous chapters, such as those concerning ambiguity, assumptions, and evidence, you will detect huge amounts of important missing information. This chapter tries to sensitize you even more to the importance of *what is not said* and to serve as an important reminder that we react to an incomplete picture of an argument when we evaluate only the *explicit* parts. We thus devote this chapter to an extremely important additional question you must ask to judge the quality of reasoning: What significant information is omitted?

 Critical Question: ***What significant information is omitted?***

THE BENEFITS OF DETECTING OMITTED INFORMATION

You should remember that almost any information that you encounter has a purpose. In other words, its organization was selected and presented by someone who hoped that it would affect your thinking. Every statistic, for example, is chosen, organized, and used to achieve a purpose. Whose purpose? The purpose of whichever human is sharing the statistic with you. Hence, your task is to decide whether you wish to be an instrument of the chosen purpose. Often that purpose is to persuade you.

Advertisers, teachers, politicians, authors, speakers, researchers, bloggers, and parents all organize information to shape your decisions. Thus, those trying to persuade you will almost always try to present their position in the strongest possible light. So, when you find what you believe to be persuasive reasons—those gold nuggets for which you are prospecting—it's wise to hesitate and to think about what the author may *not* have told you, something that your critical questioning has not yet revealed.

By *significant omitted information*, we mean information that would affect whether you should be influenced by a speaker's or writer's arguments—information that *shapes the reasoning*! Interspersed throughout this chapter will be examples of reasoning that are not very convincing, not because of what is said but because of what is omitted. Study the examples carefully and notice how in each case the failure to look for omitted information would have resulted in your making a premature and potentially erroneous judgment.

 Attention: *Significant omitted information is missing information that shapes the reasoning.*

不完整的论证在所难免

不完整的论证在所难免，主要有以下几点原因。第一，时空的限制。论证不完整是因为立论者没有足够的时间去组织这些论证，他们也没有不受限制的空间或时间来呈现他们的理由。

第二，我们大部分人注意力持续的时间都很有限，并且如果信息长得没完没了，我们就会感到厌倦。因此，立论者常觉得有必要将他们的信息尽快传达给受众。广告和社论都反映了这两个因素。例如，社论的字数都有一定的限制，论证必须既引人入胜又切中要害。为此，社论作者不得不点到为止，让人心痒难搔。电视评论员更是出名地会将极其复杂的问题弄得听上去十分简单。他们的时间有限，无法提供足够精准的信息，而你需要精准信息才能形成一个合理的结论。

第三，进行论证的人掌握的知识总是不完全的。还有第四个省略某些信息的原因：作者就想欺骗你。广告商知道他们省略了关键的信息。如果他们描述了产品中含有的所有化学成分或廉价的组成部分，那么你可能就不会购买他们的产品了。

为什么省略信息变得这样肆无忌惮？最后一个原因是那些试图给你提建议或想要说服你的人的价值观、信仰和态度常常和你的并不相同。因此，可以预料，他们的论证会受到不同的假设引导，这些假设和你对同样的问题可能提出的假设可能完全不同。批判性思维者看重好奇心和合理性，而那些力图说服你的人常常想要打消你的好奇心，鼓励你依靠不理智的情绪反应来做出选择。

一个特定的视角就像一副马戴的眼罩。眼罩让马心无旁骛，全神贯注于正前方的道路。但是，一个人看问题的视角也像马所戴的眼罩那样，阻止了他去关注某些特定的信息，而这些被他忽略的信息对那些从不同参照系进行论证的人而言可能至关重要。演员马特·达蒙所扮演的角色在电影《谍影重重3》里表达了对这个问题的理解："一个东西到底是什么样，主要取决于你坐在什么地方观察，这真有意思。"

展示 11-1　不完整的论证出现的原因

✓ 时空对论证产生的限制
✓ 由于注意力集中时间的限制，论证必须尽快完成
✓ 立论者的知识总是不全面
✓ 论证常常是为了欺骗
✓ 立论者常常与你有不同的价值观、信仰和态度

THE CERTAINTY OF INCOMPLETE REASONING

Incomplete reasoning is inevitable for several reasons. First, there is the limitation imposed by time and space. Arguments are incomplete because communicators do not have forever to organize them, nor do they have unlimited space or time in which to present their reasons.

Second, most of us have a very limited attention span; we get bored when messages are too long. Thus, communicators often feel a need to get their message across quickly. Advertisements and editorials reflect both these factors. For example, editorials are limited to a specific number of words, and the argument must both be interesting and make the author's point. Editorial writers, therefore, engage in many annoying omissions. Television commentators are notorious for making highly complicated issues sound as if they are simple. They have very little time to provide the degree of accurate information that you will need to form a reasonable conclusion.

A third reason for the inevitability of missing information is that the knowledge possessed by the person making the argument will always be incomplete. A fourth reason is an outright attempt to deceive. Advertisers *know* they are omitting key bits of information. If they were to describe all the chemicals or cheap component parts that go into their products, you would be less likely to buy those products.

A final important reason why omitted information is so prevalent is that the values, beliefs, and attitudes of those trying to advise or persuade you are frequently different from yours. You can expect, therefore, that their reasoning will be guided by different assumptions from those you would have brought to the same question. Critical thinkers value curiosity and reasonableness; those working to persuade you often want to extinguish your curiosity and to encourage you to rely on unreasonable emotional responses to shape your choices.

A particular perspective is like a pair of blinders on a horse. The blinders improve the tendency of the horse to focus on what is directly in front of it. Yet, an individual's perspective, like blinders on a horse, prevents that person from noting certain information that would be important to those who reason from a different frame of reference. Actor Matt Damon's character shows an understanding of this important point in *The Bourne Ultimatum*: "It's funny how different things look, depending on where you sit."

EXHIBIT 11-1 Reasons for Incomplete Reasoning

✓ Time and space imposes limitations on arguments.
✓ Arguments must be given quickly due to limited attention spans.
✓ The arguer will always have incomplete knowledge.
✓ Arguments often attempt to deceive.
✓ The arguer often will have different values, beliefs, and attitudes from yours.

帮你识别省略信息的问题

怎么才能识别省略的信息？首先你得提醒自己，不管支撑某一判断或观点的理由乍一看多么吸引人，你都必须再仔细看看，以寻找那些省略的信息。

你怎么去找，到底希望自己找到些什么？首先你得提出一些问题来帮你判断还需要哪些额外的信息，然后再提出一些问题来找出那些信息。

你可以利用很多种问题来识别相关的省略信息。有些你已经学会使用的问题能够凸显出这些信息。此外，为了帮你确定可能被其他批判性问题所忽视的省略信息，我们为你准备了一个清单，上面列举了一些重要的省略信息类型，同时列出一些提问的例子来帮助你发现它们。

你是否理解为什么广告词如"五个医生当中就有四个同意""全天然""不含脂肪""低淀粉含量""对你的心脏有好处""头号领先品牌""美国糖尿病协会认证"以及"不添加防腐剂"可能都是准确的，但同时又因省略信息而误导消费者？

我们需要知道确切的数字

你可能还记得，要理解特定的比较，我们必须要对可能的数值的范围和比例情况有一点了解。假设我们被告知美国人在世界幸福量表上得分比丹麦人要高。因为"幸福"常常是人们的福祉的代名词，甚至对有些人来说就是"人生的目的"，所以美国人的相对幸福感可以作为一个信号，指导丹麦和其他国家组织经济和社会习惯。但是我们在得出这样的结论之前，还需要大量的其他信息。

这个幸福量表上的刻度范围是多少？回答者可以填写的数字范围是 1～2，还是 1～100？美国和丹麦的回答者分数的确切差距到底是多少？想一想，如果平均分数差距很小，相比于平均分相差 50 分的情况，我们处理这个分数差距的方式会有多大不同。当我们看到与数字相关的信息时，一定要记住问一下具体的数字。

这个幸福量表的例子展示了一个"省略信息"的重大问题。我们使用的很多词都暗示了大小、范围或者比例，但是除非我们细究这些话暗含的具体信息，我们很容易被引入歧途。这里的关键是：如果一个字或词可以用具体的单位来测量，那么我们在做决定之前就应该问一问这些量度单位的具体数值。

QUESTIONS THAT IDENTIFY OMITTED INFORMATION

How do you identify omitted information? First you have to remind yourself that regardless of how attractive the reasons supporting a particular decision or opinion may initially seem, you need to take another look in search of omitted information.

How do you search, and what can you expect to find? You ask questions to help decide what additional information you need, and then ask questions designed to reveal that information.

You can use many kinds of questions to identify relevant omitted information. Some questions you have already learned to ask will highlight such information. In addition, to help you determine omitted information that might get overlooked by other critical questions, we provide you with a list of some important kinds of omitted information and some examples of questions to help detect them.

Do you see how advertising phrases like "four out of five doctors agree," "all natural," "fat free," "low in carbs," "good for your heart," "number one leading brand," "ADA approved," and "no added preservatives" may all be accurate but misleading because of omitted information?

BUT WE NEED TO KNOW THE NUMBERS

You may remember that to make sense of certain comparisons we must know something about the range and proportions of possible values. Suppose we are told that the United States scores higher on a worldwide Happiness Scale than do the citizens of Denmark. Because "happiness" is often the name for human well-being or, even for some people, "the purpose of life," the relative happiness of the United States could serve as a signal for how Denmark and other countries should organize their economy and social habits. But we need so much information before moving forward to any such conclusion.

What is the range of scores on the Happiness Scale? Did respondents fill out a survey that permitted answers from 1 to 2, or from 1 to 100? What was the precise difference in the scores of U.S. and Danish respondents to the scale? Notice how differently we would process the meaning of difference in scores if the average scores were almost exactly identical or if they were 50 units apart. When we see information that has a numerical dimension associated with it, we must always ask for the specific numbers.

This Happiness Scale example is an illustration of a large and significant "missing information" problem. Many of the words we use imply size, scope, or proportion, but unless we ask for the specific information implied in these words, we are easily led astray. The point here is, when a word or phrase is measurable in specific units, ask for the specifics of those units before making a decision.

寻找常见类型的重要信息的一些线索

1. 常见的反驳论证
   a. 反对的人会提供什么样的理由?
   b. 有没有研究和所说的研究相冲突?
   c. 有没有支持论证的对立面的例子、证词、备受尊敬的权威人士提供的观点或者类比被省略?
2. 遗漏的定义
   a. 如果关键词用另一种方式定义,这个论证会有怎样的不同?
3. 遗漏的价值偏好或者视角
   a. 不同的价值观会不会产生对这一论题的不同思考方式?
   b. 从与发言者或写作者不同的价值观出发,会产生怎样的论证?
4. 论证中所说的"事实"的来源
   a. 这些"事实"的来源是什么?
   b. 事实断言是否由出色的研究或可靠的信息来源支撑?
5. 用来获得事实的程序细节
   a. 有多少人完成这个问卷调查?
   b. 调查中的问题是怎样措辞的?
   c. 调查对象是否有大量的机会提供与使用这些回答的人所报告的不同的答案?
6. 收集或组织证据的替代性技术
   a. 访谈研究得到的结果和书面问卷调查得到的结果可能有怎样的不同?
   b. 实验室试验会不会产生更可靠、更丰富的结果?
7. 遗漏的或不完整的数字、图表、表格或者数据
   a. 如果数据包含早期或者后来的证据,看起来会不会不一样?
   b. 作者有没有故意"拉长"图表让差异显得更大一些?
8. 省略的结果,不管是正面还是反面结果,是短期还是长期结果,是提倡的还是反对的结果
   a. 论证有没有遗漏提议的行动所带来的重要的正面或负面后果?代价是什么?好处又是什么?
   b. 我们是否需要知道行动对下列任何一个领域的影响:政治、经济、社会、生物、精神、健康或环境?
9. 当为特殊的预测技巧进行辩护时省略掉关于预测失败或者预测失误的信息
   a. 当"通灵巫师"或"直觉主义者"推销他们的特异能力时,我们需要追问他们的预测被证明未成真的概率有多少。

Clues for Finding Common Kinds of Significant Information

1. Common counterarguments
   a. What reasons would someone who disagrees offer?
   b. Are there research studies that contradict the studies presented?
   c. Are there missing examples, testimonials, and opinions from well-respected authorities, or analogies that support the other side of the argument?
2. Missing definitions
   a. How would the arguments differ if key terms were defined in other ways?
3. Missing value preferences or perspectives
   a. Would different values create a different approach to this issue?
   b. What arguments would flow from values different from those of the speaker or writer?
4. Origins of "facts" referred to in the argument
   a. What is the source for the "facts"?
   b. Are the factual claims supported by competent research or by reliable sources?
5. Details of procedures used for gathering facts
   a. How many people completed the questionnaire?
   b. How are the survey questions worded?
   c. Did respondents have ample opportunity to provide answers different from those reported by the person using the responses?
6. Alternative techniques for gathering or organizing the evidence
   a. How might the results from an interview study differ from written questionnaire results?
   b. Would a laboratory experiment have created more reliable and informative results?
7. Missing or incomplete figures, graphs, tables, or data
   a. Would the data look different if they included evidence from earlier or later years?
   b. Has the author "stretched" the figure to make the differences look larger?
8. Omitted effects, both positive and negative and both short and long term, of what is advocated and what is opposed
   a. Has the argument left out important positive or negative consequences of a proposed action? What are the costs? What are the benefits?
   b. Do we need to know the impact of the action on any of the following areas: political, social, economic, biological, spiritual, health, or environmental?
9. Omission of prediction failures, or misses, when arguing for special prediction skills
   a. When "psychics" or "intuitionists" promote their special abilities, we need to ask how often their predictions have proven to be untrue.
   b. We need to know the frequency of prediction failures, as well as the frequency of successes, of economists, financial advisers, sports gamblers, and political pundits before concluding that they have special expertise.

b. 我们需要知道经济学家、理财顾问、体育彩票参与者和政治权威人士预测失败的概率以及成功的概率，然后才能得出结论说他们拥有特殊的才干。

为了确保你能强烈感受到细究具体数字的必要性，这里再给你举几个例子，这些陈述应该能激发你这样说："可是我需要知道更具体的数字。"

1. 如果你上大学，那么在 23 岁时你就更有可能找到一份工作。
2. 任何每天喝下 60 毫升酒的人，都能改善自己休息放松的能力。
3. 在大学时有个来自不同文化环境的室友，可以减少你前往其他国家旅行时生病的概率。

不管你在什么时候读到或看到一个论证，其中使用诸如"更大""更多""更快""更瘦""在……之后"（多久以后）之类的概念，以及任何表示大小或范围的其他概念，都要养成习惯，意识到细究具体数字的必要性。当你遇到这些需要具体数字的概念时，一定要意识到，在这种情况下"慢思考"是多么重要。你需要使用"系统 2 思维"来获得具体数字方面的信息，需要这些信息来做一个审慎的决定：应该相信什么或应该做什么。

负面视角的重要性

还有一种类型的省略信息异常重要而又常常被人忽视，我们想在这里特别强调一下：被提倡的行动（如使用一种新型药物、建设一所大型的新学校或者减税）带来的潜在负面效果。我们在这里强调负面效果，因为通常情况下这些行动的提议都发生在支持者宣称他们的意见非常棒的语境下，他们宣称这些行动能带来很大的好处，例如某个医学难题的大幅解决、更美丽的容貌、更多休闲时间、更多教育机会、更长的寿命，更多或更好的商品。但是，大部分行动都在有广泛的正面影响的同时也有广泛的负面影响，我们需要问一问下面的问题：

- 社会的哪一部分并没有从提议的行动当中受益？谁蒙受了损失？蒙受损失的人对此有什么话说？
- 提议的行动对权力分配有什么影响？
- 这个行动对我们的健康有什么影响？
- 这个行动怎么影响我们的人际关系？怎么影响我们与自然环境之间的关系？

对上述每一个问题，我们都不要忘了问一声："这个行动潜在的长期负面效果是什么？"

⚠ **注意**：在考虑省略的信息时，要记住问一问："这个行动潜在的长期负面效果是什么？"

To make certain you have a strong sense of the need to ask for specific numbers, here are a few more examples of statements that should prompt you to say, "But I need to know the specific numbers."

1. If you go to college, you will be more likely to have a job when you are 23.
2. Anyone who drinks two ounces of alcohol each day will be improving his ability to relax.
3. Having a college roommate from a culture different from yours reduces your chances of getting sick when you travel to other countries.

Try to develop the habit of being sensitive to the need for specific numbers whenever you read or see an argument that uses concepts like "greater than," "more than," faster," "thinner," "after" (how long after?), and any other concept that comes in sizes and scope. When you encounter these "concepts that require specific numbers," be aware of how important slow thinking is in this instance. You need to use System 2 thinking to gather the numerical information you require to make a careful decision about what to believe or do.

THE IMPORTANCE OF THE NEGATIVE VIEW

There is one type of omitted information that is so important to identify and so often overlooked that we want to specifically highlight it for you: the *potential negative effects* of actions being advocated, such as the use of a new medication, the building of a large new school, or a proposed tax cut. We stress the negative effects here because usually proposals for such action occur in the context of backers heralding their benefits, such as greater reduction of a certain medical problem, better appearance, more leisure, more educational opportunities, increased length of life, and more and/or improved commodities. However, because most actions have such widespread positive *and negative* impacts, we need to ask the following questions:

> - Which segments of society do not benefit from a proposed action? Who loses? What do the losers have to say about it?
> - How does the proposed action affect the distribution of power?
> - What are the action's effects on our health?
> - How does the action influence our relationships with one another? With the natural environment?

For each of these questions, we always have to ask, "What are the potential *long-term negative effects* of the action?"

 Attention: *Remember to ask, "What are the potential long-term negative effects of the action?" when considering omitted information.*

为了展示提出这些问题的用处，让我们思考下列问题：建设一所面积大的新学校会带来什么可能的负面效果？你有没有考虑到以下问题？

- 破坏环境。例如，建造这样一所新学校会不会涉及砍伐一片森林？失去栖息地会对当地的野生动物造成的影响？
- 教育质量的变化。如果新学校将其他学校有经验的教师或者天资聪颖的学生都吸引过去了怎么办？如果新学校吸纳了大部分拨付给当地学校的教育经费，而让其他学校享受不到同样的资金怎么办？
- 对房地产价格的影响。如果这座学校的办学质量相对于全国标准来说并不怎么样，那么这会对周边社区房屋的房地产价格产生怎样的影响？
- 增加税负。新学校的资金是怎么筹集的？如果新学校是一所公立学校，那么新学校的开张可能会导致当地社区不动产税的税率增加，这样才能支撑学校的运转。

这样的问题能让我们在追随被提倡的行动的浪潮时停下来思考一下。

如果省略的信息还是找不到

要求别人提供重要的缺失信息并不能保证别人给你一个让你满意的回答。很可能你探究的问题根本就无法回答。不要绝望！你已经尽力了。你要求得到这些做出决定之前必需的信息，现在你必须决定在找不到缺失的信息时是否还有可能得出结论。我们早先曾警告过你，论证的过程从来都是不完整的。如果你认为只要有信息仍然找不到，你就不能做出决断，那么这只会阻止你形成任何观点。

使用这个批判性问题

在你想到一个论证中存在缺失的信息后，应该怎么做呢？第一个符合逻辑的反应就是去寻找这个信息，但是通常情况下你会遇到一定的阻力。作为批判性思维者，你的选择是表达你对这个存在信息缺失的论证的不快，不断搜寻你所需要的信息，或者在这个论证比其他的论证好的基础上，谨慎地同意这个论证。

To illustrate the usefulness of asking these questions, let's reflect upon the following question: What are some possible negative effects of building a large new school? Did you think of the following?

- *Destruction of the environment.* For example, will building a new school involve the removal of a wooded area? How will the local wildlife be affected by the potential loss of habitat?
- *Shifts in quality of education provided.* What if the new school attracts skilled teachers or gifted students away from other schools? What if the new school absorbs a significant amount of the funds available to schools, depriving other schools of the same funds?
- *Effects of property values.* If the school does not do well in comparison with national standards, how will this affect the property values of the houses in the surrounding community?
- *Increased tax burden.* How will the new school be funded? If the new school is a public school, the opening of the new school could result in an increase in property taxes for the local community to help support the new school.

Such questions should give us pause before jumping on the bandwagon of a proposed action.

OMITTED INFORMATION THAT REMAINS MISSING

Just because you are able to request important missing information does not guarantee a satisfactory response. It is quite possible that your probing questions cannot be answered. Do not despair! You did your part. You requested information that you needed to make up your mind; you must now decide whether it is possible to arrive at a conclusion without the missing information. We warned you earlier that reasoning is always incomplete. Therefore, to claim automatically that you cannot make a decision as long as information is missing would prevent you from ever forming any opinions.

USING THIS CRITICAL QUESTION

Once you have thought about the existence of missing information in an argument, what should you do? The first logical reaction is to seek the information. But usually you will encounter resistance. Your options as a critical thinker are to voice your displeasure with the argument in light of the missing information, keep searching for the information that you require, or cautiously agree with the reasoning on the grounds that this argument is better than its competitors.

像批判性思维者那样写作和发言

正如我们在第 10 章中指出的那样,省略信息是论证过程中不可避免的一部分,作为写作者和发言者,我们将选择在论证中省略哪些信息,包含哪些信息。为了在我们自己的论证中体现批判性思维,针对省略的信息,我们要牢记以下几点:

1. 我们的论文应该尽量精确严谨。在为我们的论点写论文时,我们要确保没有省略的信息会影响到读者对我们的主张的理解能力。
2. 我们应该对自己的论点有足够的信息,知道自己在这个问题上的立场。尽管我们的论证不可避免地会省略一些信息,但在我们决定支持一个结论之前,我们要收集尽可能详尽的信息,以对自己的信念做出明智的决定。
3. 在构建自己的论证时,我们必须进行选择性的过滤。我们要仔细检查那些可以被包含进来的原因和证据,以确保重要信息不被遗漏。
4. 如果我们必须遗漏重要的信息,应该在探索反驳论证时处理这个遗漏。

在构建我们自己的论证时,我们可以通过精炼和聚焦论点来帮助我们的读者。在第 2 章和第 3 章的写作和发言部分,我们讨论了紧扣论点的重要性,以帮助我们的论点保持聚焦。在这一章,我们的重点是论点的严谨精炼。严谨精炼可以通过定义和缩小论点来实现。一个泛泛而谈或云笼雾罩的论点常常会遗漏掉那些能帮助读者或听众理解论证的信息。论点中遗漏掉的信息也会造成混乱,并有可能降低我们论证的有效性,因为我们认为自己支持或反对的内容可能不是读者或听众从我们的论证中推断出的。在我们得出结论之前,我们必须考虑哪些信息可能被省略掉了。如果遗漏的信息影响到我们论证的有效性,那么我们要在论点或引言中包含这些信息。

为了进一步说明这一点,下面举个例子。我们要完成一篇演讲稿的写作任务,回答"大学应该对每个人都免费吗"这个问题。假设我们确实相信大学应该为所有人提供免费教育,那么我们的结论将是"大学应该对所有人免费"。我们已经完成了写作聚焦的论点的第一步,但现在我们还必须问一问:"结论中遗漏了哪些信息?这些信息可能会影响我们论证的清晰度和说服力吗?"

我们可以通过查看论点中的关键术语来开始我们的检查过程。有什么信息可以帮助澄清论点中的关键术语吗?以"大学"为例,乍一看,其含义似乎很清楚,但我们指的是私立大学还是和公立大学?我们指的是所有的大学、社区学院以及文理学院吗?既包括经过

WRITING AND SPEAKING AS CRITICAL THINKERS

As we noted in Chapter 10, omitting information is an inevitable part of the argument process, and as a writer and speaker, we will be making choices about which information to omit and which to include. For the purposes of critical thinking in our own arguments, we want to keep these things in mind with regards to omitted information:

1. Our thesis should be as precise as possible. When writing the thesis for our arguments, we want to make sure that there is no omitted information that will affect our audience's ability to understand our claim.
2. We should have enough information about our thesis to feel that we know where we stand on the issue. Even though our argument will inevitably omit information, before we decide to support a conclusion, we want to gather as much information as possible about it so that we can make an informed decision about our own beliefs.
3. As we construct our argument, we have to engage in selective filtering. We want to carefully look over the possible reasons and evidence we could include to ensure that essential information is not left out.
4. If we must leave out important information, we should address this omission in our exploration of counterarguments.

When constructing our own arguments, we can help our audience greatly by a precise and focused thesis. In the writing and speaking section for Chapters 2 and 3, we discussed the importance of writing our thesis in response to an issue in order to help our arguments stay focused. In this chapter, our focus is on precision in the thesis. Precision can be achieved by defining and narrowing our thesis.

A broad or vague thesis often omits the information needed to help guide our reader or listener through our argument. Omitted information in the thesis also creates confusion and may detract from our argument's effectiveness because what we think we are arguing for or against may not be what our reader or listener infers. Before we proceed with a conclusion, we must consider what information might be omitted. And, if that omitted information will impact the effectiveness of our argument, we want to include that information in the thesis or introduction.

An example will help clarify. We are given the assignment to write a speech that answers the issue, "Should college be free for everyone?" Let's say that we do believe college should be free for all. So our conclusion is, "College should be free for everyone." We've done the first step of writing a focused thesis, but now we have to ask, "What information is omitted from that conclusion that might affect the clarity and persuasiveness of our argument?"

We can start our examination process by looking at the key terms in our thesis. Is there any information that might help clarify the key terms in our thesis? How about "college"? Seems clear at first glance, but do we mean private

学术机构认证的院校也包括未经学术机构认证的院校吗？既包括线上授课的也包括现场面授的吗？"免费"这个词又如何理解？必须有人支付教授和工作人员的薪酬，支付校舍和场地的费用，那么在这种情况下，"免费"是什么意思？税赋收入会承担一切费用吗？学生们享受免费学费但要支付宿舍、膳食和文娱等费用吗？这些问题帮助我们看到被省略的信息，以及为什么这种省略会为我们的读者或听众带来迷惑或含混不清。为使论点清晰准确，我们需要做一些修改。在继续阅读之前，试着自己另起炉灶拟一个论点，将我们提出的定义问题纳入考虑。

作为一个写作者或发言者，我们很容易故意遗漏我们知道与我们的结论不一致的信息，认为这样的策略会提高我们论证的效率。也许我们有这样的想法：读者或听众不会知道这些信息，那么我为什么要告诉他或她呢？但这是草率的想法，它不会得出有效的、强有力的论证。虽然这种论证可能会让我们对自己坚定的观点感到满意，但它们很快就会被批判性读者发现其漏洞百出、破绽尽现，无助于达到我们说服受众接受合理主张的目的。作为批判性思维者，我们要做的不是试图用任何可能的手段去说服读者或会话者，而是尽力使自己的论点得到充分有力的论证，不需要玩弄诡计就能令人信服。

为了尽可能在论证中包含重要信息，我们应该列出支持和反驳这个立场的论点。这一练习让你能够大致看到一个有效的、强有力的论证可能需要包含的内容。它还将帮助我们形成支持我们结论的高质量的理由和证据，并帮助我们加深对需要应对的反驳观点的理解。为使这份清单尽可能网罗更多我们论证过程中需要的重要信息，我们需要在思考和研究论题时提出一些批判性的问题。使用第292页的图表"寻找常见类型的重要信息的一些线索"可以帮助我们列出清单。并不是对每一个结论都要按照这个清单把所有线索都询问一遍，但是清单给我们准备了一个很好的起点。

让我们看看上面提供的结论的改写版本：

> 公立大学应该向州内所有的学生提供免学费教学。

请注意，新结论添加了重要信息，以缩小我们所说的"教育"和"免费"的范围。现在，让我们列出一个非常简短的支持和反驳论点的清单，这样我们就可以延伸我们对省略信息及其对我们论证的影响的讨论了。

> **支持：**
> 向州内所有学生提供免学费教学将缩小不平等差距。
> 人口受教育程度的提升可以使政府运行得更好，因此我们都将从教育更多的人中受益。

and public college? Do we mean all universities and community colleges and liberal arts schools? Accredited and nonaccredited? Online and face to face? And, how about the term "free"? Someone has to pay the professors and staff and pay for buildings and grounds, so what does "free" mean in this context? Will taxes pay for everything? Will students have free tuition and pay for things like dorms, food, and recreation programs? These questions help us to see the information that has been omitted and why that omission might cause confusion or a lack of clarity for our readers or listeners. To increase precision, our thesis is going to need some revision. Before reading on, try on your own to generate an alternative thesis that takes into account the definitional issues we've raised.

It is easy as a writer or speaker to intentionally leave out information that we know does not agree with our conclusion, thinking that such a strategy will increase our argument's effectiveness. Maybe, the thinking goes, the reader or listener won't know about that information, so why should I tell him or her? But this is sloppy thinking and will not lead to good, strong arguments. And while such arguments might make us feel good about our staunchly held opinions, they will quickly be seen as severely flawed by a critical reader and will not serve our purpose of convincing our audience of a well-reasoned claim. As critical thinkers, we are not trying to convince the reader or speaker using any means possible; we are trying to develop strong arguments that do not need trickery to be compelling.

To increase our likelihood of including important information in our arguments, we should make a list of the support and the counterarguments for that position. This exercise provides an overview of what might need to be included in a good, strong argument. It also will help us generate quality reasons and evidence that support our conclusion, as well as help us understand the counterarguments we will need to address. To increase the likelihood that the list will cast a wide net for possible information essential to our argument, we want to ask some critical questions as we think and research the issue. Using the chart on page 293, "Clues for Finding Common Kinds of Significant Information," can help us to make our list. Not all these clues will be necessary to ask for every conclusion, but the list gives us a good place to start.

Let's look at a rewritten version of the conclusion we offered above:

Public universities should offer free tuition to all in-state students.

Notice that our new conclusion adds important information to narrow the scope of what we mean by "education" and "free." Now, let's make a very brief list of support and counterarguments so we can extend our discussion of omitted information and its effect on our arguments.

Support:

Providing free tuition to all in-state students will narrow the inequality gap.

Governments function better with an educated population, so we will all benefit from educating more of our population.

反驳：
免交学费会增加每个人的税收，不管他们是否上大学。

即使学费免交，也不能保证人们会真的上大学，因为仍然有很多教育相关费用是不免交的。

为了发现遗漏的信息，我们需要问一些关于支持和反驳的问题。这里有几个我们在能够构建一个有效的论证之前需要回答的问题的范例：

1. 有证据表明不平等与教育有关吗？
2. 有证据表明免交学费会提高大学的入学率吗？
3. "受教育人口"和"运行得更好"是什么意思？
4. 就算学费全免，谁仍会被排除在大学门槛之外？
5. 有证据表明我们在大学接受的教育实际上提高了政府的效率吗？

从清单中我们可以看出，对于许多问题，我们都需要进一步的信息。如果我们得到了问题中的这些信息，我们可能会对论证做出不同的决定：免交学费真的像我们最初认为的那样有益吗？如果是的话，对谁有益？我们必须决定哪些遗漏的信息对论证至关重要，或者哪些可以省略掉。

此外，请注意，清单中既有支持论点的问题，也有反驳论点的问题。人们很容易想到一个结论的所有优点，而忽略缺点。确保你的清单不要过于积极。起初，强调积极的一面似乎是一件好事：我们的论点有这么多积极的方面，这显然是绝佳的立场！但是如果这个论点真的没有消极的方面，它就不是一个论点了，很可能也不值得写作或谈论了。

识别支持和反驳以及检查遗漏信息的步骤迫使我们检查甚至放弃我们的信念，因为我们发现了从未考虑过的关键信息。而第二步，即选择性过滤，在某些方面更困难。在收集到所有我们可能想要包含的信息之后，我们就必须决定哪些信息是最重要的，哪些是次重要的，哪些是不重要的。我们还必须适当考虑这样一个事实：我们不想让我们的读者厌烦得要死。有可能我们会以吸引受众的名义加入一些无关紧要的内容。

记住，我们永远不会拥有所有的信息。有些问题是无法回答的。我们没有时间去研究完美无缺的论点具备的其他要素。但我们想要确保大多数批判性读者提出的问题都得到解决。

Counterarguments:

Free tuition will increase taxes for everyone regardless of whether they attend college.

There is no guarantee that people will actually attend universities if tuition is free because there are still many costs associated with an education that won't be free.

To uncover the omitted information, we need to ask questions about our support and counterarguments. Here is a sample of just a few of the questions that need to be answered before we can construct an effective argument:

1. Is there any evidence to show that the inequality gap is related to education?
2. Is there any evidence to show that free tuition will increase attendance in college?
3. What is meant by "educated population" and "function better"?
4. Even if tuition were free, who would still be excluded from attending college?
5. Is there evidence that the kind of education we get at a university actually increases the effectiveness of governments?

As we can see from our list, there are lots of questions for which we need further information. If we had the information that these questions raise, we might make a different decision about our argument: Is free tuition really as beneficial as we originally thought, and if so, for whom? We have to decide which omitted information is essential to our argument or can be left out.

Moreover, note that our list has both support *and* counterarguments. It's easy to think of all the good things about a conclusion and leave out the bad. Make sure your list is not overly heavy on the positive. At first, emphasizing the positive might seem like a good thing: Our argument has so many positives, it's clearly the best position! But if the argument really has no negative sides to it, it's not an argument and is probably not worth writing or speaking about.

While the step of identifying support and counterarguments and checking for omitted information forces us to examine and perhaps even discard our beliefs as we identify key information that we hadn't previously considered, the second step, selective filtering, in some ways is even more difficult. Once we've gathered all the information that we might want to include, we have to decide which information is most essential, which is just somewhat essential, and which isn't essential at all. We also have to give due consideration to the fact that we do not want to bore our readers to death. It's possible that we may want to include something inessential in the name of engaging with our audience.

Remember, we will never have *all* the information. Some questions are not answerable. And we just don't have the time to research other elements of a perfect argument. But we want to make sure that the questions that most critical readers ask have been addressed.

思维体操

⚠ **批判性问题：有什么重要信息被省略了？**

下面每篇文章都有重要的信息缺失了。将你要问这些文章作者的问题列出来。解释对于每篇文章，为什么你所寻找的信息在你设法判断这篇论证的价值时非常重要。

第一篇

研究显示大学生特别容易变肥胖。说得更具体一点，最近一项持续长达十年的研究揭示，大学生的肥胖率一直呈上升的趋势。2002 年，肥胖的大学生比例为 25.4%，到 2012 年，肥胖的大学生比例达到 30.2%。研究人员指出，大学生肥胖现象变得越来越普遍的主要原因是缺乏获得健康食物的渠道以及饮酒量的增加。

第二篇

克隆技术有助于医疗领域取得很多积极的突破。如果我们适当地发展克隆技术，人们就无须因为缺少器官捐助者而死亡了。有了克隆技术，研究人员可以为那些急需做器官移植手术的人人工培育新器官。此外，因为这些器官是用病人自己的人体组织克隆出来的，所以根本不会再出现病人的身体排斥移植器官的情况。克隆的器官可以在没有头颅的身体中进行培育，这样就不会牵涉到以一个人的死亡为代价来拯救另一个人的生命。克隆的另一个优点是它将有助于人们战胜疾病。通过克隆产生的某些蛋白质可以用来治疗糖尿病、帕金森症和囊性纤维化等疾病。

第三篇

标准化考试常常声名狼藉，但最近的研究表明，标准化考试实际上可以帮助学生在学校取得成功。最近的一项研究调查了美国各地的 500 所高中。一些学校被要求在学年期间参加一次标准化考试，而另一些学校则被要求在学年期间参加两次标准化考试。参加两次考试的学校的毕业率比只参加一次考试的学校高 20%。与只参加一次考试的学校的学生相比，参加两次考试的学校的学生升入大学的比例高出 30%。这些发现清楚地表明了标准化考试的好处。

PRACTICE EXERCISES

 Critical Question: ***What significant information is omitted?***

In each of the following examples, there is important missing information. Make a list of questions you would ask the person who wrote each passage. Explain in each case why the information you are seeking is important to you as you try to decide the worth of the reasoning.

Passage 1

Studies have shown that college students are vulnerable to obesity. More specifically, a recent research study conducted over a 10-year period has revealed that obesity rates among college students are on the rise. In 2002, 25.4 percent of college students were obese, and in 2012, 30.2 percent of college students were obese. Researchers have suggested that the main reasons for the prevalence of obesity among college students is the lack of access to healthy foods and the increase in alcohol consumption.

Passage 2

Cloning technology can lead to many positive breakthroughs in the medical field. If we were to adequately develop cloning technology, there would no longer be a need for people to die because of a lack of organ donors. With cloning, researchers could artificially develop new organs for people in need of transplants. Plus, because these organs would be cloned from the person's own tissues, there would be no chance of her body rejecting the transplanted organ. The cloned organs can be made in bodies that lack a head, and thus would not involve a "death" in order to save a life. Another advantage of cloning is that it can help fight diseases. Certain proteins produced by clones can be used to fight diseases such as diabetes, Parkinson's, and cystic fibrosis.

Passage 3

Standardized testing often gets a bad name, but recent research shows that standardized testing may in fact help students succeed in school. A recent study looked at 500 high schools throughout America. Some of the schools were asked to take one standardized test during the school year, and the other set of schools was asked to take two standardized tests during the school year. The schools that took two tests had a graduation rate 20 percent higher than the schools that took only one test. The schools that took two tests also had 30 percent more students go on to college compared to the one-test schools. These findings clearly indicate the benefits of standardized testing.

给个提示

第一篇

　　结论：大学生特别容易肥胖。

　　理由：最近的一项研究发现大学生肥胖的证据，肥胖的原因是他们缺乏健康食物和饮酒增多。

有没有什么信息被省略了？在哪些其他方面（比如社会经济阶层和生活压力上），大学生和社会其他人群相比显得比较特殊，可能让他们更容易变肥胖？这些研究结果有没有在其他的研究当中发现？研究的参与人员是怎么选择的？例如，自愿参加这项研究的人是不是和随机选择的样本有所不同，因此会限制概括的范围？

第二篇

　　结论：克隆可以带来积极的医学效果。

　　理由：1. 克隆可以用于器官移植。

　　　　　2. 克隆可以用来帮助人类战胜某些疾病。

　　首先，这个论证呼吁我们去追求一项新技术——人体克隆，而且只援引了它的好处。作者省略了可能存在的不利之处。我们需要综合考虑其有利之处和不利之处。克隆人体器官可能带来什么严重的副作用？克隆器官是不是和正常器官一样稳定？克隆技术会给人类的决策带来什么样的正面和负面作用？如果人们知道新器官可以被任意培育出来取代现有的器官，他们会不会对身体和器官不再那样精心呵护？克隆技术的可得会不会导致人们为了不可告人的目的而滥用，以制造完整的克隆人？人们会不会克隆出自己，让地球上现有的人口负担又加重一层？或许这项技术的优点大于它的缺点，但是我们需要明白地知道它的优缺点，这样才能判断这个结论的价值。

　　此外，让我们进一步看看和研究相关的缺失信息。你有没有发现这篇文章没有引用过一项研究？实际上，论证没告诉我们的是，在美国，根本没有发生过一例将克隆技术应用于人体的实验。因此，所有这些有关克隆技术的优点的讨论都只是假设而已。真实的实验会不会证明这些假设的优点确实有可能存在？我们根本不知道。

Sample Responses

Passage 1

CONCLUSION: *College students are vulnerable to obesity.*

REASON: *A recent research study found evidence of obesity among college students due to lack of healthy foods and increased alcohol consumption.*

Is there any omitted information? In what other ways, such as socioeconomic class and life stressors, do college students differ from the rest of the population that may make them more vulnerable to obesity? Have these results been found in other studies? How were participants selected? For example, do volunteers for such studies differ from a random sample, thus limiting the generalization?

Passage 2

CONCLUSION: *Cloning can provide positive medical benefits.*

REASONS: 1. *Clones can be used for human transplants.*
2. *Clones can be used to help combat certain diseases.*

First, this reasoning urges us to pursue a new technology—human cloning—and cites only its advantages. The writer omits possible disadvantages. We need to consider both advantages and disadvantages. What serious side effects might result from using cloned organs? Are cloned organs as stable as regular organs? What positive and negative effects might cloning technology have on human decision making? Would people be less likely to take care of their bodies and their organs if they knew that new organs could be grown to replace their current ones? Would the availability of the technology lead people to misuse cloning to produce complete human clones for an insidious purpose? Would people clone themselves, helping add to the burden already placed on Earth by the current population? The advantages of the procedure may well outweigh the disadvantages, but we need to be aware of both in judging the merits of the conclusion.

Furthermore, let's look at the missing information regarding the research. Did you notice that no research has been cited? In fact, the argument fails to tell us that no tests on human cloning have occurred in the United States. Therefore, all of the discussion on the benefits of cloning is hypothetical. Would actual research prove the hypothetical benefits to be possible? We do not know.

第 12 章

能得出哪些合理的结论

| 学 习 目 标 |

1. 意识到二分式思维的危险。
2. 培养灰度思维,而非黑白思维。

到了这个阶段,你已知道该怎样"披沙拣金"——把坚实的理由和牵强的理由区分开来。请思考下面这个论证:

> 大型企业花大量时间和金钱对孩子进行广告轰炸。儿童节目中精心设计了各种商业广告,竭力向孩子推销最新款的玩具,告诉孩子只有得到这些新玩具,他们才能活得幸福快乐。向孩子打广告这种行为简直令人发指,应该被宣布为非法。孩子根本不能客观评价他们看到的广告,向孩子打广告实际上是给他们的父母戴上了紧箍,让他们要么对孩子说"不",使得孩子不高兴,要么对孩子的各种要求有求必应,最终宠坏孩子。

你是否应该呼吁本地的议员宣布针对儿童的广告为非法的?假设你检查过作者的理由,发现它们都很确凿可信。还有没有其他的结论可能和作者这个结论一样与这些理由契合无间?本章将会提出几个可能存在的备选结论。换句话说,理由本身并不会引导出一个唯一

CHAPTER

12

What Reasonable Conclusions Are Possible?

LEARNING OBJECTIVES

1. Become aware of the dangers of dichotomous thinking.

2. Develop grey thinking rather than black-and-white thinking.

By this stage, you know how to pan for intellectual gold—to distinguish stronger reasons from weaker ones.

Consider the following argument:

Large corporations spend far too much time and money advertising to children. Children's programming is riddled with commercials trying to sell them the latest toy, telling the children they will not be happy unless they have it. The practice of advertising to children is horrendous and should be illegal. Advertising to children, who cannot critically evaluate the ads they see, puts a strain on parents to either say "no" to their children and have them get upset, or to give in to their children's demands, ultimately spoiling the children.

Should you urge your local congressperson to criminalize advertisements to children? Suppose you checked the author's reasons and found them believable. Are there other conclusions that might be as consistent with these reasons as the author's conclusion? This chapter suggests several possible

的结论。它们好像会带领我们走向一个可靠的结论,但它们常常为不止一个结论提供基础。

你很少会遇到只能从理由中合理推断出一个结论的情况。在第9章中,我们讨论过替代原因的重要性,那里的重点是,对一个特定的结果来说,可能会存在不同的原因。本章则主要关注可以从同一套理由中推断出的多个备选结论。

因此,你必须确保最终采纳的结论最合乎情理,和你的价值偏好最吻合一致。如果你发现了其他的备选结论,你就能更好地从一系列可选的结论中发现那个最说得通的结论。

⚠ **批判性问题:可能得出什么合理的结论?**

二分式思维:妨碍我们考虑多种可能性

重大的问题很少能用简单的"是"或绝对的"不是"来回答。当人们习惯用非黑即白、非是即否、非对即错、非正即误式的方式来思考问题时,他们就是在使用二分式思维。这种类型的思维往往假设一个可能存在多种答案的问题只有两个可能的答案。持这种思维方式的人总是看到和提到一个问题的两个方面,好像天下所有问题都只有两面一样,这种习惯对我们的思维具有毁灭性的破坏效果。

前面我们讨论"虚假的两难选择"谬误时也遇到过这种二分式思维。这种谬误,以及一般的二分式思维,都因为过度限制我们的视野而破坏了论证的进程。我们以为考虑了两个选项之后就万事大吉,因此忽略了很多其他的选项,从而错过从其他选项中进行选择可能会带来的积极后果。

二分式思维者常常比较僵化,容不得异议,因为他们不能理解语境对特定答案的重要性。为了更清楚地看到这一点,请你想象一下下面的情形:

你的室友请你帮他构思一下生物学的论文。这份论文要设法解答以下问题:"科学家是否应该继续从事干细胞研究?"在他看来,这份论文需要他选择一个"应该"或"不应该"的立场,然后再加以论证。

你已经知道二分式思维可以通过限定结论的条件,通过将结论放到具体的语境中来加以避免。这种限定的过程需要你对任意一个结论提出以下问题:

1. 结论在什么时候是准确的?
2. 结论在什么地方是准确的?
3. 结论为什么或为了什么目的是准确的?

alternative conclusions. In other words, reasons by themselves do not lead to only one conclusion. They seem to take us to a reliable conclusion. But they often provide a basis for more than one conclusion.

Very rarely will you have a situation in which only one conclusion can be reasonably inferred. In Chapter 9, we discussed the importance of rival causes. The point was that there are different possible causal bases for a particular causal conclusion. This chapter, however, focuses on the alternative *conclusions* that are all possible outcomes from a single set of reasons.

Consequently, you must make sure that the conclusion you eventually adopt is the most reasonable and the most consistent with your value preferences. Once you find alternative conclusions, you will be better prepared to discover a stronger conclusion from among the enlarged number of options.

 Critical Question: **What reasonable conclusions are possible?**

DICHOTOMOUS THINKING: IMPEDIMENT TO CONSIDERING MULTIPLE CONCLUSIONS

Very few important questions can be answered with a simple "yes" or an absolute "no." When people think in black or white, yes or no, right or wrong, or correct or incorrect terms, they engage in *dichotomous thinking*. This type of thinking assumes there are only two possible answers to a question that actually has multiple potential answers. This habit of seeing and referring to *both* sides of a question as if there are only two has devastatingly destructive effects on our thinking.

We encountered dichotomous thinking earlier when we discussed the either-or fallacy. This fallacy, and dichotomous thinking in general, damages reasoning by overly restricting our vision. We think we are finished after considering two optional decisions, thereby overlooking many options and the positive consequences that could have resulted from choosing one of them.

Dichotomous thinkers often are rigid and intolerant because they fail to understand the importance of context for a particular answer. To see this point more clearly, imagine this situation:

Your roommate asks you to help plan her biology paper. The paper is to address the question, "Should scientists pursue stem cell research?" In her mind, the paper requires her to defend a "yes" or "no" position.

You have learned that dichotomous thinking can be avoided by qualifying conclusions, by putting them into context. This qualification process requires you to ask about any conclusion:

1. *When* is it accurate?
2. *Where* is it accurate?
3. *Why* or *for what purpose* is it accurate?

然后你开始将这个过程应用到这篇论文中。

当你解释说在某个特定的时间、特定的环境里，为了最大化地实现某种价值或目标，人们应该允许干细胞研究，你的室友因而变得越来越泄气时，你会不会觉得惊讶？他本来是在找"应该"或"不应该"，而你则提供了一个复杂的"这取决于……"的思路。

僵化的二分式思维限制了你决定和选择的范围。更糟糕的是，它过度简化了复杂的情况。采用二分式思维的人很容易就变成一条糊涂虫。下一节展示了二分式思维带来的局限性。

灰度思维：两面还是多面

几乎我们能想象到的任何重要问题都不只有两个可能的答案。因此，非黑即白的思维就像正反思维一样，用一种简化了的形式掩盖了一个问题可能出现的回应方式的复杂性，这种简化了的形式隐匿了一个问题可能的答案的多样性。

例如，假设我们想知道是否应该主修化学。我们需要背景信息来回答这样的问题，需要很多背景信息。我们是否有足够的能力来胜任化学工作者的工作？我们是否发现自己一想到化学，就心旌荡漾？有什么其他的选择方案？化学领域是否有抵制高级人工智能设备出现的工作机会？与化学工作者的岗位需求相比，主修化学的人数量如何？那些将把我们培养成化学工作者的人技术有多娴熟？在化学业界，他们有相互联系的网络吗？在我们回答是否应该主修化学这个问题之前，这些只是我们需要探索的问题中的几个例子。

灰度思维是一种反思型思维，致力于探索用"是"和"不是"之外的其他方式回答复杂的问题。非黑即白的思维看似迅捷而令人满意，但当我们对黑白答案的混合体视而不见时，我们就误解了通往对后果的明智考虑和可能的成就的路径。

在我们一起看看几个可能存在多种结论的论证之前，首先让我们确保你已经体会到，多数重要的争议都有可能会出现大量的结论。我们先来看一个在美国历久弥新的问题：

> 美国应不应该在其他国家进行维和行动？

乍一看，这个问题以及很多类似的问题好像是在寻求一个"是"或"不是"的答案。然而，一个附加限定条件的"是"或"不是"往往才是最佳的答案。用"也许"或者"这取决于……"进行回答的优点，就是它迫使你承认，你所知道的一切还不足以给出确定无疑的回答。但你在避免一个确定答案的同时，也形成了一个不确定的决定或观点，需要进

You then begin to apply this process to the paper assignment.

Would you be surprised by your roommate's growing frustration as you explained that at certain specified times, in certain situations, to maximize particular values or objectives one should allow stem cell research? She's looking for "yes" or "no"; you provided a complicated "it depends on ..."

Rigid, dichotomous thinking limits the range of your decisions and opinions. Even worse, it overly simplifies complex situations. As a consequence, dichotomous thinkers are high-risk candidates for confusion.

The next section illustrates the restrictive effects of dichotomous thinking.

GREY THINKING: TWO SIDES OR MANY?

Almost no important question we can imagine would have only two possible answers. Black-and-white thinking, therefore, like pro-and-con thinking, is masking the complexities of possible responses to a question in a simplified form that hides the robust array of possible answers to a question.

For example, suppose we wondered whether we should major in chemistry. We need context to answer a question like that, and lots of it. Do we have the kind of intelligence that would permit us to function as a chemist? Do we find ourselves energized by the very thought of chemistry? What are the other options like? Are there jobs in the field of chemistry that will resist the emergence of smart artificial intelligence devices? How many other people major in chemistry compared to the probable demand for chemists? How skilled are those who would train us to be a chemist? Do they have a network of contacts in the world of chemistry? Those are just a few of the contingencies we would need to explore before we answered the question of whether we should major in chemistry.

Grey thinking is reflective thinking devoted to exploring the options to "yes" and "no" as answers to complex questions. Black-and-white thinking is deceptively quick and satisfying, but when we do not see the blending of the white and the black answers, we have misunderstood the route to a wise consideration of consequences and possible achievements.

Before we look at several arguments in which multiple conclusions are possible, let's make sure you appreciate the large number of conclusions that are possible with respect to most important controversies. Let's look at a question that is always alive in the United States.

> Should the United States engage in peacekeeping in other countries?

At first glance, this question and many like it seem to call for a "yes" or "no" answer. However, a qualified "yes" or "no" is often the best answer. The advantage of "maybe" or "it depends on" as an answer is that it forces you to admit that you do not yet know enough to make a definite answer. But at the same time you avoid a definite answer, you form a tentative decision or opinion that calls for commitment and eventual action. It's wise to seek additional

一步投入精力和最终行动。寻找额外信息以增强对你的观点的支撑力度是明智的做法，但是到了一定程度你就必须停止搜寻并做出决定，即使你愿意为之辩护的最有力的回答是"是，但是……"。

问问你自己，在回答这个关于美国干涉其他国家的问题时，可能会得出什么结论。自然，简单地回答"是"或"不是"就可以得出两个结论。还有没有其他的结论？当然，还有很多结论！让我们来看看这个问题的一部分可能的答案。

注意，我们给每种情况都添加了一个必要条件，这样才能证明这个结论。在没有任何数据或定义的前提下，这五个结论中的任何一个都可能是最合理的。而它们只是这个问题诸多可能的结论中的一小部分。

美国应不应当涉足其他国家的和平行动？

1. 应该，当这个国家和美国有千丝万缕的联系时，例如沙特阿拉伯。
2. 应该，如果美国要被看成唯一的超级大国，有责任维护世界和平。
3. 应该，如果美国的角色定位是维护和平而不涉及打一场战争。
4. 应该，当美国的海外经济利益面临危险的时候。
5. 不应该，美国国内亟待处理的问题已经多不胜数，根本不应该在其他国家浪费时间。

条件句的生产力

如果你回头重温一下本章中讨论的所有备选结论，就会注意到每个备选的结论都有可能成立，因为我们缺少了某些信息、定义、假设，或者分析这些理由的人的参照系。所以，我们可以谨慎地使用条件句来创造多种结论。在条件句中，我们陈述一个假设的条件，目的是帮助我们得出某个特定的结论。注意，使用条件句让我们能得出某个结论，而不用假装自己对某个具体争论比实际知道得更多。对于批判性思维者来说，一个非常重要的价值观就是保持谦卑。经常使用条件句和谦卑这个重要的价值观是一致的。

当你在结论前使用条件句时，你就指出了这个结论是建立在你所不确定的特定断言或假设的基础上。为了帮助你理解我们的意思，请看看下面这些可能会作为结论的引导的条件陈述句。

1. 如果税收减免政策是针对那些低收入人群，那么……

information that would improve the support for your opinions, but at some point you must stop searching and make a decision, even when the most forceful answer you are willing to defend is a "yes, but ..."

Ask yourself what conclusions would be possible in response to the question about U.S. intervention in other countries. Naturally, a simple "yes" or a "no" answer would be two possible conclusions. Are there others? Yes, there are many! Let's look at just a few of the possible answers to the question.

Notice that in each case we added a condition necessary before the conclusion can be justified. In the absence of any data or definitions, any of these five conclusions could be most reasonable. These five are just a few of the conclusions possible for the first question.

Should the United States Engage in Peacekeeping in Other Countries?

1. Yes, when the country is intricately tied to the United States, such as Saudi Arabia.
2. Yes, if the United States is to be perceived as the sole superpower responsible for maintaining world peace.
3. Yes, if the U.S. role is to be limited to keeping peace and does not involve fighting a war.
4. Yes, when our economic interests abroad are at stake.
5. No, the United States has enough domestic problems to handle such that we should not spend time in other countries.

PRODUCTIVITY OF IF-CLAUSES

If you go back over all the alternative conclusions discussed in this chapter, you will notice that each optional conclusion is possible because we are missing certain information, definitions, assumptions, or the frame of reference of the person analyzing the reasons. Consequently, we can create multiple conclusions by the careful use of *if-clauses*. In an if-clause, we state a condition that we are assuming to enable us to reach a particular conclusion. Notice that the use of if-clauses permits us to arrive at a conclusion without pretending that we know more than we actually do about a particular controversy. The frequent use of if-clauses is consistent with the value of humility that is so important for critical thinkers.

When you use if-clauses to precede conclusions, you are pointing out that your conclusion is based on particular claims or assumptions about which you are uncertain. To see what we mean, look at the following sample conditional statements that might precede conclusions.

1. If the tax cut is targeted toward those at the lower end of the economic spectrum, then ...

2. 如果一部小说包含一个极易辨认的正面人物，一个一望而知的反面人物，以及一段扣人心弦的高潮，那么……

3. 如果汽车制造商能制造出油耗更低的汽车，那么……

使用条件句特别有助于为评价型的论证（例如评价音乐、艺术、大学或者总统演说的质量如何的论证）找到合理的结论，因为这些论证需要我们对使用什么标准来进行评价选定立场。

条件句为你提供了多种结论，在对争论做出评判之前，你应该先评估一下这些结论。它们也拓宽了可能的结论范围，从中你可以挑选出自己的立场。

识别多个结论能解放思维

如果逻辑、事实或者研究可以自圆其说，我们就会用特定的方式来学习。我们的任务就是请教另一个人，也许是老师，让他来告诉我们应该持有的信念。具体说来，我们会去寻找逻辑和事实昭示的那唯一一套信念。

虽然我们对于逻辑和事实怀有极大的敬意，但我们不能过分夸大它们作为形成结论的先导的价值。它们只能引导我们走到一定地步，然后我们就不得不借助逻辑和事实提供的帮助，自己走完通往确定看法的剩余道路。

要借助逻辑和事实为我们提供的帮助，第一步就要寻找可能存在的与我们所知的逻辑和事实相一致的结论。这种寻找让我们获得了极大的解脱，让我们从上文勾勒的那种刻板僵化的学习模式中解脱出来。在认识到种种可能存在的结论之后，我们每个人都会体验到个人选择的力量得到增强的那种激动。

小结

理由很少会只指向一种结论。在评估一套理由以后，你还要判断什么结论和争议中最好的理由最为契合。为了避免在寻找最好的结论时出现二分式思维，你可以使用前面提到过的"什么时候""什么地方"和"为什么"等问题来为这些结论添加限制性的语境。

为结论添加的限制将会让你远离二分式思维。条件句就是表达这些限制所需的技术。例如，让我们再来看一下本章一开始提到的关于限制针对儿童的广告的论证。什么样的备选结论可能与给出的理由相一致？

　　　　作者的结论：针对儿童的广告应该被判定为非法。

2. If a novel contains an easily identifiable protagonist, a clear antagonist, and a thrilling climax, then …

3. If automakers can make cars that are more fuel efficient, then …

Generating if-clauses is especially helpful in determining reasonable conclusions for evaluative arguments, such as those evaluating the quality of music, art, colleges, or a president's speech, because such arguments require taking a position on what criteria to use for making the evaluations.

If-clauses present you with multiple conclusions that you should assess before making up your mind about the controversy, and they also broaden the list of possible conclusions from which you can choose your own position.

THE LIBERATING EFFECT OF RECOGNIZING ALTERNATIVE CONCLUSIONS

If logic, facts, or studies were self-explanatory, we would approach learning in a particular manner. Our task would be to have someone else, a teacher perhaps, provide the beliefs that we should have. Specifically, we would seek that single identifiable set of beliefs that logic and facts dictate.

While we have tremendous respect for logic and facts, we cannot unduly exaggerate their worth as guides to forming a conclusion. They take us only so far; then we have to go the rest of the way toward belief, using the help that facts and logic have provided.

A first step in using the help that facts and logic provide is the search for possible multiple conclusions consistent with logic and the facts as we know them. This search liberates us in an important way. It frees us from the inflexible mode of learning sketched above. Once we recognize the variety of possible conclusions, each of us can experience the excitement of enhanced personal choice.

SUMMARY

Very rarely do reasons fit just one conclusion. After evaluating a set of reasons, you still must decide what conclusion is most consistent with the best reasons in the controversy. To avoid dichotomous thinking in your search for the strongest conclusion, provide alternative contexts for the conclusions through the use of "when," "where," and "why" questions.

Qualifications for conclusions will move you away from dichotomous thinking. If-clauses provide a technique for expressing these qualifications.

For instance, let's take another look at the argument at the beginning of this chapter for restricting advertisements aimed at children. What alternative conclusions might be consistent with the reasons given?

AUTHOR'S CONCLUSION: *Advertisements aimed at children should be illegal.*

备选的结论：1. 如果企业被当作人一样对待，那么它们也有言论自由的权利，其中就包含了广告权。因此，它们打广告的权利不应该受到限制。

2. 如果可以证明儿童评价不了他们所看到的东西，因此会深受他们看到的广告的影响，那么针对儿童的那些广告理应被判定为非法。

3. 如果提议的立法旨在限制针对儿童的广告的内容，那么政府不应该判定这类广告非法，而应该在规范针对儿童的广告的内容方面扮演积极的角色。

根据作者提供的理由，很多其他的备选结论都有可能成立。如果我们不考虑这些备选结论，不把它们当成形成自己看法的基础，那么我们决策的质量就会大大缩水。

思维体操

⚠ **批判性问题：** 可能得出什么合理的结论？

请找出以下每个论证中可以从理由得出的不同结论。

第一篇

为很多人提供伙食并不是一件容易的事，但是学校食堂应该尽量满足不同口味的人的饮食需求。整个校园的学生都在异口同声地抱怨，不仅抱怨食堂饭菜的质量，而且抱怨食堂饭菜千篇一律，无可选择。其实食堂需要做的只是提供种类繁多的饭菜来取悦更多的学生而已，这样就能让更多的学生选择在食堂就餐，而不用去校外觅食。学校后勤服务如果没有每天提供种类丰富的伙食选择，就是没有尽到为学生服务的职责。

ALTERNATIVE CONCLUSIONS:
1. *If corporations are to be treated as persons, then they have a right to free speech that includes advertisements; thus, their right to advertise should not be limited.*

2. *If it can be demonstrated that children are unable to assess what they view, and thus are heavily influenced by the advertisements they see, then advertisements aimed at children should be illegal.*

3. *If the purpose of the proposed legislation is to limit the content of advertisements aimed at children, then the government should not make such ads illegal, but rather take a more proactive role in regulating the content of advertisements aimed at children.*

Many additional alternative conclusions are possible in light of the author's reasons. We would shrink the quality of our decision making if we did not consider those alternative conclusions as possible bases for our own beliefs.

PRACTICE EXERCISES

 Critical Question: **What reasonable conclusions are possible?**

For each of the following arguments, identify different conclusions that could be drawn from the reasons.

Passage 1

Feeding large numbers of people is not easy. However, dining halls on campus should try to accommodate a larger variety of tastes. Students across campus consistently complain not only about the quality of food but also about the lack of selection they find in the dining halls. All that the dining halls need to do is offer a wider range of food to better please more students, and thus keep more of them eating on campus as opposed to eating off campus. Dining services are failing in their duty to the students when they do not provide a large selection of food options every day.

第二篇

我最近发现教堂可以不用缴税。这个例外违背了美国宪法中要求宗教和国家相分离的条款。通过给教堂免税这样的措施，政府其实是在经济层面上支持宗教。对教堂减免税收迫使美国人支持宗教，即使他们反对身边的宗教信条。教堂不应该再继续享受任何这类的免税待遇。

第三篇

特朗普总统最近宣布了一项计划，该计划将使所有上社区大学的人享受免交学费的待遇。这种对教育的补贴使我们离美国的资本主义根基更远了。此外，如果让更多人能有途径上大学，那么进入技术工人行业的人会越来越少。美国已经面临技术工人短缺的紧迫问题，而这个项目只会使短缺更加严重。

给个提示

第一篇

结论：学校食堂没有充分履行为学生提供膳食的职责。

理由：1. 学生对饭菜质量不满意。

2. 每天提供的饭菜选择范围有限。

3. 更多选择会让学生心情舒畅，让他们愿意留在学校用餐。

要练习这项特殊的批判性思维技能，我们首先需要假设其理由站得住脚。如果我们接受这些理由，认为它们都很可靠，那么我们也能合情合理地推断出下面这些结论：

如果食堂服务的目标是要提供种类繁多的饭菜，同时还要确保每天用餐结束后浪费的饭菜最小化，那么他们目前为学生提供的饭菜选择就并没有让学生失望。

如果食堂的目标是让校园饭菜的价格维持在最低水平，而提供一份丰富的菜单会导致饭菜的价格上涨，那么他们对学生就不算没有尽职尽责。

注意，备选的结论和原始结论中对食堂的负面描述相比，对食堂的评价完全不同。

Passage 2

I recently discovered that churches are exempt from taxes. This exemption is a violation of the separation of church and state required by the U.S. Constitution. By providing churches with tax exemptions, the government is financially supporting religion. A tax break for churches forces Americans to support religion, even if they oppose the religious doctrine at hand. Churches should no longer receive any of these tax exemptions.

Passage 3

President Trump recently announced his plans for a program that would make community college free for anyone who attends. This subsidizing of education takes us even further away from America's capitalist roots. Furthermore, by giving more people access to college, fewer and fewer people will go into skilled labor trades. America is already experiencing a shortage of skilled laborers, and this program would only worsen the shortage.

Sample Responses

Passage 1

CONCLUSION: *Dining services are not doing an adequate job of providing food on campus.*

REASONS:
1. *Students are upset about the quality of the food.*
2. *There are not enough options provided every day.*
3. *More options would keep students happy and keep them eating on campus.*

To work on this particular critical-thinking skill, we need to assume that the reasons are strong. If we accept these reasons as reliable, we could also reasonably infer the following conclusions:

If dining services' goal is to provide a wide selection of food while ensuring the least amount of wasted food at the end of the day, then they are not letting students down with the current selections offered to students.
　　If dining services aim to keep the price of on-campus food down, and a more expansive menu would cause an increase in prices, they are not failing in their duty to students.

Notice that the alternative conclusions put dining services in quite a different light compared to the negative portrayal they received in the original conclusion.

第二篇

　　结论：教堂不该再继续享受免税待遇。

　　理由：1. 对教堂的免税政策违背了美国宪法中规定的政教分离的条款。

　　　　　2. 对教堂的免税政策迫使美国人支持宗教，即使他们反对身边的宗教信条。

基于这些理由，我们可以得出好几个合理的结论。

（1）如果美国宪法和这些免税政策之间产生了冲突，也许我们需要通过司法解释来修订宪法，就像法庭调整其他规定来适应当代的需求一样。

（2）第二个理由可能会导致我们得出这样的结论：我们需要在学校里提高民众的教育水平。为什么？因为民主的立法过程的本质就是公民总要为某些他们作为单独个体不情愿去做的事情缴税。如果国会说我们需要一支庞大的军队，那么公民就需要缴税来支撑这支庞大的军队，不管我们作为个体愿不愿意支持国防开支。

Passage 2

CONCLUSION: *Churches should no longer receive tax exemptions.*

REASONS: 1. *Tax exemptions for churches are a violation of the separation of church and state required by the U.S. Constitution.*

2. *Tax breaks for churches force Americans to support religion, even if they oppose the religious doctrine at hand.*

On the basis of these reasons, we could infer multiple reasonable conclusions:

(1) If there is a conflict between the Constitution and these tax exemptions, perhaps we need to modify the Constitution via legal interpretations just as the courts adjust other rules to adapt to contemporary needs.

(2) This second reason can lead to the conclusion that we need to improve civics education in our schools. Why? Because the nature of a democratic legislative process is that citizens are always going to be paying taxes for some things that they as individuals do not want. When the Congress says we need a huge military, then citizens pay their taxes to support that large military regardless of whether we as individuals support the scope of defense spending.

第 13 章

干扰批判性思维的障碍

| 学 习 目 标 |

1. 分析批判性思维的多重障碍。
2. 认清提出批判性问题导致他人不适、快思考、信念固着、答非所问、自我中心和一厢情愿的想法的危险。

即使你已经学会批判性思维,并且计划用它来塑造一个更有思想的自己,接下来,在使用你已掌握的批判性思维的道路上,你也仍会碰到许多障碍,就好像你在开车时遇到减速带。之所以称它们为"减速带",是因为:(1)只要你放慢速度,你就可以跨过它们;(2)不管你有没有注意到,它们就在那里;(3)即使你注意到了它们,它们也照样会干扰你的前进。

不过,了解潜在的问题是解决问题的第一步。因此,最后这一章旨在提醒你,在成为一个批判性思维者的道路上,该注意哪些让你放慢速度的减速带。如果你了解妨碍批判性思维的障碍,你就可以努力让自己在这些威胁出现时更加谨慎。

正确提问带给人的不快

第 1 章里提到过,批判性思维是一项社交活动。想和别人打成一片,我们首先要愿意提出一些正确的问题,以便了解其他人的看法。一定要记住,并不是每个人都乐于见到自

CHAPTER 13

Speed Bumps Interfering with Your Critical Thinking

LEARNING OBJECTIVES

1. Analyze the multiple obstacles to critical thinking.
2. Identify the dangers of the social discomfort of asking critical questions, fast thinking, belief perseverance, answering the wrong question, egocentrism, and wishful thinking.

Even if you learn critical thinking and plan to use it to create a more thoughtful you, there are still many speed bumps on the road to making use of the critical thinking you have learned. We call these obstacles "speed bumps" because (1) you can overcome them when you just slow down, (2) they are there whether you are aware of them or not, and (3) once you are aware of them, they still exist to interfere with your progress.

But knowledge of potential problems is a first step in defeating them. Consequently, this final chapter is devoted to making you aware of the speed bumps that slow our progress toward becoming critical thinkers. If you know the obstacles that interfere with critical thinking, you can work to push yourself to be more cautious when these threats make their appearance.

THE DISCOMFORT OF ASKING THE RIGHT QUESTIONS

As we learned in Chapter 1, critical thinking is a social activity. To engage with others, we must be willing to ask the right questions to understand the points of view of others. It is important to remember that not everyone is comfortable

己的论证被质疑。所以第一个障碍就是，你提出正确的问题可能使他人产生的不适。

作为被提问对象，被人不断提出批判性的问题，可能会让人感觉自己好像站在法庭的证人席上接受盘问。随着提出的问题不断增多，被提问对象可能会渐渐感到不快，甚至会觉得自己受到了威胁，结果就是他可能发脾气，甚至不愿再继续交谈。他可能不习惯于解释支持其论点的各种理由，或者他为什么要坚持那些理由。虽然你认为提出这些问题对你和其他谨慎思考者而言是必不可少的步骤，但这并不意味着被提问者也觉得有这个必要。

我们得时刻注意所提的问题给被提问者带来了怎样的影响。批判性思维者如果不够小心谨慎，就有可能会因为给周围的人带来不必要的不快而伤感情、失去朋友。所以，为了维系社会关系，我们必须要了解自己的谈话对象，在使用批判性思维的时候注重策略。

思考过快

我们时刻都在思考。我们的大脑几乎总在不间断地处理周围世界的信息。无论我们在决定早晨出门该穿哪件衬衫，还是在决定该信仰哪一个宗教，我们都在不断地思考。

我们的大脑能够很快思考。但幸运的是，我们的大脑还有另外一种能力，一种被卡尼曼称为"慢思考"的能力。后一种思考方式，或者说"系统2思维"，正是本书关注的重点。

所谓慢思考，就是使用我们的大脑来吸收和理性评估别人与我们交流的内容。如果必须要用两个字来概括本书传递的信息，那就是在你思考重大事情的时候一定要"放慢"。

相反，我们的"系统1思维"只是根据手边既有的一点信息，不做任何深刻、全面的思考就仓促做出决断。如果不对我们所做的决定进行慢速、有条理的思考，那么我们犯错的概率就会大大增加。

但是，我们也无须绝望。"系统2思维"有能力否决"系统1思维"所做的决定。我们的任务就在于训练"系统2思维"不要依赖"系统1思维"来做决定。如果依赖"系统1思维"固然容易，可以节省很多精力，使我们无须对自身的认识做大量分析和评估的工作。但是，如果依赖"系统1思维"，我们就会牺牲准确和智慧以换取速度。我们要培养的习惯，是常常问一问自己："我到底依赖什么来支持我的信念和结论？"

刻板印象

在我们接触任何主题之前，我们都带有一定的信念或思维习惯。当我们形成刻板印象时，我们就会断言，因为某个人是特定集体中的一员，所以他肯定具有一系列明确的特征。

having his arguments questioned. So speed bump number 1 is the discomfort of asking the right questions in light of the probable reaction of others.

Being on the receiving end of critical questions can make someone feel as if he is being questioned on the witness stand in a courtroom. As more questions are asked, he may feel uncomfortable or even threatened. As a result, he may become angry or refuse to continue talking. He may not be accustomed to explaining reasons that support his arguments or why he supports those reasons. Just because you see asking these questions as essential to your and the other person's careful thinking does not mean he sees the activity in a similar way.

We have to be aware of how our questions affect the people we are interacting with. If critical thinkers are not careful, they can unnecessarily damage or lose relationships due to the discomfort of those around them. Therefore, in the interest of preserving relationships, we must know our audience and use our critical thinking diplomatically.

THINKING TOO QUICKLY

We are thinkers. Our minds process our world pretty much nonstop. Whether we are deciding which shirt to put on in the morning or which religion to believe, we are constantly thinking.

Our brains can think rapidly. But fortunately for us, our brains have another capacity, one that Kahneman calls "slow thinking." This second form of thinking, or System 2 thinking, is the focus of this text.

Slow thinking is the use of our brain to absorb and evaluate rationally what others are communicating. If you had to summarize the message of our text in two words, it would be "SLOW DOWN" when you are thinking about things important to you.

Our System 1 thinking, on the other hand, makes snap judgments based on what little information is available without any deep, conscious thought. Without slow, methodical thinking about the judgments we make, there is a lot of room for error.

However, there is hope. System 2 thinking has the ability to overrule the judgments made by System 1. Our task is training our System 2 to not rely on System 1. Relying on System 1 is easy, and it saves us from having to put in the work of analyzing and evaluating our perceptions. On the other hand, by relying on System 1, we are sacrificing accuracy and wisdom for speed. The habit we want to form is asking ourselves, "What am I relying on as the support for my beliefs and conclusions?"

STEREOTYPES

You approach any topic with certain preliminary beliefs or habits of mind. When we stereotype, we allege that because a person is a member of a particular group, he must have a specific set of characteristics.

刻板印象会拙劣地取代"慢思考"。下面,我们举几个小例子:

1. 有面部毛发的男人都很睿智。
2. 胖子活得都很快乐。
3. 日本人都很勤劳。
4. 年轻人办事比较草率。
5. 女人最适合做秘书工作。
6. 领取社会福利的人都比较懒惰。

上述 6 个例子都在假装告诉你特定类型的人的某些重要特征。如果我们相信这些刻板印象,我们就会失去强势批判性思维所需的开放、包容的精神,从而难以持开放的态度去接触这些人和他们的思想。此外,一旦某一问题或争端牵涉这些人,我们心里马上就有了成见。那些刻板印象一定会提前介入,抢在理性分析之前占据先机。

每个人都值得我们尊重,每个人所提出的论点都值得我们加以注意。刻板印象之所以会挡住批判性思维的道,是因为它们总是让人试图绕开客观评价的艰难过程。要成为一名批判性思维者,我们要成为充满好奇心和开放心态的表率,而刻板印象会阻断我们的道路,让我们无法仔细考虑别人所说的话。它总是早早关闭我们的思维,导致我们忽略有价值的信息。

背叛我们的思维习惯

虽然我们的认知能力无穷无尽,但有一系列思维习惯会限制和背叛我们。这些认知偏见经常会左右我们,除非我们约束住它们,逼迫它们乖乖"就范"。它们引导我们得出一些结论,但只要我们充分利用批判性思维的技能,就绝不会接受这些结论。本章只会介绍这些思维习惯中的一小部分,不过,只要你能透彻地理解并尽量抵制这些思维习惯,就能大幅提高你今后得出的结论的质量。

晕轮效应

所谓晕轮效应,是指我们有这样一种倾向:先认识到一个人身上所具有的一个积极或消极特征,然后就把这一特征和这个人的其他一切都联系起来。

我们对一个人的认识,会决定我们接收和评价其观点的方式。如果有一个人在生活的某个方面有一技之长,我们就会在大脑里给他"涂上一层光环",以为他在生活的其他方面肯定也有过人之处。这样一来,我们就特别容易接受他的观点。

例如,有个名人天生有副好嗓子,并且特别乐善好施。然后,我们在听到她因为染上

Stereotypes are poor substitutes for slow thinking. Here are a few examples:

1. Men with facial hair are wise.
2. Overweight individuals are jolly.
3. Japanese are industrious.
4. Young people are frivolous.
5. Women make the best secretaries.
6. Welfare recipients are lazy.

All six of these illustrations pretend to tell us something significant about the quality of certain types of people. If we believe these stereotypes, we will not approach people and their ideas with the spirit of openness necessary for strong-sense critical thinking. In addition, we will have an immediate bias toward any issue or controversy in which these people are involved. The stereotypes will have loaded the issue in advance, *prior to* the reasoning.

Each person deserves our respect, and her arguments deserve our attention. Stereotypes get in the way of critical thinking because they attempt to short circuit the difficult process of evaluation. As critical thinkers, we want to model curiosity and openness; stereotypes cut us off from careful consideration of what others are saying. They cause us to ignore valuable information by closing our minds prematurely.

MENTAL HABITS THAT BETRAY US

Our cognitive capabilities are numerous, but we are limited and betrayed by a series of mental habits. These cognitive biases push and pull us, unless we rope and tie them to make them behave. They move us in the direction of conclusions that we would never accept were we exercising the full range of critical-thinking skills. While this section touches on only a few of them, understanding and trying to resist the ones we discuss will make a major contribution to the quality of your conclusions.

Halo Effect

The halo effect refers to our tendency to recognize one positive or negative quality or trait of a person, and then associate that quality or trait with everything about that person.

The perceptions we have of people shape how we receive and evaluate their arguments. If someone is skilled in one aspect of her life, we place a halo on her in our minds. We assume that she must be skilled in other areas of her life. Consequently, we are overly open to her arguments.

For example, a famous celebrity has an incredible singing voice and gives large amounts of money to charity. We are then surprised to learn that she is going to rehab for a drug addiction. We have over-exaggerated the

毒瘾而要到戒毒所接受强制治疗时，会感到特别震惊。我们过高估计了这个名人的高尚品行。因为晕轮效应的存在，我们想象这个名人在生活的每个方面都很高尚，甚至连她的思维方式都包括在内。

同样地，如果有人做了什么让我们觉得比较讨厌的事情，我们就会觉得这人方方面面都让人讨厌，他的论证会被我们自动屏蔽掉。哪怕这个人要说的话我们连一个字都还没听到，我们就已仓促对这人到底是好是坏做出了判断，然后基于这样的"快思考"，我们就开始对他的观点做出回应。

信念固着

我们在与所有人谈话时都背负着大量的"包袱"。我们拥有无数的社会经历，这些经历已在某种程度上塑造了我们；我们每个人都拥有各种梦想，这些梦想引导我们的所见所闻；我们每个人都拥有一定的文化传统，这些传统促使我们以特定的方式思考。一句话，我们总是带着各种观点"上路"。让我们回到淘金式思维的比喻，在你将淘金盘放到沙子里之前，你就认为自己的盘里有了金子。你的信念之所以有价值，只因为它们是你的。你想要坚守这些信念是可以理解的，你已经倾注了大量的心血，使这些观点成为你生命中不可或缺的一部分。

这种对于个人信念的坚持不改或绝不言弃的倾向是批判性思维的一个巨大障碍。它使我们从一开始就心存偏见，不愿更改，只偏爱自己当前的观点和结论。

如果我偏好让民主党候选人当选市长，不管我的理论根据多么薄弱，我都会坚决抵制你让共和党候选人上台的提议。我如果承认自己之前的判断有缺陷的话，心里就会觉得特别不舒服。这种对个人信念的过分忠诚正是"确认偏误"的一个重要根源。所谓"确认偏误"，就是指我们总是倾向于只把可以确认我们既有信念的证据当成可靠证据。这样一来，信念固着就会导致弱势批判性思维。

信念固着的一部分原因是我们对自身能力的过分自信。我们总是以为自己的牌技高、文法好、管理时间能力强，即使面对由合理的评估得出的结果，我们都不相信。这种不幸的思维习惯也让我们总以为自己周围的人都偏执得不可理喻，只有自己不偏不倚。我们总是自以为我们看到的才是真实的世界，其他人都戴着雾蒙蒙的有色眼镜来看世界。我们最大的偏见可能正是认为只有我们自己没有偏见，而那些和我们意见不同的人都心怀偏见！

要抵制信念固着的倾向，我们需要记住，强势批判性思维需要我们认识到所有的判断都是暂时的，或者与语境相关。我们决不可因坚信某件事而故步自封，不愿去寻找更好的答案。正如著名科学家弗朗西斯·培根在 1620 年所指出的那样："当我们改变自己的想法，接受一个更好的观点时，我们应为此感到自豪，我们抵制住了诱惑，没有死心塌地维护自己长期以来的信念。这样的思想转变应当被视为一种罕见力量的反映。"

goodness of the celebrity. Because of the halo effect, we have assumed that the celebrity is good in every aspect of her life, probably including even her thinking.

Similarly, when someone does something we regard as awful, we think he is awful in all regards; we are closed off to his arguments. Even before we listen to a word such a person offers, we make a snap decision about whether that person is a good or bad person. Then we react to his opinions based on that fast thinking.

Belief Perseverance

We enter all conversations with a huge amount of baggage. We have already had numerous experiences that have shaped us in some way; we each have dreams that guide what we see or hear; we each have cultural traditions that push us to think in certain ways. In short, you start with opinions. To return to the panning-for-gold metaphor, *before* you even dip your pan into the gravel you think you have gold in the pan. Your beliefs are valuable because they are yours. Understandably, you want to hold on to them. You have invested a lot of yourself in making those opinions part of who you are.

This tendency for personal beliefs to stick or persevere is a major obstacle to critical thinking. We are biased from the start of an exchange in favor of our current opinions and conclusions.

If I prefer the Democratic candidate for mayor, regardless of how shallow my rationale is, I may resist your appeal on behalf of the Republican candidate. I might feel bad about myself if I were to admit that my previous judgment had been flawed. This exaggerated loyalty to personal beliefs is one of the sources of *confirmation bias*, our tendency to see only that evidence that confirms what we already believe as being good evidence. In this manner, belief perseverance leads to weak-sense critical thinking.

Part of what is going on with belief perseverance is our exaggerated sense of our own competence. We consistently tend to rate ourselves as more skilled at poker, grammar, and time management than any reasonable assessment would be able to find. This unfortunate habit of mind is probably responsible also for our sense that we are living in the midst of incredibly biased people, while we are unbiased. We tell ourselves that we see things as they are, while others look at the world through foggy, colored lens. Our biggest bias may be that *we* are not biased, but those with whom we disagree are!

To counter belief perseverance, it's helpful to remember that strong-sense critical thinking requires the recognition that judgments are tentative or contextual. We can never permit ourselves to be so sure of anything that we stop searching for an improved version. As the famous scientist Francis Bacon pointed out in 1620, *when we change our minds in light of a superior argument, we deserve to be proud that we have resisted the temptation to remain true to long-held beliefs. Such a change of mind deserves to be seen as reflecting a rare strength.*

可得性启发法

与"系统1思维"联系在一起的懒惰,部分在于我们习惯于依赖自己已有的信息,而不是依赖做出更好的决定所必需的信息。获取和处理额外的信息需要付出时间和精力。所谓可得性启发法,指的是我们反复使用的心理捷径,即只根据我们手边最容易获得的信息来形成结论。

假如有人问你,恐怖主义和饥饿哪个对人类安全的威胁最大。你听得最多的说法是什么?哪个问题有好几个大型政府机构都致力于减小其影响?你是不是会说"恐怖主义"?那你就大错特错了。事实上,每天有超过60 000人死于饥饿和不安全的饮用水,与之相比,只有几个人死于恐怖主义。这个信息对于我们做出应致力于攻克哪些难题的决策至关重要。

下面,我们再举一个类似的可得性启发法的例子。对人的生命威胁最大的是什么,是疟疾还是暴力?你脑海里浮现出什么样的画面?你可以回想一下你在新闻里见到的有关这两个致命现象的报道数量,再考虑一下专门控制疟疾和控制暴力行为增长的政府工作人员的数量,你一定也记得随时爆发的各类战争。那么,到现在为止,你也许能猜出哪一个是更加致命的"敌人"了?没错,正是疟疾。每年死于疟疾的人数远远高于死于身体暴力的人数,大概高出33个百分点。

这种可得性启发法和另一个有害的思维习惯——近因效应紧密相连。作为我们思维基础的最容易得到的信息,往往是我们最新看到的那部分信息。例如,尽管乘飞机是最安全的出行方式,但在出现一场空难以后,常常有很多乘客几个月都不敢乘飞机出行。单单一场空难在他们的思维中所起的作用,就要远远大于系统的安全数据所揭示的真相:其实让他们铭记在心的空难出现的概率极低。

答非所问

如果我们想要成为一个善于思考和判断的人,就要竭力避免一个不好的思维习惯,这一思维习惯让我们无法与他人有效沟通。经常有人问我们一个问题,我们立刻不假思索地回答他们,往往怎么容易就怎么回答,也不管回答得对不对。也许我们根本就答非所问。我们在无意中用自己的问题替代了别人的问题。

举个例子,如果有人问:"斯蒂芬·库里是不是历史上最佳篮球运动员?"你对下面的这个回答怎么看:"我在某个地方读到过,斯蒂芬·库里自己说他在个人职业生涯里输掉超过300场职业篮球赛。"有没有人问过斯蒂芬·库里输掉过多少场篮球赛?如果库里输掉过300场球赛,我们会想知道其他"最佳篮球运动员"的候选人一共输掉过多少场比赛。

Availability Heuristic

Part of the laziness associated with System 1 thinking is that we naturally rely on the information we possess, instead of information we need to make a better decision. Obtaining and processing additional information requires time and energy. The availability heuristic refers to the mental shortcut we use again and again of forming conclusions based on whatever information is immediately available to us.

Suppose someone asked you whether terrorism or starvation is the biggest threat to human safety. Which do you hear the most about? Which problem has several huge governmental agencies working to reduce its effects? Did you say "terrorism"? You would be wrong by only a factor of several thousand percent. Only a handful of people die from terrorism when compared to the more than 60,000 who die each day from starvation and unsafe drinking water. This information is crucial to shaping what problems we decide to attack.

Here is another example of the availability heuristic along the same lines. What is the biggest threat to human life: malaria or violence? What pictures come to mind? Think about the number of instances you have witnessed on news reports of these two deadly phenomena. Consider the number of public employees whose job it is to halt the growth of malaria and violence. Remember the huge number of wars occurring at any time. By now, you can guess what is the more deadly foe. Right, malaria. There are more annual deaths by malaria than from physical violence by approximately 33 percent.

The availability heuristic is closely related to another harmful mental habit, the *recency effect*. What is immediately available as a basis for our thinking is often the most recent piece of information we have encountered. For example, even though flying in an airplane is extremely safe, many travelers refuse to fly for a few months after an airplane crash. A single crash plays a larger role in their thinking than do the systematic safety statistics that reveal how unusual that remembered crash was.

Answering the Wrong Question

Part of our failure to communicate effectively with one another is an unfortunate mental habit that we must fight to avoid if we are going to be a skilled critical thinker. Often when someone asks us a question, we provide an immediate automatic answer that comes easily to mind and fail to respond to the question that was asked. We give an answer to the wrong question. We unconsciously substitute our question for the one we were asked.

Consider this example. Is Steph Curry the best basketball player ever? What would you think about this answer: "I read somewhere that he said he had personally lost more than 300 professional basketball games"? Did anyone ask how many games Steph Curry had lost? If Curry lost 300 games, we wonder how many games the other candidates for "best player ever" have lost.

你可以测试一下自己，看能否快速发现在最近一期《滚石》杂志的某场采访中起此种作用的"减速带"。基思·理查兹在被问及他和米克·贾格尔之间的宿怨是否已经了结时，他回答说：

> 米克和我都是专业歌手。我们所做的一切都是为了做好自己的音乐。

很显然，如果理查兹真的要回答这个问题，那就得大费周折，绝非一句话可以说清。此外，他可能有一个答案，但他不愿在《滚石》这样的媒体上公开，所以，他的思维很快就从这个答案上溜走了。

一个批判性思维者应该从这里学到，只要有人给出一个驴唇不对马嘴的答案，这种行为就会将大家的注意力从开始讨论的问题上引开。然后，它开始引出另一场全新的讨论。不管怎样，"慢思考"都很困难，如果有人不让我们专注于一个问题，那么我们提出有效的批判性问题的能力更会大受影响。

自我中心

在回顾这些"减速带"的时候，我们可以思考一下，为什么这么多"减速带"都源自同一个位置。我们总是自我陶醉并忠于自己。所谓自我中心，是指相对于其他人的经历和观点，我们赋予自我的世界以中心地位。对我们来说，自己胃里一时半会儿的空空如也，常常比这个星球上每天有超过 35 000 人因饥饿而死这个事实显得更加万分火急。我们总觉得自己的经历、自己的观点、自己的需求排在第一位，世界围着它们转，或者说至少应该围着它们转才好。

其实，不妨换一种视角来回顾每一个"减速带"，像关注自己的思维那样去关注其他人的思考方式，这是非常好的"系统 2 思维"练习方法。你需要非常投入地去关注很多与你截然不同的人的生活。你需要倾听他们，一遍遍地问他们："那么，你是不是说……呢？你之所以这样说，是不是因为……呢？"你不得不进入他们的大脑，查看一下那里有没有什么强烈的偏见，导致他们得出那样的结论。

以信念固着为例。如果我们换一种视角，那么其他人的信念会获得同样的尊重，并且会和我们自己的信念一样，得到我们全面的了解。我们开始真正用心倾听各种各样的理由和看法。而在此之前，一旦意识到它们并非来自跟我们同一阵营的人或观点相近的人，我们立刻就会将它们拒之门外。如果我们意识到我们持有的信念中有多少仅仅因为它们是我们自己的信念而被相信，结果一定会让我们感到大为震惊。

我们在进行论证或者评价其他论证的时候，常常会忘记我们所面对的对象，我们沉浸在自己的知识世界里，沉浸在我们所知道的解决方案里。这正是我们的自我中心在发挥作用。

See whether you can quickly see this speed bump at work in a recent *Rolling Stone* interview. Keith Richards was asked whether the feud between him and Mick Jagger was over. Richards replied:

> Mick and I are professionals. We do what is necessary to make our music.

Apparently, it would have just been too much work for Richards to answer the question he was asked. In addition, his mind probably quickly ran away from an answer he preferred not to give in a public forum like *Rolling Stone*.

The point for a critical thinker is that whenever anyone provides an answer to a question that was not asked, that behavior diverts attention away from where the discussion began. Instead, it starts an altogether different discussion. Slow thinking is very difficult anyway, and when someone does not permit us to focus on a single question, our ability to ask effective critical questions is sharply reduced.

EGOCENTRISM

When you review the speed bumps, consider how so many of them have their source in the same location. We are fascinated by and loyal to ourselves. *Egocentrism* refers to the central role we assign to our world, as opposed to the experiences and opinions of others. The temporary emptiness of our own pantry is often much more compelling to us than the fact that more than 35,000 people starve to death each day on our planet. We think our experiences, our opinions, and our needs somehow move the world or at least deserve to move it.

Indeed, it would be a good System 2 exercise for you to review each of the speed bumps from a perspective in which you pay as much attention to the thinking of others as to your own. You would need to be very engaged with the lives of many people quite different from you. You would need to listen to them and ask them again and again, "So, is this what you are saying, and is this why you are saying it?" You would be forced to get inside their heads to see whether there is some strong basis for the conclusions they have.

Take belief perseverance as an example. With our new perspective, the beliefs of others get the same respect and as thorough a hearing as we give our own. We start to really hear at a deep level various reasons and perspectives that would otherwise be dismissed immediately once we recognized they were not coming from people in our immediate tribe or family of opinion. It is frightening to realize how many things we believe just because the belief is ours.

When we make arguments or when we evaluate arguments, we often forget our audience as we focus on what we know and what we know how to do. Our egocentrism is at work.

在与那些并未系统学习过批判性思维技能、不了解其重要性的人交往时，我们尤其要提醒自己注意这些交往对象的存在。

批判性思维者与其他人一样，也时刻在和知识的诅咒做斗争。所谓知识的诅咒，是指我们无法回想起我们不具备现在掌握的知识时的情况。

当我们忘掉知识的诅咒带来的危险时，我们可能会发现自己和其他人之间的对话听起来就像是《生活大爆炸》里谢尔顿和佩妮之间的对话：

> 谢尔顿：我需要你帮忙解决一个符号学上的问题。
>
> 佩妮：你说什么？
>
> 谢尔顿：符号学，就是研究符号和象征的学问，它是和语言学有关的一个哲学分支。
>
> 佩妮：好了，亲爱的，我知道你以为你在解释自己的话，但其实你等于什么也没说。

谢尔顿的自我中心阻碍了任何理性对话的可能，如果他多想一想佩妮是个什么样的人，也许他们之间的对话就有可能继续下去。

一厢情愿：批判性思维最大的障碍

2005年，斯蒂芬·科尔伯特曾提醒我们"相信内心而非真相"这种危险的思维习惯。如果一个人宁愿相信那些他希望是真的概念或事实，而不愿相信那些业已证明为真的概念或事实，他就在忠于内心而非真相。我们希望这个世界具备一些特点，希望世间事能够更公平、更和气、更有成效。但很多人并不去想这样的升平世界是否接近现实，而是自创一套信念来迎合这个虚幻的世界。我们希望什么是真的，索性就宣布它是真的。我们希望商品的标签简单易懂并且名副其实，所以买起东西来毫不犹豫，相信买到的商品百分百会印证标签上的文字。

这样一来，我们要的是事实符合我们的信念，不愿让我们的信念与事实契合。我们相信，你肯定能发现其中的问题。仅仅因为我们认为很多事情不应该是现在这个样子，我们就相信它们确实不是这个样子。一旦我们意识到自己身上存在这样的倾向，就要不断地问自己："我相信这件事是真的是因为我希望它是真的，还是有确凿的证据证明它是真的？"否则，我们可能会自取其辱，说出的话就像哈利·波特在《哈利·波特与混血王子》里说的那样，完全受"系统1思维"的控制。

Being aware of our audience is especially important when interacting with those who have not learned the skills and importance of critical thinking.

Critical thinkers, like everyone else, struggle with the *curse of knowledge*. The curse of knowledge is that we cannot recall what it is like when we did not know what we now know.

When we forget about the dangers of the curse of knowledge, we may find our conversations with others sound like that of Sheldon and Penny in *The Big Bang Theory*:

> Sheldon: I need your help in a matter of semiotics.
>
> Penny: What?
>
> Sheldon: Semiotics, the study of signs and symbols as a branch of the philosophy related to linguistics.
>
> Penny: Okay, honey, I know you think you are explaining yourself, but you're really not.

Sheldon's egocentrism is getting in the way of any rational conversation that might have been possible had he thought more about who Penny was.

WISHFUL THINKING: PERHAPS THE BIGGEST SINGLE SPEED BUMP ON THE ROAD TO CRITICAL THINKING

In 2005, Stephen Colbert reminded us of the dangerous mental habit of *truthiness*. A person is loyal to truthiness when he prefers concepts or facts he wishes to be true, rather than concepts or facts known to be true. We wish for the world to have certain characteristics. Things could be much more fair and kind and productive. But in place of wondering about whether such a world is even close to reality, many of us just form beliefs to match our make-believe world. What we wish to be true, we simply declare *is* true. We want the product label to be honest and straightforward. So we buy with little hesitation, believing that the product is precisely reflecting the words on the label.

That way, the facts conform to our beliefs rather than fitting our beliefs to the facts. We are sure you can see the problem here. Because we think that things should be different than they are, we believe that indeed they are different. Once we recognize this tendency in ourselves, we need to keep asking, "Is that true because I want it to be true, or is there convincing evidence that it's true?" Otherwise we will embarrass ourselves by saying something like Harry Potter says in *Harry Potter and the Half-Blood Prince* in a fit of System 1 thinking:

> Harry: It was Malfoy.
>
> Professor Minerva McGonagall: That is a very serious accusation, Potter.
>
> Professor Severus Snape: Indeed. Your evidence?

哈利：肯定是马尔福干的。

麦格教授：波特，这个指控可非同小可。

斯内普教授：说的是。波特，你有什么证据？

哈利：我就是知道。

斯内普：你……就是……知道？（语带讥讽）波特，你的天赋又要让我大吃一惊了。

一厢情愿的思维之所以会如影随形，是因为我们否认的思维模式常常出现。我们不知不觉地与事实抗争，竭力在冷冰冰的现实之外强化我们对升平世界的幻想。我们对共同面对或单独面对的各种问题产生的焦虑和恐惧情绪构成了一道防护墙，让我们看不清自己身处其中的现实世界。

想一想，我们生平多少次反复听到某国的元首宣布，他们进行中的战争很快就会结束，胜利指日可待。但结果总证明：这些预测都不过是空洞的诺言。战争或许会永无休止地打下去，或许本国部队获胜的前景并不明朗，想到要面对这样残酷的现实会让人觉得难以承受。所以，我们的思维干脆将它抹掉了。

有种一厢情愿的思维方式叫作奇迹式思维。对于一些事情，在科学还不能提供令人信服的解释时，人们就会依赖奇迹来解释它们的来龙去脉，或是设法用奇迹来控制科学无法掌控的事物。让我们一起听听动画片《辛普森一家》里的顽童巴特·辛普森是怎样打击奇迹式思维的吧。

马格：好了，孩子们，把你们的信都交给我。我会把这些信寄到圣诞老人在北极的工作室去。

巴特：得啦，这世上只有一个肥佬会不断给我们带礼物来，他的名字才不叫圣诞老人呢。

当人们无法理解或改变一个处境时，奇迹式思维往往最能大显身手。在强大需求的驱动下，任何对人生的随机性或事件偶发性的信念都让人觉得难以接受并被抛到一边；与之相反，奇迹式的因果关系则带给人希望。某个人或某个新思想可能会让一切峰回路转。只要听从政治候选人许下的各种诺言，就会梦想成真。我们之所以相信他们，并不是因为他们的宣言有什么可靠的证据，而是因为我们渴望相信他们。

治疗一厢情愿的良药，就是积极主动地使用本书教授的批判性思维方式。在我们通往批判性思维的道路上，"减速带"会不时出现，它们是我们生命的一部分。我们无法忽视它们，但只要我们不缺好奇心，只要我们发自内心地尊重批判性思维的各项原则，我们就一定能够顺利地跨越这些障碍。

Harry: I just know.

Severus Snape: You ... just ... know? (sarcastically) Once again, you astound me with your gifts, Potter.

Wishful thinking has staying power because of the frequency of our denial patterns. Quite unconsciously, we fight with the facts, trying to reinforce visions of the world that are rosy beyond the bounds of reality. Anxieties and fears about the problems we face together and individually serve as a protective shield against seeing the actual world in which we live.

Think of how frequently over the course of your life you will hear leaders of nations declare that the war they are fighting will soon be over, and victory will be won. But such predictions usually turn out to be hollow promises. To have to face the fact that the war may go on and on or that it will not result in a clear victory for the home team is just too painful to consider. So the mind erases it.

A form of wishful thinking is magical thinking. People tend to rely on magic as a causal explanation for things that science has not acceptably explained, or to attempt to control things that science cannot. Listen as Bart Simpson deflates magical thinking:

Marge: Alright kids, hand me your letters. I'll send them to Santa's workshop up at the North Pole.

Bart: Oh, please. There's only one fat guy who brings us presents, and his name ain't Santa.

Magical thinking tends to be greatest when people feel most powerless to understand or alter a situation. In the face of great need, any belief in the randomness or accidental aspects of life is set aside as grim and replaced with the promise of magical causal relationships. Somebody or some new idea will make everything wonderful. Simply listen to the promises of political candidates. We believe them not because of any evidence for their claims, but because we so much want to believe them.

The antidote to wishful thinking is active use of the critical questions taught in this text. Speed bumps will always be in the way of our critical thinking; they are part of us; we cannot ignore them, but we can surely resist them with curiosity and a deep respect for the principles of critical thinking.

FINAL WORDS

Critical thinking is a tool. It does something for you. In serving this function, critical thinking can perform well or not so well. We want to end this text by urging you to get optimal use of the attitudes and skills of critical thinking that you have worked so hard to develop.

最后的话

批判性思维是一个工具，它能助你一臂之力。在实现这一功能的时候，批判性思维可能让你如虎添翼，也可能让你面临阻碍。你已耗时费力地苦练了批判性思维能力，在本书的最后，我们鼓励你最大限度地利用批判性思维的态度和技能。

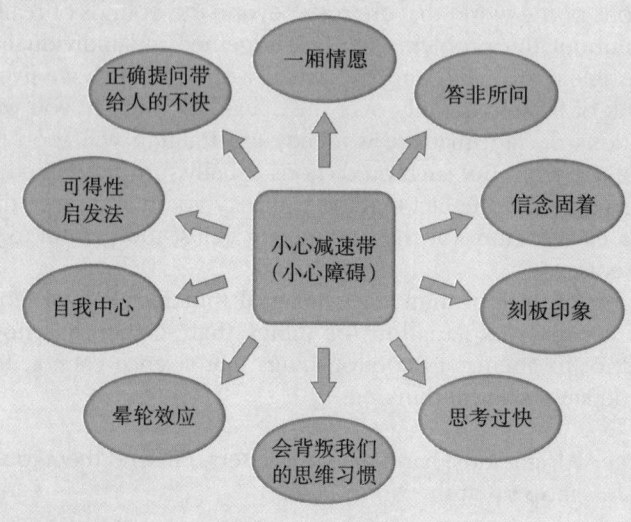

批判性思维的障碍

本书花费大量篇幅为你打造批判性思维的"技能库"。在第1章中，我们指出批判性思想者的基本价值观是自主决断、好奇心、谦恭有礼和对好的论证的尊重。

要按照这些价值观生活并付诸行动，就需要培养一定的思维习惯。这些习惯既不是天生的，也不易培养。我们强烈建议你经常进行自我评估："在生活中，我是不是运用了所学的批判性思维技能？"为帮助你做到这一点，我们提供了一份对批判性思维者特有习惯的简单描述。

一个批判性思维者：

1. 博览广识，涉猎广泛，为从多角度理解假设提供基础；
2. 以理由和证据作为决定的依据；
3. 以愿意接受的姿态对待他人的信念，但以质疑的态度来确认该信念是否有强有力的支撑的；
4. 迫使他或她自己寻求和尊重对其主张的真相的多种思考方法。

Chapter 13 Speed Bumps Interfering with Your Critical Thinking   341

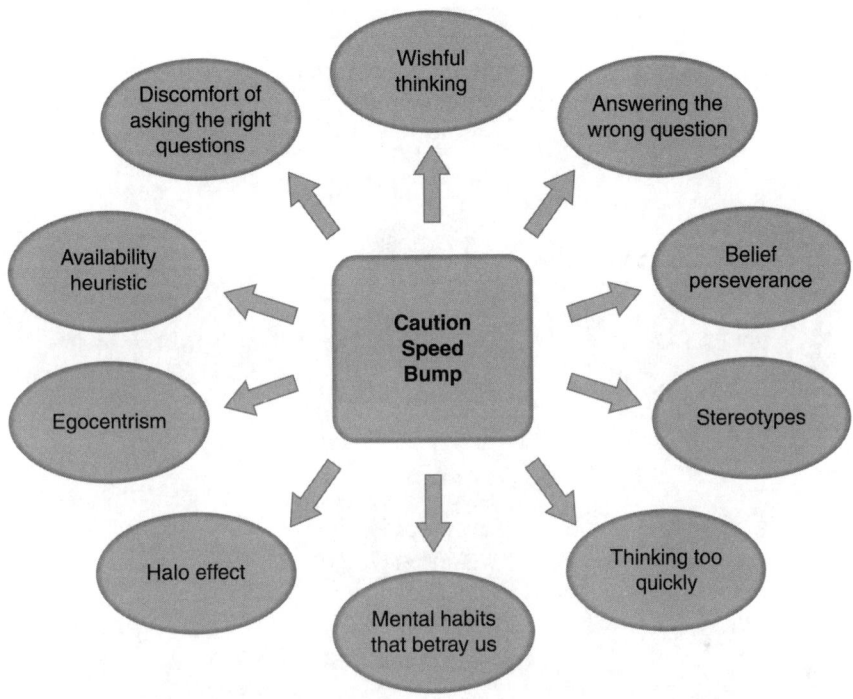

Speed Bumps to Critical Thinking

This text has spent a lot of time building your repertoire of critical-thinking "skills." In Chapter 1 we pointed out that the primary values of a critical thinker are *autonomy, curiosity, humility,* and *respect for good reasoning.*

To live so as to act on those values requires the development of certain habits of mind. These habits are neither natural nor easy to nurture. Our challenge to you is frequent self-assessment: "Am I living so as to use the critical-thinking skills I have learned?" To help you do so, we provide a brief description of the habits that distinguish your identity as a critical thinker.

A critical thinker

1. reads a broad range of materials to provide a base for comprehending assumptions from multiple perspectives;
2. depends on reasons and evidence as the basis for decisions;
3. approaches beliefs with a willingness to agree, but uses a questioning attitude to ensure the belief has strong support; and
4. forces him- or herself to seek and respect multiple perspectives about the truth of any claim.

How can you give others the sense that your critical thinking is a friendly tool, one that can improve the lives of the listener and the speaker, the reader and

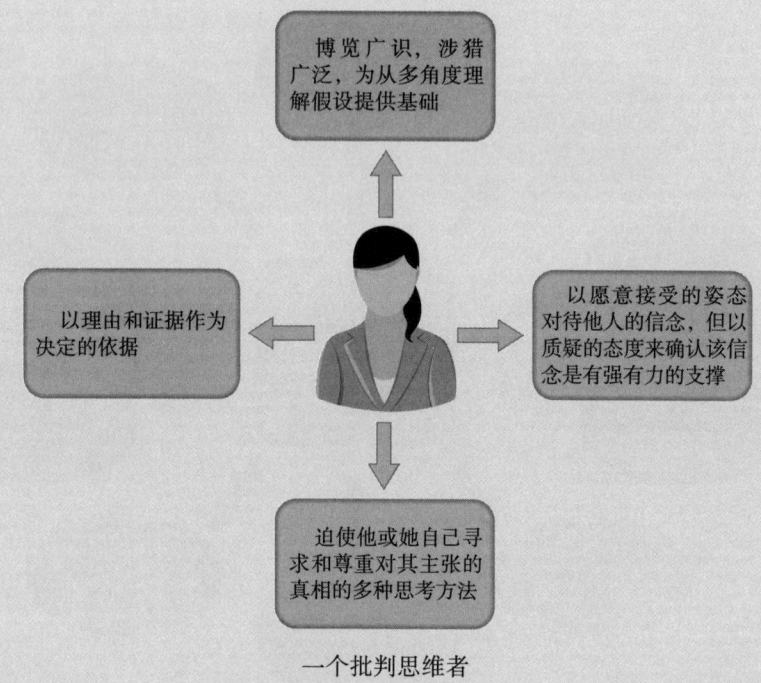

一个批判思维者

你怎样向别人传达你的批判性思维是一种友善的工具，它可以改善发言者和听众、写作者和读者的生活质量？和其他批判思维者一样，我们也一直在试图回答这个问题。我们发现最有用的一个办法是大声把你的批判性问题说出来，表现出你对它充满了好奇。如果总是摆出一副"哈哈，我可逮着你的一个错了"的态度，将会对批判性问题的有效性产生最致命的影响。

作为临别赠言，我们想鼓励你真正投入论题中。批判性思维不是只开花不结果的业余爱好，只能在教室里摆摆架子，在考试时临阵磨枪，或者在要显摆你智力超群时拿出来充充门面。它是通情达理的人采取联合行动的坚实基础。信念固然很奇妙，但是信念的回报寓于我们随后的行为之中。在你发现一个问题的最佳答案之后，请依据这个答案采取行动。让批判性思维成为创造一个你为之自豪的身份认同的基础。让批判性思维转化为你自己和你所在的集体的行动。

期待有朝一日，我们能从你学会的这一切之中大大受益。

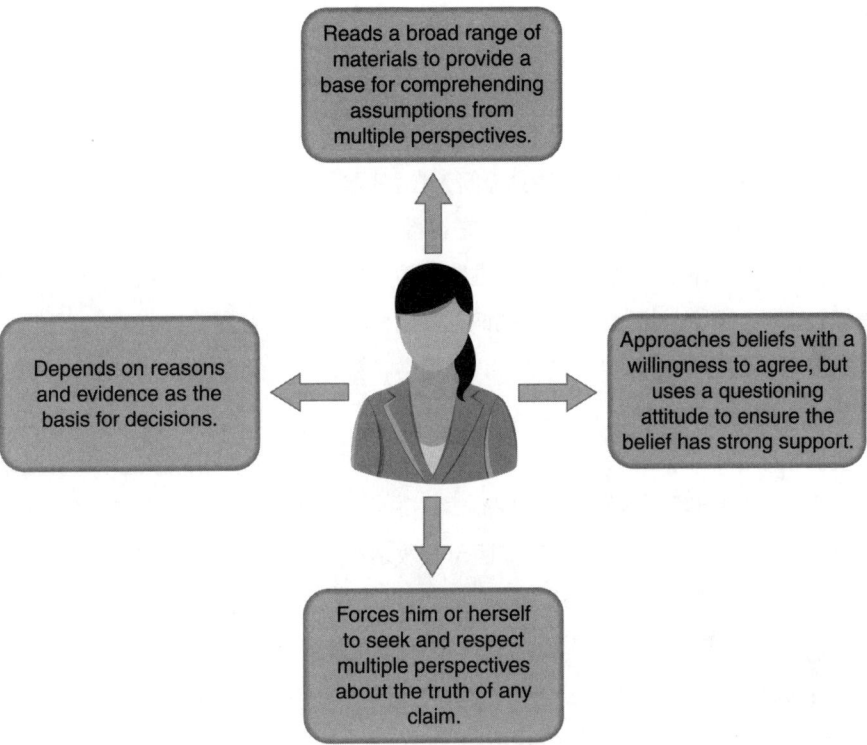

A Critical Thinker

the writer? Like other critical thinkers, we are always struggling with this question. But the one strategy we find most useful is to voice your critical questions as if you are curious. Nothing is more deadly to the effective use of critical thinking than an attitude of, "Aha, I caught you making an error."

As a parting shot, we want to encourage you to engage with issues. Critical thinking is not a sterile hobby, reserved only for classrooms, for taking exams, or for showing off your mental cleverness. It provides a basis for a partnership for action among the reasonable. Beliefs are wonderful, but their payoff is in our subsequent behavior. After you have found the best answer to a question, act on that answer. Make your critical thinking the basis for the creation of an identity you can be proud of. Put it to work for yourself and for the community in which you find yourself.

We look forward to benefiting from what you have learned.

译后记

作为一本学术通俗读物，《学会提问》在美国学界享誉已久，自从1978年第1版刊行以来，迄今已有12个版次陆续面世，它在批判性思维训练领域的权威性、可信度可见一斑。它在紧扣批判性思维培养和发展方面殚精竭虑、不遗余力，不仅保留了短小精悍的传统篇幅，而且不断注入了最新的研究证据。因此，正如作者所说，它可以"被广泛应用于法律、英语、制药学、哲学、教育学、心理学、社会学、宗教文以及各类社科课程，同时还被普遍应用于中学课堂里。"

我们生活在一个面对百年未有之大变局的新时代，每时每刻都被这个世界嘈杂、喧嚣、泛滥成灾的信息所冲击，大到国际政治、经济、军事形势，小到个人日常生活的决策。我们如果缺乏独立思考和判断，就只能照单全收地接受别人兜售的观点，那样我们就失去了一个独立个体存在的意义和价值。中国古代亚圣孟子曾经说过："尽信书，不如无书。"(《孟子·尽心下》) 古希腊哲学家苏格拉底和人讨论问题时，常用诘问法，又称苏格拉底法 (the socratic method)。苏格拉底认为一切知识均从疑难中产生，越求进步疑难越多，疑难越多进步越大。因此，在日常生活、学习和工作中，我们应该重视批判性思维的培养，学会始终拥有审慎的态度、质疑的精神和探究的意识，养成独立思考和求实创新的习惯。

本书的作者通过探索论证的组成部分（问题、结论、原因、证据、假设和语言），传授富有理性、逻辑性和批判性地提问、思考、判断和解决问题的方法。全书叙述深入浅出，立论严谨精确，例证贴近实际。这是一本授人以渔的打开想象翅膀之书，是一座让我们能够砥节砺行、行思致远的前行的灯塔。医生诊断得再精确，也无法替我们救急，我们要想培养自己独立思考、创新作为的能力，有必要依靠自己的大脑，激发批判性思维，提出批判性问题，才能保证自己不会自欺欺人和人云亦云。

《学会提问》第12版的翻译分工如下：全书的翻译由许蔚翰博士完成，吴礼敬博士对各个章节的翻译进行了统稿，最后全书的审译由许俊农教授负责。本书的出版要感谢机械工业出版社相关编辑辛勤、细致而专业的工作，也要感谢上海英硕文化传播公司的鼎力帮助。由于能力有限，翻译错误在所难免，敬请各位方家读者批评指正。

许蔚翰　谨识
2020年10月26日